ITABC

NASA

The Reconstruction of Archaeological Landscapes through Digital Technologies

Proceedings of the 1st Italy-United States Workshop, Boston, Massachusetts, USA, November 1-3, 2001

Sponsored by CNR (National Research Council - ITABC, Institute of Technologies Applied to Cultural Heritage, Roma, Italy) and Boston University's Center for Archaeological Studies, Center for Remote Sensing, and the Department of Archaeology (USA) with the collaboration of NASA (National Aeronautics and Space Administration). This workshop was made possible by grants from the J. M. Kaplan Fund, U.S.A. and the National Research Council, Italy

Edited by

Maurizio Forte
P. Ryan Williams

BAR International Series 1151
2003

Published in 2016 by
BAR Publishing, Oxford

BAR International Series 1151

The Reconstruction of Archaeological Landscapes through Digital Technologies

ISBN 9781841715216 paperback
ISBN 9781407325552 e-format
DOI https://doi.org/10.30861/9781841715216
A catalogue record for this book is available from the British Library

BAR Publishing is the trading name of British Archaeological Reports (Oxford) Ltd. British Archaeological Reports was first incorporated in 1974 to publish the BAR Series, International and British. In 1992 Hadrian Books Ltd became part of the BAR group. This volume was originally published by Archaeopress in conjunction with British Archaeological Reports (Oxford) Ltd / Hadrian Books Ltd, the Series principal publisher, in 2003. This present volume is published by BAR Publishing, 2016.

BAR
PUBLISHING

BAR titles are available from:

BAR Publishing
122 Banbury Rd, Oxford, OX2 7BP, UK
EMAIL info@barpublishing.com
PHONE +44 (0)1865 310431
FAX +44 (0)1865 316916
www.barpublishing.com

Italy-United States Workshop:
"The Reconstruction of Archaeological Landscapes through Digital Technologies"
Boston, Massachusetts, USA, November, 1-3, 2001

This workshop is made possible by grants from the J. M. Kaplan Fund, U.S.A., and the National Research Council, Italy.

Sponsored by CNR (National Research Council - ITABC, Institute of Technologies Applied to Cultural Heritage, Roma, Italy) and Boston University's Center for Archaeological Studies, Center for Remote Sensing, and the Department of Archaeology (USA) with the collaboration of NASA (National Aeronautics and Space Administration).
Organized by Kathryn Bard (BU), Farouk El Baz (BU - CRSA), Maurizio Forte (CNR-ITABC),
Salvatore Garraffo (CNR-ITABC), P. Ryan Willams (Field Museum), James Wiseman (BU).
Co-chairs: Farouk El Baz (BU - CRSA), Maurizio Forte (CNR-ITABC), James Wiseman (BU).

Promoting Collaboration between Italy and the United States

Editors: **Maurizio Forte**, **Patrick Ryan Williams**
Co-editors: **Farouk El Baz**, **James Wiseman**

Graphic: Bartolomeo Trabassi
editing and layout: Sofia Pescarin

“The Reconstruction of Archeological Landscapes Through Digital Technologies”

Contents

THE CONFERENCE

Introduction

The organizers, CNR-ITABC (Rome) and Boston University's (B.U.) Center for Archaeological Studies, Center for Remote Sensing, and Department of Archaeology, represent a high level of competence in the fields of territorial studies and advanced digital technologies, mainly with reference to landscape archaeology. CNR-ITABC is by definition an institution for inter- and multi- displinary research, and therefore its research includes the fields of applied technology, archaeology, architecture, geology, geophysics, chemistry, archaeometry, and computer applications.

B.U.'S archaeological units are internationally respected for their innovative educational and research programs. The B.U. Center for Remote Sensing is one of the most important research centers for remote sensing in the U.S. and works in close collaboration with B.U.'s Center for Archaeological Studies and Departments of Archaeology, Geography, and Earth Sciences.

Furthermore, there is an official agreement for scientific cooperation between CNR-ITABC and B.U.'s Center for Remote Sensing (signed 20 October, 2000 by the contributing scientists, Dr. Farouk El-Baz and Dr. Maurizio Forte, in order to conduct scientific research, education, and to co-organize workshops on GIS and remote sensing in archaeology.

Aims

Diachronic and visual analysis for the classification of the archaeological landscape is a key concept in understanding the complex evolution of the anthropic factors for settlement patterns, use of resources, and social, political, and economic relationships which connect the study of the past with the present. Because of the complexity of this approach in our research, input from different fields is essential, including archaeology, the environmental sciences, GIS, remote sensing, geology, anthropology and the other social sciences. The main goal of this workshop is the reconstruction of the archaeological landscape, with special reference to the application of GIS, spatial analyses and remote sensing. Another key element will be the evaluation, valorization and representation of the archaeolgical landscape as an important and rich cultural resource of the territory (CRM, Cultural Resource Management). In fact, this is a valuable contribution in terms of the restoration/reconstruction of the relationships between the social/cultural history and the collective memory.

"The map is not the territory..." (Bateson), and, in digital language, GIS is not the landscape, because it cannot represent a complex dataset, such as a landscape. In recent scientific literature the relationships between the landscape and cognitive archaeology have been studied, especially in two significant contributions: *The ancient mind* (Renfrew and Zubrow 1994[1]) and *Semiotics of Landscape: Archaeology of Mind* (Nash 1997[2]). Post-processualist archaeology is also studying GIS spatial applications, for reconstructing past landscapes (mainly "resource landscapes") using viewshed and cost surface analyses (on the last, see Van Leusen 1999). Since the late 1980s a significant number of these spatial applications have been undertaken, specifically with cognitive aims. The issue is very complex because it involves methodological, epistemological and technological aspects: How can we represent archaeological landscapes? How is the landscape perceived? What is the difference between the real and imaginary landscape? How is it possible to describe an archaeological landscape? The complexity of the issues is evident: for analyzing, describing and representing an ancient landscape, spatial techniques, epistemological, archaeological and anthropological approaches are needed.

Furthermore, we think that for a reconstruction of the cognitive landscape an integrated methodology is needed consisting of digital spatial technologies (GIS, remote sensing, virtual reality) and epistemological discussions. The conditions for interpreting an archaeological landscape mainly depend in large part on many factors of visual representation. In fact, visibility, perception, cognition and communication represent the main factors of knowledge of the landscape, especially the ancient landscape. A distinction between perception and cognition has been mentioned by M. Van Leusen (Van Leusen 1999[3]), in which perception is "the simple act of being aware of the landscape..." and cognitive archaeology is "the study of past ways of thought as inferred from materials remains" (Renfrew and Zubrow 1994). In Nash (1997), the "landscape, comprising natural features, is a series of stories that are constructed through time and space and formulate a series of histories; not history". Thus, for the reconstruction of cognitive archaeological landscapes the following questions and topics are suggested:

---

[1] Renfrew C., Zubrow E. (eds.) 1994, *The Ancient Mind: elements of cognitive archaeology*, Cambridge UP.
[2] Nash G. (ed.) 1997, *Semiotics of Landscape: Archaeology of Mind*, Oxford.
[3] Van Leusen M. 1999, *Viewshed and Cost Surface Analysis Using GIS* (Cartographic Modelling in a Cell-Based GIS II)," in New Techniques for Old Times, Computer Applications and Quantitative Methods in Archaeology, Proceedings of the 26th Conference, Barcelona, 1998, bar 757.

- What does the landscape consist of *(now)?*
- How do we perceive the landscape and what do we perceive *(now through our minds)?*
- What did the ancient landscape consist of *?*
- How did ancient people perceive their landscape *(through the ancient minds)?*
- How can we perceive the ancient landscape *(now, but through the perspective of ancient minds and mental maps).*

Topics

- Archaeological landscapes of power relationships
- CRM, cultural heritage and archaeological parks
- 3-D, GIS and virtual reality
- Geomorphology, pedogenesis and landscape evolution
- Genetic algorithm and GIS
- Multispectral classifications
- Radar classifications
- New high resolution satellites
- 3-D models and cognitive models
- Paleoenvironmental analyses
- General perspective on the use of remote sensing in terms of landscape archaeology

The contribution of Italian scholars will mainly focus on GIS, remote sensing and desktop virtual reality for the reconstruction of the archaeological landscape. Other contributions will include valorization of the territory and "musealization" of archaeological parks through models of CRM research.

The contribution of American scholars will focus on digital and multi-spectral classification of the landscape (spectral, radar, and multi-resolution), new satellite programs, and technical aspects of fieldwork, such as the analysis of soils and strata (natural and anthropic), and archaeological interpretation.

Finally, it is important to stress that in the composition of the workshop members, we have included all levels of professionals, from post-graduate research fellows to full professors. We believe that this workshop also provides an important opportunity for young scientists to continue their scientific education with internationally renowned scientists.

An additional goal of this international workshop is to provide more informal periods of discussion for the involved scientists, to compare different archives of data. Finally, from these discussions we hope to see the creation of projects generated by new international cooperation.

Sponsoring Participants

Farouk EL-BAZ
Center for Remote Sensing
Boston University
675 Commonwealth Ave.
Boston, MA 02215
617/353-5081
farouk@crsa.bu.edu

Salvatore GARRAFFO
CNR-ITABC
NATIONAL RESEARCH COUNCIL
Institute for Technologies Applied to Cultural Heritage
Via Salaria km.29,300, c.p.10
00016 Monterotondo St., Roma
06/90672.382
salvatore.garraffo@mlib.cnr.it

Compton Jim TUCKER
Biosphysic Sciences Branch
NASA/Goddard Space Flight Center
Greenbelt, MD 20771
301/614-6644
compton@gsfc.nasa.gov

James WISEMAN
Center for Archaeological Studies
Boston University
675 Commonwealth Ave. #347
Boston, MA 02215
617/3538-1670
jimw@bu.edu

PROGRAM

October 31, Wednesday

Participants arrive

Accommodation: Howard Johnson, 1271 Boylston St., Boston, MA 02215; Radisson Hotel, 777 Memorial Drive, Cambridge, MA 02139.

6:00-8:00 p.m. Opening Reception

November 1, Thursday

Welcome (Chair J. Wiseman) 9.30 a.m.

Opening Remarks

- by Jon Westling, President of Boston University

Introductory Comments

- by Farouk El Baz, James Wiseman (Boston University), Salvatore Garraffo (National Research Council, Italy).

Presentation of the workshop

- by Maurizio Forte, Ryan Williams (CNR-ITABC, Boston University)

SESSION 1 (10:20-12:15 a.m.), J. Wiseman, presiding

- Keynote speaker: Michael THOMAS (NASA)

NASA, Earth Science, and Archaeology

Coffee break (11:00-11.30 a.m.)

- **Compton Jim TUCKER** (NASA/Goddard Space Flight Center)

SESSION 2 (13:30-17:45 p.m.), F. El Baz, presiding

- **Carlo M. MARINO** (Dipartimento di Scienze dell'Ambiente e del Territorio, University of Milan)

"Hyperspectral Airborne Remote Sensing as an Aid to a Better Understanding and Characterization of Buried Elements in Different Archaeological Sites"

- **Tom SEVER, Daniel IRWIN** (NASA-Marshall Space Flight Center/GHCC, Universities Space Research Association NASA-Marshall Space Flight Center/GHCC)

"Recent Advances In Maya Studies Using Remoteley Sensed Data"

- **Stefano CAMPANA, Riccardo FRANCOVICH** (Dipartimento di Archeologia e Storia delle Arti, University of Siena)

"Landscape Archaeological Prospections in Tuscany: Research Methods, Settlement Patterns and Cultural Resource Management"

Coffee Break (15:45-16:15 p.m.)

- **Ron BLOM, Douglas COMER, Brian BYRD, Andrew YATSKO** (NASA/JPL)

"Locating Archaeological Sites Using AIRSAR Derived Data"

November 2, Friday

SESSION 3 (9:00-12:30 a.m.), M. Forte, presiding

- **Patrick Ryan WILLIAMS** (Field Museum of Natural History)

"Hydraulic Landscapes and Social Conflict in Middle Horizon Peru"

- **Armando DE GUIO** (Dipartimento di Archeologia, University of Padua)

"Dirt Roads to Power on the Surface of Complexity : the CISAS Experience"

Coffee break (10:30-11:00 a.m.)

- **Ezra B. ZUBROW** (Department of Anthropology, SUNY Buffalo)

"The Archaeologist, the Neural Network, and the Random Pattern: Problems in Spatial and Cultural Cognition"

- **Maurizio FORTE** (CNR-ITABC, Rome)
"Remote Sensing, GIS and Virtual Reconstruction of Archaeological Landscapes"

SESSION 4 (13:30-17:30 p.m.), S. Garraffo, presiding

- **Elizabeth STONE** (Department of Anthropology SUNY Stony Brook)
"The Role of the Tigris in Ancient Mesopotamian Settlement"

- **Cinzia PERLINGERI, Rosario PERLINGERI, Monica FOCCILLO** (IUO, Istituto Universitario Orientale, Naples)
"Real Space Beyond Solid Models: Spatial Metadata in Ethnoarchaeology".

Coffee Break (15:45-16:15 p.m.)

- **Andrea MANZO, Cinzia PERLINGERI** (IUO, Istituto Universitario Orientale, Naples)
"Genetic Programming and Archaeology"

- **Rodolfo FATTOVICH** (IUO, Istituto Universitario Orientale, Naples)
"From Artifact to Landscape: A Theoretical Approach to a Simulated Reconstruction of Historical Processes in Ancient Ethiopia"

- **John KANTNER** (Georgia State University, Department of Anthropology and Geography)
"Digital Technologies and Prehistoric Landscapes in the American Southwest"

November 3, Saturday

SESSION 5 (9:00-12:00a.m.), T. Sever, presiding

- **Kathryn A. BARD, Michael C. DI BLASI, Magaly KOCH** (Boston University), **Rodolfo FATTOVICH, Andrea MANZO, Cinzia PERLINGIERI** (IUO, **Istituto Universitario Orientale, Naples**), **Maurizio FORTE,** (CNR-ITABC, Rome), **Stefano TILIA** (TREERRE srl, Rome)
"The BU/IUO Archaeological Project at Bieta Giyorgis (Aksum), Ethiopia: Archaeological Results and Computer Applications"

- **Sheldon SANDLER** (Department of Archaeology, Boston University)
"Archaeological Subsurface Site Reconstruction Using Computer Processing of GPR Data"

Coffee Break (10:15-10:45 a.m.)

- **Paolo MOZZI** (Dipartimento di Geografia, University of Padua)
"Geo-Landscapes, Cultural Landscapes, Archaeo-Landscapes: Integrated Dynamics and Digital Analysis"

- **Francisco ESTRADA BELLI** (Vanderbilt University)
"The Power of GIS and Remote Sensing: Multi-Scalar Spatial Modeling and Analysis of Settlement Data in SE Pacific Coastal Guatemala and the Southern Maya Lowlands"

SESSION 6 (13:30-17.30), P. R. Williams, presiding

- **Eileen VOTE** (Department of Computer Science, Brown University)
"A New Methodology for Archaeological Analysis: Using Visualization & Interaction to Explore Spatial Links in Excavation Data"

- **Frederic LEYMARIE** (SHAPE Lab, Brown University)
"Progress Report on Shape Modeling of 3D Archaeological Artifacts and Fragments with a Discussion of Potential Applications in Digital Archiving"

- **Conclusions and plenary session**

PROCEEDINGS

Welcoming Remarks

Conference on "The Reconstruction of Archeological Landscapes Through Digital Technologies"

Jon Westling

November 1, 2001

The remote sensing of the remote past is perhaps the most important scientific frontier in archaeology, and we are privileged to have at this workshop some of the scholars who are leading the way. Archaeologists have always sought to make the remote a little more familiar. The task for this workshop is to help make the various techniques of remote sensing a more familiar part of both fieldwork and interpretation.

In reconstructing the lives of ancient peoples, we face numerous difficulties. The first, and perhaps most persistent, is the scarcity of human records. Instead, we find ourselves examining ruins—some quite elaborate, others but ambiguous traces of human actions.

Archaeologists, of course, are not the only people to be intrigued by the physical remains of the human past. Those remains have often excited the imagination of poets and amateurs, who devise startlingly clear pictures of what must have been—and, of course, never was.

In 1762 the Scots poet James Macpherson presented what he claimed to be a translation of third century manuscripts of the work of the legendary Celtic poet, Ossian. Macpherson's fraud was suspected by some, including Samuel Johnson, but accepted—indeed celebrated—by many, who found in it exactly the image of the ancient Celts that they had expected. Today, we are not likely to be moved by Macpherson's vision of the Celts in lines such as:

> Her hair flew slowly on the blast,
> that rushed unfrequent from the hill.
> Not every experiment in remote sensing, it appears, is successful.

More attentive to the evidence at hand was Josiah Priest who in 1833 published his *American Antiquities and Discoveries in the West Being an Exposition of the Evidence that an Ancient Population of Partially Civilized Nations Differing Entirely from Those of the Present Indians Peopled America Many Centuries Before its Discovery By Columbus*. Priest's book offered what many Americans thought was a highly plausible account of the landscape—the mounds and other ancient earthworks scattered throughout the Ohio and Mississippi Valleys—and his book went through numerous editions. Unfortunately, some of the details of his work have not been confirmed by later investigators. The Roman fort that Priest found at Marietta, Ohio, for example, has succumbed to more mundane explanations, and few now follow Priest in characterizing Native Americans as descendants of the Scythians and ancient Israelites.

Trapped in webs of our own making, we are always at least a little susceptible to conjectures about the past that turn out to be airy nothings. Nor is it just romantics like James Macpherson or fantasists like Josiah Priest who mislead us. Much scholarship about ancient peoples bears an uncomfortable resemblance to the spirit of Macpherson's and Priest's conjectures. Unrestrained by respect for fact and unhinged by speculative theory, some ruminations on the human past sound a lot like "the blast, that rushed unfrequent from the hill," if that blast had been written in academic jargon and leaned heavily on the passive voice.

The program of your conference is a testament to the extraordinary scientific tools available to archeologists today. Ground Penetrating Radar sifts through centuries of habitation to reveal traces that may cohere in the landscape of one age. High-resolution optical sensors, such as those deployed on the IKONOS satellite, can pierce layers of development to reveal swamps that the ancient Mayans may have farmed.

Satellites and electronic imaging open up entirely new fields of data for archeologists, historians, and anthropologists, but present new epistemological problems as well. Macpherson and Priest conjured visions from suggestive scraps of the past; we, on the other hand, are inundated with detail of the physical landscape, much of which is irrelevant or misleading. Contemporary archeologists need to sift through their databases with the kind of patience and skepticism that kept their predecessors hunched over trays of sand for months on end.

Furthermore, the pursuit of the past is hampered today by obfuscations our ancestors could not have imagined. Theoretical fads blow up on the academic landscape like sudden dust storms, clouding the real evidence of the past with vague terms drawn from the lexicon of literary criticism. I would be delighted if there were a satellite that could penetrate opaque prose, or a computer program that could pick insights from the debris of abandoned constructs, but we must still rely on the kind of sound judgment that guided the great scholars of the past.

This conference presents a splendid opportunity to consider ways in which scholars of the human sciences and experts in digital technology can collaborate, and I am deeply grateful to you for joining a multidisciplinary effort of great promise. I would also like to thank the J. M. Kaplan Fund and the National Research Council of Italy for so generously supporting your endeavors. I believe this conference has the potential to make an important contribution to our understanding of the landscapes of civilizations hidden in the depths of time.

Preface

Proceedings of the first Italy-USA Workshop on "*The Reconstruction of Archaeological Landscapes Through Digital Technologies*" (Boston, Nov. 2001) are published in this volume.

First goal of this workshop was, as it is self-evident by its title, the checkout of the opportunities given by the most recent technologies, firstly ICT, to reconstruct archaeological landscapes by means of interdisciplinary and multidisciplinary approach to actual case studies; also methodological and epistemological aspects of this approach were discussed.

The second goal, no less important than the first, was to study and to appreciate archaeological landscapes as a cultural resource of the territory, understood as the place of all the human activities, where during and through the time cultural and historical phases have been developed: i.e. to study the archaeological landscapes as factor of cultural and social identity.

The papers here published, that are devoted to achieve the above mentioned goals, are the work of more or less young scientists, of different academic rank, and show an ever growing attention to the exploitation of ICT, not merely oriented to the data management, but as a decisive support to produce new historical knowledge in the field of archaeological and topographic research.

These proceedings, whose publication has been affected by some delay in comparison with its expected date, offer, to conclude, a contribution to the implementation of new cognitive models in archaeology. They will be surely discussed, together with other recent studies, on the occasion of the second Italy-USA Workshop that will take place at Rome next November 2003.

Rome, May 12, 2003

Salvatore Garraffo
Director of Institute of Technologies
Applied to the Cultural Heritage
CNR- Montelibretti (Rome)

The Joint Archaeological Project at Bieta Giyorgis (Aksum, Ethiopia) of the Istituto Universitario Orientale, Naples (Italy), and Boston University, Boston (USA): Results, Research Procedures and Preliminary Computer Applications

Kathryn A. Bard[1], Michael C. DiBlasi[1], Magaly Koch[1], Livio Crescenzi[2], A. Catherine D'Andrea[3], Rodolfo Fattovich[4], Andrea Manzo[4], Cinzia Perlingieri[4], Maurizio Forte[5], M. Scott Harris[6], Gerald H. Johnson[7], Stefano Tilia[8], Bartolomeo Trabassi[5]

[1]Boston University, Boston, USA
675 Commonwealth Avenue, Boston, MA 02215 -617/358-1662 - fax: 617/353-6800
kbard@bu.edu; mdib@bu.edu

[2]Soprintendenza Archeologica per il Lazio, Roma, Italy

[3]Simon Fraser University, Burnay, BC, Canada

[4]Istituto Universitario Orientale, Napoli, Italy
P.zza S. Domenico Maggiore, 12 - 80134 Napoli
fattovich@spinweb.it; ybeyene@iuo.it

[5]CNR - ITABC, Roma, Italy
Via Salaria, km.29,300, c.p.10,
Monterotondo St., 00016, Roma. Tel. +39.06-90672.721
maurizio.forte@mlib.cnr.it

[6]Carolina Coastal University, Conway, SC, USA

[7]The College of William and Mary, Williamsburg, VA, USA

[8]TREERRE s.r.l
Via Agri 11, Roma
stilia@tin.it

Abstract. Boston University (BU), Boston (USA) and the Istituto Universitario Orientale, Naples (Italy) are conducting a joint research project at Bieta Giyorgis near Aksum in Tigrai (northern Ethiopia) under the direction of Kathryn Bard (BU) and Rodolfo Fattovich (IUO). The project, in progress since 1993, includes research in archaeology, paleoethnobotany, archaeozoology, ethnoarchaeology, history, geology/geomorphology, palynology, as well as systematic topographic mapping and conservation. At present, the project involves the application of advanced digital technologies for a detailed reconstruction of the archaeological landscape, analysis and classification by GIS and remote sensing, and interpretation and presentation of the results through virtual reality and visual information systems. The innovative aspect of the project is a multidisciplinary approach, starting with the acquisition of the data on the ground and, then, creating predictive classified maps of the geomorphologic and anthropic landscape, including all paleoenvironmental factors. In this paper the major results of the project will be outlined, and the main computer applications to data recording and analysis will be presented.

1. Introduction

Aksum in Tigrai (northern Ethiopia) was the capital city of a powerful kingdom, which dominated the southern Red Sea in the 1st millennium AD, and was an important commercial partner of the Roman and Byzantine empires.[1] The town is located about 22 km to the west of Adwa, at an average elevation of 2200 m (Figure 1). The hills of Bieta Giyorgis and Mai Qoho dominate the present town to the northwest and east, respectively. Some hills, with an average elevation ranging between 2406 m and 2289 m, form a crown delimiting a roughly circular plain, gently sloping from north to south and southeast, about 10 km in diameter. From this plain a number of seasonal rivers drain to the Mareb-Gash valley to the north, Shirè to the west, Simien to the south, and Tembien to the southeast. Thus, the region of Aksum can be described as a fertile plain delimited by a crown of hills, forming a natural border, and connected to the surrounding regions through a radial system of river valleys. Aksum occupies

[1] Munro-Hay 1991; Brakmann 1992; Burstein 1997; Phillipson 1998.

the central area of the plain, while Bieta Giyrogis and Mai Qoho occupy the northeastern sector of it (Figure 1).[2]

This region corresponds to the archaeological area of Aksum, which can be divided into three main zones with a different density of sites: A) Core zone, including the town of Aksum, the hills of Bieta Giyorgis and Mai Qoho, and the plateau of Abba Liqanos. B) Peripheral zone, including the sloping plain to the south and north of Aksum. C) Marginal zone, including the hills, which surround the plain of Aksum.[3]

Despite surveys and excavations have been conducted at Aksum since the early 19th century,[4] systematic off-site and landscape archaeological investigations[5] began only in 1997 as part of the research program of the Joint Archaeological Expedition of the Istituto Universitario Orientale (IUO), Naples (Italy), and Boston University (BU), Boston (USA) at Bieta Giyorgis.[6]

In this paper the major results of the IUO/BU project will be outlined, and the fieldwork procedures and main computer applications to data recording and analysis will be presented. The aim of the paper is to briefly explain how the project was designed at the different stages of implementation from the collection of data in the field to the reconstruction of the ancient landscape.

2. The Project

2.1 Description of the project

Figure 1: Satellite image with a general view of the region of Aksum (Landsat/Spot)

[2] See Fattovich , Bard, Petrassi & Pisano 2000.
[3] Fattovich , Bard, Petrassi & Pisano 2000.
[4] See Brandt & Fattovich 1990; Fattovich 1992.
[5] For an introduction to off-site archaeology and landscape archaeology see e.g., Miller & Gleason 1994; Ashmore & Knapp 1999.
[6] See Fattovich , Bard, Petrassi & Pisano 2000.

The IUO/BU project, in progress since 1993 under the direction of Rodolfo Fattovich (IUO) and Kathryn Bard (BU), includes research in archaeology, paleoethnobotany, archaeozoology, ethnoarchaeology, history, geology, geomorphology, palynology, as well as systematic topographic mapping and conservation. An important goal of the project is to investigate the origins and development of Aksum within its environmental setting.[7] The project was initially aimed at testing the hypothesis, based on traditional Ethiopian sources, that Bieta Giyorgis was an area of early development at Aksum.[8] At present, the project involves the application of advanced digital technologies for a detailed reconstruction of the archaeological landscape: analysis and classification by remote sensing and GIS, and interpretation and presentation of the results through virtual reality and visual information systems.[9]

A relevant aspect of the project is a multidisciplinary approach, starting with the acquisition of the data on the ground and, then, creating predictive classified maps of the geomorphological and anthropic landscape, including all paleoenvironmental factors.

Eight field seasons have been conducted in 1993, 1994, 1995, 1996, 1997, 1998, 2000, and 2001. In 1996 the University of Cagliari (Italy) conducted a preliminary geophysical survey at Bieta Giorgis in collaboration with the IUO/BU project under the direction of Prof. Roberto Balia and Antonio Vernier. In 1999 fieldwork was suspended because of the conflict at the Ethiopian-Eritrean border.[10]

2.2 Area of investigation

The area under investigation encompasses the upper part of Bieta Giyorgis, a hill dominating Aksum to the northwest, and covers a surface of about 4 sq. km.

The occurrence of ancient sites on the top of the hill was already recorded in 1906 and 1974 by German, Italian, and American expeditions. These sites included an elite residential area (Ona Nagast) in the southwestern sector of the hill, a cemetery with stelae (Ona Enda Aboi Zewgè) in the central sector, and a quarry and two Aksumite churches (BGS, BGI) on the southern slope of the hill. Two rock inscriptions were recorded to the

[7] See Fattovich 1997; Bard., Coltorti, Diblasi, Dramis and Fattovich, 2000. Funding for this project was provided by the Ministry of Foreign Affairs, Rome; the National Council for the Research (CNR), Rome; the Ministry for the University, and Scientific and Technological Research (funds 60%), Rome; the National Geographic Society, Washington DC.
[8] See Fattovich & Bard 1993, 1994.
[9] See Forte, Bard, Fattovich, Focillo, Manzo & Perlingieri 2001
[10] See Bard 1995, 1998; Bard & Fattovich 1993a, 1993b, 1995, 1997; Bard, Di Blasi, Fattovich, Manzo, Perlingieri & Crescenzi 1996; Bard, DiBlasi, Fattovich, Manzo, Perlingieri, D'Andrea & Chaix 1998; Bard, Fattovich, Manzo & Perlingieri 1997; Fattovich 1994, 1995; Fattovich & Bard 1993, 1994, 1995, 1996, 1997a, 1997b, 1998; Ogundiran 1995; Balia & Vernier 1997.

northeast of Ona Nagast, and on the southeastern slope of Bieta Giyorgis, respectively (Figure 2).[11]

Beginning in 1993, the IUO/BU Expedition conducted excavations at the sites of Ona Nagast (ON) and Ona Enda Aboi Zewgè (OAZ).

Ona Nagast is located on a slope about 700 m to the west of OAZ. The site is over 12 ha in area. Part of the ancient settlement is covered by a modern village, and to the west of the site is a large rock-cut cistern of unknown age, most likely ancient (Ela Nagast). According to local traditions, this site was the residence of the earliest Aksumite kings.

Ona Enda Aboi Zewgè extends over an area of about 10 ha, and is partly covered by a modern village. Over 100 stelae are visible on the surface, including roughly hewn monoliths and ones with rounded tops, and are up to 10 m high.

In 2000, the full-coverage systematic survey of the top and slopes of Bieta Giyorgis hill identified many sites, including rock-shelters, open air sites, stele quarries, and sites with rock inscriptions. Four sites were also selected for test excavation: Baati Asba (BA), a rock-shelter on the northwestern slope of the hill; Gunda Nebri (GN), an open air site on the northeastern ridge of the hill; Tukul Emeni (TE), an open air site in the northeastern sector of the hill; and Guadguad Agazien (GA), an open air site on the northern edge of the hill.

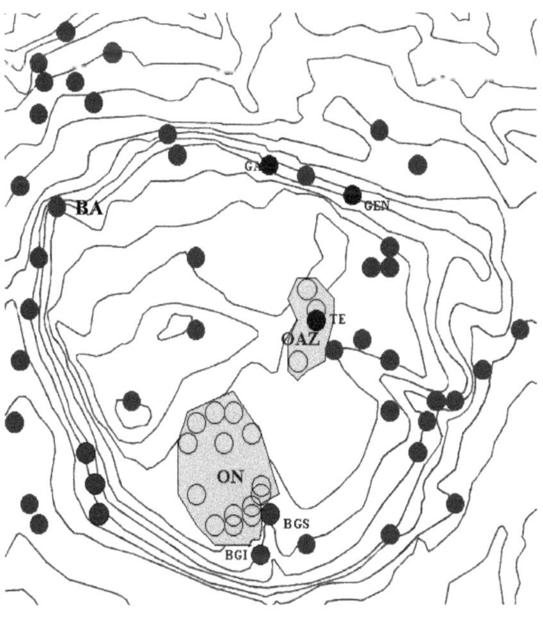

Figure 2: 2-D map with site location (ON, OAZ, TE, BA, GN, GA)

[11] See Littmann, Krencker & von Lüpke 1913, II/1: 125-127; Michels 1994; Ricci 1974, 1990; Ricci & Fattovich 1988; Drewes & Schneider 1967: 96; Bernard, Drewes & Schneider 1991: 292.

Preliminary investigations of features not associated with sites on the top of the hill were conducted during the 2000 and 2001 field seasons. These investigations were aimed at recording possible ancient man-made features for a reconstruction of the ancient rural landscape at Bieta Giyorgis.

2.3 Results

Major results of the expedition were:
1) To establish a precise cultural sequence of Aksum, and a firmer chronology for the phases of development of the kingdom.
2) To outline the subsistence economy in Aksumite times.[12]
3) To provide more information about the net of external contacts of the kingdom, and its changes through the time.[13]

Four main culture history periods were identified: 1) *Pre-Aksumite Period*; 2) *Proto-Aksumite Period*; 3) *Aksumite Period*; 4) *Post-Aksumite Period.* The Aksumite Period has been further divided into four phases: Aksumite 1 (Early Aksumite), Aksumite 2 (Classic Aksumite), Aksumite 3 (Middle Aksumite), and Aksumite 4 (Late Aksumite).

Some late prehistoric evidence was recorded at Baati Asba, where pottery has been collected that exhibits the range of variation found in the Late Stone Age/Pastoral Neolithic of East and Northeast Africa. It is unrelated to Pre-Aksumite, Aksumite, and Post-Aksumite pottery traditions.

Pre-Aksumite evidence (ca. 700 - 400 BC) has been recorded at OAZ, ON, GN, and sites 6, 9, 44, 52, 56. This evidence is mainly related to ceremonial areas. Faunal remains from ON include cattle and caprines.

Proto-Aksumite evidence (ca. 400 - 150 BC) has been recorded at OAZ, ON, and possibly sites 1, 4, 19, 23, 36, 44, 51. The Proto-Aksumite evidence includes man-made stone platforms associated with roughly hewn stelae and pit-graves at OAZ, and some traces of a monumental building at ON. Plant remains from domestic refuse at ON include free-threshing wheat, emmer wheat, barley, lentil and flax. Faunal remains include cattle, caprines, and dog.

Evidence dating to Aksumite times (ca. 150 BC - AD 700) occurs at OAZ, ON. Two churches (BGS and BGI), stele quarries, and most surveyed sites also date to this period. A lithic workshop dating to Aksumite 1 times was also identified in the western sector of the site.[14] Plant remains from Aksumite contexts at Bieta Giyorgis include emmer wheat, free-threshing wheat, barley, tef, lentil,

[12] Plant remains were examined by A. C. D'Andrea. Animal remains were examinred by L. Chaix.
[13] Imported materials were examined by A. Manzo, see Manzo 1996, n.d.
[14] Usai 1997.

grape, and linseed. Animal remains include domestic cattle and sheep/goats.

Post-Aksumite evidence (from ca. AD 700) has been recorded at OAZ, ON, BGS, BGI, GA, and sites 8, 9, 12, 13, 14, 15, 18, 23(?), 24, 26(?), 31, 35, 37, 39, 46, 47, 50, 51, 52, 54, 57, 61, 62, 63, 64. At present, the Post-Aksumite evidence can be described as scattered compounds and possibly two churches. Post-Aksumite sites have produced free-threshing wheat, barley, probable tef, linseed, and finger millet. Domestic cattle and sheep/goats occur in the deposits of this phase.

Finally, imported ceramics and glass point to contacts with central Sudan and Nubia from the 4th century BC to the 8th century AD, the Mediterranean regions from the 1st century BC to the 7th century AD, and the Muslim regions of the Near East since the 7th century AD.

3. Fieldwork procedures and data analysis

3.1 Fieldwork procedures

The excavations were conducted in "Excavation Units" (EU), 10x10 meters in area, in order to combine an extensive horizontal excavation with a stratigraphic excavation.[15] Each unit was divided into a grid of 25 squares, 2x2 meters in area, in order to maintain a topographic control of the finds. The 2x2 meters squares were recorded by means of an alpha-numerical matrix with A to E along the west-east axis, and 1 to 5 along the north-south axis (from A1 to the NW to E5 to the SW). Each EU was called with the abbreviated name of the site (e.g., Ona Nagast = ON) and Roman numerals (e.g., ON I, ON II, etc.).

Horizontal excavations in each EU were systematically recorded with plans at a 1:50 scale in order to show the spatial relationship among the single features at the different levels. Plans of specific details of the excavation were drawn at a 1:20 scale.[16]

Stratigraphic excavations were conducted in conformity with the "Stratigraphic Unit" (SU) procedure suggested by E. Harris.[17]

"Soil strata" (i.e., discrete deposits resulting from the combined action of natural and cultural factors) and "features" (i.e., all man-made spatially organized structures), as well as their vertical and horizontal relationship, were regarded as the basic components of the stratigraphic sequence. Soil strata over 20 cm in thickness were excavated distinguishing artificial levels, ca. 5-10 cm deep, in order to follow their possible internal stratigraphy.

[15] Excavations at the single sites were conducted under the supervision of K. A. Bard. M. C. Di Blasi, R. Fattovich , A. Manzo, C. Perlingieri, D. Usai.

[16] Plans of excavation were drawn by K. A. Bard, L. Crescenzi, M. C. Di Blasi, A. Manzo and C. Perlingieri.

[17] Harris 1989.

Each SU was described on a form recording: 1) Name of the site. 2) Excavation Unit number. 3) Grid number (2x2m squares). 4) SU number. 5) SU description. 6) Elevations at the top and bottom of the SU. 7) Type and density of artifacts and ecofacts. 8) Vertical and horizontal position in the stratigraphic sequence.

The (traditional) profiles of each wall of the trenches were traced to record the vertical correlation between strata and features.[18] The sequence of SUs in each EU was also represented with an "Harris Matrix" describing each unit in order to make easily understandable the relationship among walls, wall collapse, room fill and floors.

Both horizontal and vertical excavations were systematically recorded with photos at a different scale of detail. A monopode was used to take vertical photos of some EUs in order to generate a photo-mosaic of the excavated area.

In 2000 a systematic surface survey was conducted on foot with a grid system superimposed on an enlarged aerial photo base map of the region (1:7800). Site locations were recorded by means of GPS. Data from the survey have been complied in a GIS database for computerized analysis and interpretation of land use dynamics from the earliest periods of settlement to recent times.[19]

All artifacts and ecofacts, together with samples of charcoal for radiocarbon dating and soil for laboratory analysis, were collected and recorded. The color of the soil strata, pottery and glass was described by means of the Munsell Soil Color Charts, 1994.

Artifacts from each SU were recorded on the basis of their main attributes (category, shape, size, material, etc.), and photos were taken for a visual record of them.[20] Metal artifacts were divided into two main groups: iron and copper/bronze artifacts. A more specific identification of the metal was not possible because of Ethiopian regulations, which do not allow the export of antiquities from the country. Stone artifacts were identified in the field with the support of the geologists participating to the expedition.

The pottery was described by means of an open attribute list with ca. 50 descriptive fields, specifically created for the Aksumite ceramics, and all attributes were recorded on preprinted forms to facilitate their quantitative analysis.[21]

[18] Profiles were traced by M. C. Di Blasi, L. Crescenzi, A. Manzo, C, Perlingieri, D. Usai.

[19] The survey was conductet by M. C. Di Blasi and A. Manzo.

[20] Arifacts were described by K. Bard (minor finds), R. Fattovich (minor finds), A. Manzo (imported materials, minor finds, pottery), D. Usai (lithics). Photos of the artifacts were taken by K. A. Bard.

[21] The ceramics were studied by C. Perlingieri.

3.2 Data analysis

3.2.1 Cultural sequence

The cultural sequence at Bieta Giyorgis was established on the basis of the stratigraphic sequence in the single EUs, and the typological sequence of the ceramics.
The stratigraphic sequence in the EUs was interpreted generating two different representations along with the Harris' Matrix and side profiles:

a) Compressed profiles of the Excavation Units aimed at representing in one profile the vertical and horizontal relationships of the SUs, with the possible calibrated C14 dates of the SUs.
b) 3-D models of constructed features in order to show the different phases of construction.[22]
In such a way, the diachronic (stratigraphic) correlation of phases of wall construction, floors, room fill, and wall collapse was outlined.

The cultural "phasing" of the SUs was based on pottery occurrences (in terms of types and frequencies) in the units.

Quantitative procedures with the application of SPSS were followed in the analysis of the ceramics in order to obtain a more rigorous classification of wares and types and to establish the cultural phases.[23]

The wares were classified on the basis of the association of color and fabrics, and three main ware groups were distinguished: Orange Ware (OW), Brown Ware (BW), and Grey Ware (GW). In turn, these major wares were divided into classes on the base of the fabric characteristics and colors. Moreover any class can comprise several variants.

The attributes of all fragments and vessels were included into a database to infer their frequency and association in the single SUs. Specific sets of attributes (i.e., manufacturing technique and style) were identified for the single SU, and cultural phases were inferred from the changes in "types," and their relative frequencies in the stratigraphic sequence.[24]

A seriation of the SUs, based on coefficients of similarity/dissimilarity in the frequency of all morphological attributes of the pottery in the single SUs, was conducted and was represented with a matrix in order to test the cultural sequence in the excavation.

The phases resulting from these analyses were represented on a Harris Matrix, and were compared with the absolute chronology based on radiometric dating and typological identification of imported materials in order to identify possible discrepancies in the sequence.

[22] These models were generated with Autocad.
[23] Perlingieri n.d.
[24] See Bard, Fattovich, Manzo & Perlingieri 1997; Fattovich, Bard, Petrassi & Pisano 2000.

3.2.2. Imported materials

SPSS were used for a quantitative analysis of imported materials for a more precise reconstruction of the development of Aksumite long-distance trade.[25]

In this analysis imported ceramics, glass and stone vessels were considered as one statistic population. This approach, unusual in archaeology, is more consistent with the mental templates of ancient people, as we can infer from textual evidence.[26]

The quantitative analysis was conducted with a frequency procedure on diagnostic fragments.[27] Although classes of materials with a frequency <10% can be overestimated with this procedure,[28] this procedure was adopted to assess the trends in the single phases and areas, rather than to calculate the number of the vessels. The frequency of the materials for each phase was calculated combining different SUs in order to reduce the variability due to a functional difference.[29] The frequency of pottery and glass was always tested with the Chi-square procedure.[30] The Chi-square test suggested that the differences in the frequency of the materials in the single Sus are a consequence of sampling (always less than 5%), and have a level of confidence always higher than 95%.

The "equivalent vessel estimate" (EVE) was not used because of the great number of fragments from Bieta Giyorgis,[31] and the quantification by weight or simple potsherd counting could emphasize the presence of amphorae and big containers and underestimate the fine pottery and glass.

4. Absolute chronology

The absolute chronology was established on the basis of about 50 imported artifacts (glass and ceramics), five coins, and 35 calibrated radiocarbon dates (according to cal. 3.0).[32]

Imported ceramics and glass provided a *terminus post quem* to date the single Sus.[33] This means that each SU is contemporary or later (never earlier) than the time when the imported ceramics and glass started to be produced. This evidence suggested the following chronological range for the single periods and phases: Proto-Aksumite Period - ca. 300-100 BC. Aksumite Period – ca. 100 BC – AD 800: Early Aksumite (Aksumite 1) Phase - 100 BC

[25] Imported materials were examined by A. Manzo.
[26] See e.g., *Satyricon* by Petronius Arbiter where glass and metal vessels are distinguished only on their different resistence and value, and glass was regarded as more estetically appealing than precious metal; see Manzo nd.a.
[27] See Doran & Hodson 1975: 105; Drennan 1996: 65-73.
[28] Orton 1982.
[29] Tomber 1993: 150.
[30] Drennan 1996: 187-191.
[31] See Orton 1982; Tomber 1993: 148-150.
[32] Stuiver & Reimer 1993.
[33] Manzo 1996, n.d.

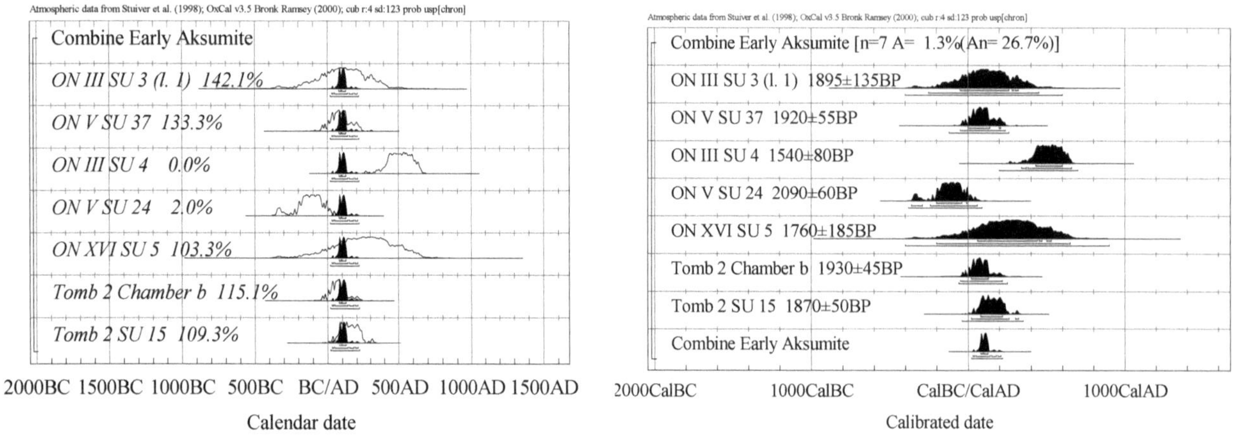

Figure 3: Example of calibrated radiocarbon dates from Bieta Giyorgis

– AD 100/200; Classic Aksumite (Aksumite 2) Phase - AD 100/200-300; Middle Aksumite (Aksumite 3) Phase - AD 300-400; Late Aksumite (Aksumite 4) Phase - AD 400-800; Post-Aksumite Period - (AD 800-1500).

Thirteen coins were collected at Ona Nagast. Ten of them could be identified, but only five were found in undisturbed Sus. These coins provided a *terminus post quem* to date the Stratigraphic Units where they were collected, and suggested that the Middle Aksumite Phase can be dated up to the mid-6th century AD, and the Late Aksumite Phase covered the late 6th – mid-7th centuries AD.

Forty-one charcoal samples from SUs ascribable to different periods and phases were processed with OxCal. 3.0, so far. One sample from a soil stratum just beneath the topsoil resulted intrusive and modern in age, and was excluded from the evaluation of the chronology.

The radiometric dates suggested the following sequence (Figure 3): Proto-Aksumite Period; 400-150 BC; Aksumite Period, 150 BC-AD 700: Early Aksumite (Aksumite 1) Phase, 150 BC-AD 150; Classic Aksumite (Aksumite 2) Phase, AD 150-350; Middle Aksumite (Aksumite 3) Phase, AD 350-550; Late Aksumite (Aksumite 4) Phase, AD 550-700. Post-Aksumite Period, from 700.

The following absolute chronology of these periods has been established on the basis of 35 calibrated radiocarbon dates, ca. 50 imported artifacts (glass beads and vessels, and amphoras), and some coins: Proto-Aksumite Period - ca. 400-150 BC; Aksumite Period - ca. 150 BC - AD 700; Aksumite 1 - ca. 150 BC - AD 150; Aksumite 2 - ca. AD 150 - 400/450; Aksumite 3 - ca. AD 400/450 - 550; Aksumite 4 - ca. AD 550 - 700; Post-Aksumite Period - from AD 700.

5. Genetic Algoritms

A preliminary application of Genetic Programming, a correlation system already tested in the area of biomedicine and genetics, to the archaeological data from Bieta Giyorgis is in progress since 1999.[34] This is the first application of this method in archaeology.[35]

Genetic Programming consists of automatic calculations to combine different classes of data, which were previously organized with statistic methods. The input of the system is formalized databases. The output is a correlation between different variables expressed in a mathematical form. In such a way, the factors affecting changes in the material culture can be better detected, and possible rules, which may excape to less sophisticated analyses, can be recognized.

The final steps of this research will be the simulation of changes in Aksumite pottery as a consequence of induced changes in the external natural and cultural environment.

6. Mapping

6.1 General topography

In a first stage of the project the survey of the area on the top of the hill including ON and OAZ was conducted with a mechanic theodolite. The EUs, as well as the contour lines and man-made landscape infrastructures were mapped at a scale 1:5000.[36] A 3-D CAD map of the area based on this preliminary survey was also created.[37]

[34] The application of Genetic Algoritms is conducted by A. Manzo and C. Perlingieri in collaboration with the CNR-IRSIP (Istituto di Ricerca sui Sistemi Informatici Paralleli), Naples (Italy).
[35] See Manzo & Perlingieri in this volume.
[36] This survey was conducted by L. Crescenzi.
[37] The survey data were elaborated by F. Cioffi.

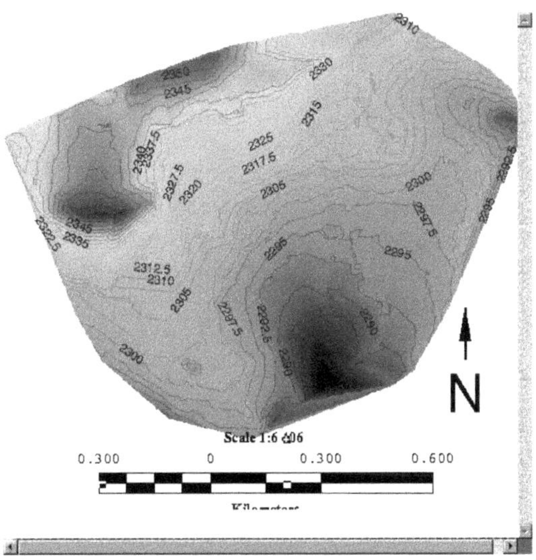

Figure 4: Surveyed area with differential GPS.

Beginning in 1997, remote sensing analyses of aerial photos and satellite images as well as GIS were used to complement the topographic base of Bieta Giyorgis in order to generate archaeological maps and predictive models of site distribution to be tested in the field.[38] These applications were initially conducted on the 1:50,000 map of the Aksum area,[39] as well as the Landsat TM (30 m of resolution) and Spot XS (20 m of resolution) satellite images, and the aerial photos (ca. 3 m of resolution) of the Aksum area. We expect to obtain an IKONOS image with high spatial resolution (1 m panchromatic, and 4 m multispectral bands) in the near future.

At present, a detailed survey of the top of Bieta Giyorgis hill with TLS and differential GPS is in progress.[40]

In 2001 a network of fixed stations was established with TLS in order to obtain a complete topographical coverage of Bieta Giyorgis hill. This procedure was integrated with a differential GPS in order to geo-referencing any point with a high degree of precision in.

Application of differential GPS was to:
1. Rectify earlier maps not obtained with differntial GPS in order to generate a new cartographic and GIS database.
2. Generate a micro-topographic DEM of the top of Beta Gyorgis hill.
3. Define training regions for spectral classifications of SPOT XS and Landsat TM satellite images.

The accomplishment of these three aims has substantially improved the archaeological investigations *in situ*.

The creation of a micro-DEM on the top of Bieta Giyorgis by a detailed survey (ca. 30,000 elevation points were recorded and interpolated) has filled the gaps of the micro-relief map at a scale of 1:25,000 and has increased the visibility of the local morphology, man-made field terraces, and some paleoenvironmental evidence.

Finally, the survey of the training regions at a 1 cm scale has provided a more precise supervised spectral classification (Figure 4).

6.3 Remote sensing

Remote sensing and pedology could further enhance existing knowledge about the landscape characteristics on and around Bieta Giyorgis.[41] The objectives of these studies included:

(1) Relating different land-cover types to the spectral signatures found with Landsat ETM+ for an area in and around Bieta Giyorgis;
(2) Identifying characteristics of the soil land-cover types found on Bieta Giyorgis;
(3) Assessment of a possible soil study in order to be able to relate soil characteristics with archaeological findings.

A Landsat Thematic Mapper TM image of 22nd of November 1984 was processed using standard techniques, such as linear stretching and filtering, to enhance the color contrast of land surface materials as well as linear (structural) features. Hardcopies of the area of interest were printed at different scales, ranging from coarse (1:150000) to fine (1:50000) scale. The purpose was to study (a) the landscape components and tectonic features at a general scale, and (b) to use the image prints for conducting a detailed survey at a local scale in Bieta Giyorgis. All image products were geo-referenced using the WGS 84 datum, and a UTM grid with a spacing of 500 m was overlaid on the images to facilitate geo-location of land-cover types later in the field.

A preliminary field survey was conducted by selecting different land-cover types (soil, rock and vegetation) in the region of Bieta Giyorgis (Figure 1). This included taking 15 land-cover types, and when a land-cover type was represented by a bare soil surface, a surface sample was collected. Furthermore, viewpoints with photos and control points were collected in order to give additional information about the respective land-cover types. All these points have been geo-referenced in order to include their position in a GIS database.

The land-cover types are represented by selecting homogenous areas of a minimum of 30 by 30 m (dimension of 1 pixel in the Landsat image). However, it was more suitable to select homogenous areas, which would include several pixels. For main land-cover surfaces, such as bare soils, rock outcrop and vegetation, the largest possible area was marked by a central reference point, and the perimeter of the respective land-cover was marked with further reference points. In order to be able to make an assessment for a future soil survey,

[38] See Fattovich, Bard, Petrassi & Pisano 2000: 39-41.
[39] This is the most detailed map of the region presently available.
[40] The survey with TLS ands differential GPS were conducted by S. Tilia and M. Forte, respectively.
[41] Remote sensing and pedological analyses were conducted by Magaly Koch and Thomas Schmid, respectively.

different soil profiles were taken at selected points related mainly to archaeological sites in the Bieta Giyorgis area. Samples were collected for future analysis.

The land-cover types selected in the field were used as training sites for the classification of a TM sub-image enclosing Beita Giyorgis. Four different classifiers were tested, namely the *Mahalanobis Distance, Minimum Distance, Maximum Likelihood, and Paralellepiped* classifier (Figure 2). The classification results in Figure 2 show that land-cover types (classes) were recognized by the classifiers with various degrees of accuracy. For instance, all four classifiers performed more or less well in identifying and mapping the spatial distribution of the eucalyptus class. However, the colluvium and red soil classes were mainly detected by the mahalanobis and minimum distance classifiers and to a lesser degree by the other two classifiers. In addition, some class confusion seems to occur between the colluvium and red soil class, indicating that the TM sensor's resolution is insufficient to spectrally and spatially separate these two land cover types. The same occurs with the dark cultivated soil class that includes pixels with shadowing effects caused by the northern escarpment zone of the hill. The central plane of Beita Giyorgis shows some characteristic light soils (high reflectance values) which are spectrally similar to the pyroclastic material (trachyte; light blue class) found in neighboring hills (e.g., Mai Qoho). Here again the classifiers differ in the way this class is mapped. The maximum likelihood and parallelepiped seemed to perform better in detecting this class than the other two classifiers. In summary, these examples show that despite TM's spatial and spectral limitations, land-cover types as identified in the fields were mapped by the classifiers with different degrees of accuracies. This suggests that a higher resolution satellite sensor such as IKONOS will without doubts be better suited for precise land-cover mapping than Landsat TM alone.

Nevertheless, the coarse-scaled TM images used in the field gave a first general description of the geologic and tectonic features of the study area and surrounding region. A preliminary assessment shows that Bieta Giyorgis forms part of an array of three hills aligned with a NE-SW direction. These hills are made up of igneous rock (plutonites) and represent denuded domes that are uncovered by erosion. Parallel to this array of hills is an important fault zone that was probably responsible for uplifting Bieta Giyorgis and the surrounding hills in relation to the hills (domes) lying SE of the fault zone (e.g., Mai Qoho). The fault zone strikes between the town of Aksum and May Koho.

In the TM image it is possible to discriminate between three types of plutonic domes:

1) not breached domes where erosion has not yet uncovered the igneous rocks, but the shape of the dome is clearly visible;
2) breached domes (Mai Qoho) where erosion has started to disclose pyroclastic rocks;
3) denuded domes (Bieta Giyorgis) where the core of the dome (syenitic intrusion) is being exposed by erosion.

All three types of dome structures are perfectly identifiable on TM and SPOT images based on their characteristic spectral response. Major fault structures are also clearly visible. Understanding the general geological setting of the region may help in locating other areas with similar geological and geomorphological characteristics as found at Bieta Giyorgis.

As previously stated for land-cover mapping a finer (more detailed) scaled image is needed. At present, we are limited to the use of TM images, but we expect to obtain an IKONOS image with high spatial resolution (1 m panchromatic, and 4 m multispectral bands) in the near future. The fieldwork conducted in the 2001 season will be useful for processing and interpreting the IKONOS data and we expect to produce a more accurate land-cover distribution map.

6.4 Geological maps

A major objective of the geological research was to identify and map the distribution of rocks and sediments on Bieta Giyorgis using field techniques and GPS.[42]
These investigations demonstrated that:

1) The rocks of Bieta giyorgis are cut by numerous normal faults and, to a lesser extent, are folded and jointed. The topography and distribution of late Cenozoic deposits are strongly influenced by the faulting. Large displacement-normal faults (> 250 m) occur on the southeastern, western, and northern perimeter of the mountain. Two major high-angle normal fault systems, one trending east-southeast and the other northeast control the orientation of the escarpments and drainage patterns on the mountain, and the movement of groundwater. For example, the early inhabitants took advantage of the jointing in order to quarry syenite, siltstone and sandstone.
2) Exposed faults and joints are rare because of deep weathering and extensive colluviation. An unnamed fault, in which contorted sedimentary rocks rest against deeply weathered jointed syenite, is exposed along the old access trail to the southeastern part of the mountain. Steeply inclined and folded sedimentary rocks crop out in the southern side of Bieta Giyorgis. Highly jointed and deeply leached syenitic rock appear to define a fault zone on the northern edge of the Ona Nagast plain.
3) Jointing is prevalent in the Dungur formation and in the uppermost parts of the Bieta Giyorgis syenite where it affects groundwater migration, strongly influences the character of weathering, and methods of quarrying by humans. In addition, the Dungur formation appears to dig in a south-southwesterly direction and is bounded on the south and northwestby normal faults that bring syenite up against the truncated Dungur.

[42] Geological Investigations were conducted by Gerald H. Johnson and M. Scott Harris with the assistance of three students (J.R. Soltess, J.G. Matkowski, J.D. Long)

4) Faulting occurred after deposition of the Dungur formation during the late Tertiary, and may be continuing.

6.5 Archaeological maps

The systematic topographic survey of the central and northern parts of Bieta Giyorgis, including Ona Enda Aboi Zewgé and Ona Nagast, was conducted in 1993 to 2000 field seasons in order to generate a basic archaeological map of top of the hill. Particular attention was paid to evidence of pre-modern agriculture terraces and stone arrangements in order to determine earlier land-use practices as well as the geological features.

The actual topographic survey was aimed at mapping the remains of the churches Ricci and Fattovich excavated in 1974 at the southeastern side of the hill, as well as all 2001 excavation units (particularly TE I, GA I, and OAZ X-XI-XII) were located and surveyed. Some excavation units opened in the previous seasons were also re-located in order to connect the former topographic survey with the 2001 one.

The relocation of sites and relevant archaeological areas with differential GPS (at a 1 cm scale) has provided a more precise map of the investigated area. The integration of maps with a differential GPS and topographic maps with TLS (at less than 1 cm scale) has provided data to generate a new archaeological map of the top of Bieta Giyorgis hill.

7. Off-site and landscape archaeology

7.1 Man-made landscape features

Man-made landscape features at Bieta Giyorgis include agricultural terraces and fields, cisterns, dams, roads and paths. These features are the connective tissue linking together the whole investigated area into one system, and may reflect temporal changes in social organization and land use patterns in the region.[43]

In 1997 all agricultural features that occur in the area between Ona Enda Aboi Zewgè and Ona Nagast were systematically mapped, in order to generate a model of ancient land use pattern to be tested in the field. Four recent cisterns that have been constructed on the Ona Nagast plain, and a possibly ancient cistern at 'Ela Nagast, were also examined.[44]

Several hundred features have been identified in the aerial photos of the area, and two main categories of features were distinguished: a) linear features (including quadrangular or polygonal structures), and b) circular structures (including circles and compacted structures).[45] Major concentrations of features were visible in the northwestern and northeastern sectors of the top of the hill. The 2000 survey demonstrated that most of these features are recent (although they may be a reuse or rearrangement of earlier structures), but others may be old or ancient structures.

The remains of two possible dams, associated with three walls, were found along the Guadguad Agazyen. They delimited a small basin on the northen edge of the hill. A man-made terrace (TI) wall was also built along the northeastern side of the basin. All these features most likely formed an elaborate system of water and soil management. These structures can be dated between the mid-1st millennium BC and mid-1st millennium AD.

An ethnoarchaeological study of land use practices on Bieta Giyorgis was initiated in 2001.[46] Four categories were identified: *d'en* (forest conservation), *wofri* (cultivated land), *wajet* (residential), and several types of *badme* (grazing lands). A sketch map illustrating the distribution of these categories was completed using aerial photos, and boundaries were mapped using GPS. In all land use categories, major plant species present were identified, with a total of 54 species (trees, shrubs, herbs) for all of Bieta Giyorgis.

7.2 Natural and cultural components of the ancient landscape

Bieta Giyorgis is particularly suitable for a reconstruction of the ancient Aksumite landscape, as the hill encompasses all features which characterized the territory of the capital-city.

7.2.1 Natural landscape

Bieta Giyorgis rises 200 to 250 m above the surrounding terrain. The hill itself is comprised of three distinct physiographic sections: the flanks, or lower steep and cliffed part; the horse-hoe shaped, gently sloping Ona Nagast plain; and the central uplands of Bieta Giyorgis. The drainage pattern on the mountain is radial and streams are intermittent.

The flanks of Bieta Giyorgis include the very steep slopes and cliffs that form the perimeter of the mountain. These slopes are littered with a discontinuous cover of large boulder colluvium.

The Ona Nagast plain is a horseshoe-shape, gently plain that surrounds the central uplands on three sides. The plain, which is about 175 m above the Aksumite plain, slopes inward toward the central upland and is bounded on its outer edge by a knobby rim.

[43] Off-site archaeological investigations were conducted by R. Fattovich and A. Manzo.

[44] See Fattovich, Bard, Petrassi & Pisano 2000: 41-43.

[45] The aerial photo interpretation was done by Mr. Antonio Solazzi, Suprintendency for the Archaeology of Latium, Rome (Italy).

[46] This investigation was conducted by A. C. D'andrea.

Figure 5: DEM of BG

The central uplands consist of a rolling terrain underlain by relatively homogeneous, but jointed, igneous rock. It stands 30 to 50 m above the surrounding Ona Nagast plain.

The preliminary results of the pollen-analytical studies suggest that in the middle to late 1st millennium BC and in the second half of the 1st millennium AD the vegetation cover on Bieta Giyorgis hill was dominated by shrubs and herbaceous plants characteristic of open vegetation and areas of human settlement. The very low arboreal frequencies in the samples (and virtual absence of tree pollen normally transported long distances by wind) suggest that trees were not common components of the vegetation pattern in the general area of Aksum.

7.2.2 Rural landscape

Man-made landscape features at Bieta Giyorgis include agricultural terraces and fields, cisterns, dams, roads and paths. Cultivated fields with associated terraces, grazing areas, roads and paths date to recent times (19th-20th centuries), but most likely maintain the basic organization of earlier land-use patterns.

Four major cultivated areas can be distinguished on the top of Bieta Giyorgis. They are located in the southeastern (between Ona Nagast and Ona Enda Aboi Zewgè), northeastern, northern, and northwestern sectors of the hill respectively. A possible evidence of ancient terraces occurs around the Guaguad Agazyen on the northern ridge of the hill. Grazing areas are located on the outcrop in the southwestern sector and in the northeastern sector of the hill. These areas are definitely recent, but most likely they maintain the location of ancient ones as they occupy areas not convenient for cultivation.

A large path, crossing the top of Bieta Giyorgis from south to north, most likely follows the direction of a possible ancient road. Another large path connects Tukul Emeni to Geza 'Agmai at Aksum. This path too, probably, follows an ancient road, as this is the shorter way to the settlement along the Mai Malahso at Aksum. These paths cross each other in the area of Tukul Emeni, at a place where a rough monolith was erected in late Pre-Aksumite times, suggesting that the monolith was a landmark.

A reservoir at 'Ela Nagast, very close to the residential area at Ona Nagast, and the two possible massive dams Guadguad Agazyen on the northern edge of Bieta Giyorgis hill can be also dated to Aksumite or Pre-Aksumite times.

Finally, stele quarries were identified on the southern and northeastern ridge of the hill.

At present, the largest concentration of houses and agricultural activity on the Ona Nagast plain occurs on the better-drained soils formed from the weathering of stratified sedimentary rocks.

7.2.3 Power landscape

The settlement pattern on the top of the hill is characterized by an elite residential settlement with evidence of at least one palace at ON, an elite cemetery with platforms associated with stelae and rock-cut tombs at OAZ, and scattered compounds along the ridge of the hill. Both ON and OAZ are located on prominent areas and were distinctive features of the local landscape. The palace at ON was occupied Proto-Aksumite to Late Aksumite times. The cemetery was used in Proto-Aksumite to Early Aksumite times.

Two small houses, dating to Aksumite times, have been excavated at Tukul Emeni and Guadguad Agazyen.

Rock-inscriptions, most likely used as property landmarks, were located at the base of the southern side of the hill.

7.2.4 Sacral landscape

Three types of ceremonial areas have been recorded on the top of the hill: 1) an open air, votive place; 2) a possible monumental temple; and 3) two Christian churches.

An open-air site, dating to late Pre-Aksumite or initial Proto-Aksumite times, has been discovered at Gunda Nebri, on the northeastern side of the hill. No ancient man-made structure was recorded at this site, but pottery was placed around and beneath two large natural boulders.

The remains of a possible monumental temple, dating to Proto-Aksumite and Early Aksumite times, were found at ON, on the most elevated area at the site. The walls with a precise southeast-northwest orientation were carefully constructed with small stones, which were regularly laid. These walls delimited seven long and narrow rooms (6.0 x 1.3 m) and three possible open areas in the northern part of the unit.

Finally, two churches, dating to the Aksumite 4 phase, were excavated on the southeastern slope of the hill.[47] One of this churches was cruciform in plan.[48]

8. Virtual models of BG

A more sophisticated "digital" project began in 1999, in collaboration with the Institute of Technologies Applied to Cultural Heritage, CNR, Rome (Italy).[49]

The project (in progress) is aimed at generating a "conceptual model" of the ancient landscape at Bieta Giyorgis and the Aksum area using the following procedures: 1) Vector and raster digitalization (maps, aerial photos, morphological features, etc.). 2) DEMs creation (micro-macro scale). 3) Satellite image classification (multispectral). 4) Integration and overlay of raster and vector data: aerial photos, satellite images, DEM, multilayered surfaces. 5) 3-D processing and VR navigation through the archaeological landscape. 6) Multispectral classification of Landsat TM and SPOT XS images. 7) GIS applications (spatial analyses). 9) 2-D and 3-D Spatial analysis with GIS applications (Arcview and Idrisi). 10) VR applications through remote sensing data and archaeological landscapes reconstructed by OpenGl applications (Figure 5).

Conclusions

The reconstruction of the ancient landscape of Bieta Giyorgis (and the Aksum region in general) is now one of the main aims of the IUO/BU Expedition, along with the outline of the cultural and environmental history of the region.[50] This reconstruction, however, is not (and cannot) be generated by the simple application of high technologies, but results from a continuos application of careful excavation and survey procedures, artifacts and ecofacts analysis, paleoenvironmental investigation, and ethno-archaeological and ethno-historical research. In regions such as Aksum, moreover, the application of high technology is particularly difficult because of logistical constraints (e.g., insufficient electrical power, lack of good roads, etc.).

At the end of the project, planned for 2003, we hope to produce a detailed database and virtual models to be used for planning an archaeological park at Bieta Giyorgis, which may integrate site and off-site archaeology and contribute to a sustainable development of Aksum.

[47] Ricci & Fattovich 1988.

[48] See Ricci & Fattovich 1988.

[49] See Forte ,Bard, Fattovich, Foccillo, Manzo & Perlingieri 2001

[50] See Bard 1997; Bard, Coltorti, Di Blasi, Dramis & Fatovich 2000.

References

ASHMORE W. & A. KNAPP, 1999. *Archaeology of Landscape*, Oxford: Blackwell.

BALIA R. & A. VERNIER, 1997. *Geophysical and Geological Studies - Scientific Expedition in the Archaeological Area of Axum* - Tigray. Technical Report to CRCCH, Addis Ababa.

BARD K.A., 1995. *New Excavations in Ethiopia*, *Context* (Boston University) 12 (1-2): 13-15.

BARD K.A., 1997. *Environmental History of Early Aksum*, in K.A. Bard (ed.), The Environmental History and Human Ecology of Northern Ethiopia in the Middle and Late Holocene, 19-28, Napoli.

BARD K.A., 1998. *Multidisciplinary Archaeology in Aksum, Ethiopia*, Context (Boston University) 13 (3-4): 16-18.

BARD K.A. & R. FATTOVICH, 1993a. *The 1993 Excavations at Bieta Ghiorghis (Aksum, Tigray)*, Africana Bullettin (Boston University), 10: 3,6.

BARD K.A. & R. FATTOVICH, 1993b. *The 1993 Excavations at Ona Enda Aboi Zaguè (Aksum, Tigray)*, Nyame Akuma, 40:14-17.

BARD K.A. & R. FATTOVICH, 1995. "The I.U.O./B.U. *Excavations at Bieta Giyorgis (Aksum): An Interim Report*, Nyame Akuma, 45: 25-27.

BARD K.A. & R. FATTOVICH, 1997. *The 1997 B.U./I.U.O. Excavations at Bieta Giyorgis, Aksum: A Preliminary Report*, Nyame Akuma, 48: 22-28.

BARD K.A., M. DI BLASI, R. FATTOVICH, A. MANZO, C. PERLINGIERI & L. CRESCENZI, 1996. *The B.U./I.U.O. Archaeological Excavations at Bieta Giyorghis (Aksum), Ethiopia: A Preliminary Report of the 1996 Field Season*, Nyame Akuma, 46: 21-23.

BARD K.A., M. DIBLASI, R. FATTOVICH, A. MANZO, C. PERLINGIERI, C.A. D'ANDREA & L. CHAIX , 1998. *The IUO and BU Archaeological Expedition at Bieta Giyorgis (Aksum): 1998 Field Season*, Nyame Akuma, 50, 1998, pp. 10-12

BARD K.A., R. FATTOVICH, A. MANZO & C. PERLINGIERI, 1997. *Archaeological Investigations at Bieta Giyorgis (Aksum), Ethiopia: 1993, 1994 and 1995 Field Seasons*, Journal of Field Archaeology, 24: 387-403.

BARD. K.A., M. COLTORTI, M. DIBLASI, F. DRAMIS & R. FATTOVICH, 2000. *The Environmental History of Tigray (Northern Ethiopia) during the Holocene: A Preliminary Outline*, The African Archaeological Review, 17 (2): 65-86

BERNARD E., A.-J. DREWES & R. SCHNEIDER , 1991. *Recuil des inscriptions de l'Éthiopie des périodes pré-axoumite et axoumite*, I-II, Paris :Académie des Inscriptions et Belles Lettres.

BRAKMANN H., 1992. *Axomis (Aksum)*, in E. Dassmann et al. (eds.), Reallexikon für Antike und Christentum, Suppl. I, Lieferung 5/6, 718-810, Stuttgart.

BRANDT S. A. & R. FATTOVICH, 1990. A *History of Late Quaternary Archaeological Research in the Horn of Africa*, in P. Robertshaw (ed.), History of African Archaeology: 95-108, London: Currey.

BURSTEIN S.M., 1997. *Ancient African Civilizations: Kush and Aksum*, Princeton: Markus Wiener Publisher.

DORAN J.E. & F.R. HODSON , 1975. *Mathematics and Computers in Archaeology*, Edimburg: Edimburg University Press.

DRENNAN R.D., 1996. *Statistics for Archaeologists*, New York and London: Plenum.

DREWES A.-J. & R. SCHNEIDER , 1967. *Documents épigraphiques de l'Éthiopie, I*, Annales d'Ethiopie, 7: 89-102.

FATTOVICH R., 1992. *Lineamenti di Storia dell'archeologia dell'Etiopia e della Somalia*, Napoli: Istituto Universitario Orientale.

FATTOVICH R., 1994. *Scavi archeologici nella zona di Aksum. D. Ona Enda Aboi Zaguè (Bieta Goyorgis)*, Rassegna di Studi Etiopici, 36: 49-55.

FATTOVICH R., 1995. *Archaeological Excavations at Bieta Giyorgis (Aksum, Tigray): A Preliminary Report of the 1994 Field Season*, Nyame Akuma, 43: 34-37.

FATTOVICH R., 1997. *Archaeology and Historical Dynamics: the Case of Bieta Giyorgis (Aksum), Ethiopia* Annali dell'Istituto Universitario Orientale di Napoli, 57 (1-2): 48-79.

FATTOVICH R. & K.A. BARD, 1993. *Scavi archeologici nella zona di Aksum. C. Ona Enda Aboi Zaguè (Bieta Giyorgis)*, Rassegna di Studi Etiopici, 35: 41-71.

FATTOVICH R. & K.A. BARD, 1994. *The Origins of Aksum: A View from Ona Enda Aboi Zaguè (Tigray)*, in H.G. Marcus (ed.), New Trends in Ethiopian Studies. Papers of the 12th International Conference of Ethiopian Studies, I: 16-25, Lawrenceville: Michigan State University.

FATTOVICH R. & K.A. BARD, 1995. "*Scavi archeologici nella zona di Aksum. E. Ona Enda Aboi Zeugè ed Ona Nagast (Bieta Giyorgis*)," Rassegna di Studi Etiopici, 37: 5-35.

FATTOVICH R. & K.A. BARD, 1996. "*Scavi archeologici nella zona di Aksum. F. Ona Nagast (Bieta Giyorgis*)," Rassegna di Studi Etiopici, 38: 71-94.

FATTOVICH R. & K.A. BARD, 1997a. "*The I.U.O./B.U. Excavations at Bieta Giyorgis (Aksum) in Tigray (Northern Ethiopia).*" Journal of Ethiopian Studies, 30 (1): 1-29.

FATTOVICH R. & K.A. BARD, 1997b. "Scavi archeologici nella zona di Aksum. G. Ona Enda Aboi Zeugè e Ona Nagast (Bieta Giyorgis," *Rassegna di Studi Etiopici*, 39: 1-22.

FATTOVICH R. & K.A. BARD, 1998. "*Scavi archeologici nella zona di Aksum. H. Ona Enda Aboi Zeugè e Ona Nagast (Bieta Giyorgis*)," Rassegna di Studi Etiopici, 41: 57-76.

FATTOVICH R., K.A. BARD, L. PETRASSI & V. PISANO, 2000. *The Aksum Archaeological Area: a Preliminary Assessnent*, Napoli: Istituto Universitario Orientale.

FATTOVICH R., C.A. D'ANDREA, M. DIBLASI, A. MANZO & YAQOB BEYENE, 2000. "*Scavi archeologici nella zona di Aksum. I. (Bieta Giyorgis*)", Rassegna di Studi Etiopici, 43: 43-61.

FORTE M., K.A. BARD, R. FATTOVICH, M. FOCCILLO, A. MANZO & C. PERLINGIERI, 2001. "*The Aksum Project (Ethiopia): GIS, Remote Sensing Applications and Virtual Reality*", in Z. Stancic e T. Veljanovski (eds.), Computing Archaeology for Understanding the Past - CAA 2000: 241-251, Oxford: British Archaeological Reports.

Harris E., 1989. *Principles of Archaeological Stratigraphy*, San Diego, CA: Academic Press.

Littmann E., D. Krencker & T. von Lüpke, 1913. *Deutsche-Aksum Expedition*, I-IV, Berlin: Druck und Verlag von Georg Reimer.

MANZO A., 1996. "Preliminary Report on the Imported and Exotic Materials from Bieta Giyorgis (Aksum, Ethiopia)", paper delivered at the 13th Biennial Conference of the Society of Africanist Archaeologists, Poznan.
n.d. "The Imported Ceramics and Glasses from Bieta Giyorgis", unpublished report.

MICHELS J.W., 1994. "*Regional Political Organization in the Axum-Yeha Area during the Pre-Axumite and Axumite Eras*," in C. Lapage (ed.), Etudes éthiopiennes, I: 61-80, Paris: Société Française des Etudes Ethiopiennes.

MILLER N.F & K.C. GLEASON , 1994. *The Archaeology of Garden and Field*, Philadelphia: Universuty of Pennsylvania Press.

MUNRO-HAY S.C., 1991. *Aksum. An African Civilisation of Late Antiquity*, Edinburgh: Edimburgh University Press.

OGUNDIRAN A., 1995. *A 1995 Archaeological Reconnaissance in Aksum, Northern Ethiopia: Perspectives on the Aksumite Settlement Geography and Material Culture*, report submitted to the Social Science Research Council, August 1995, Boston.

ORTON C., 1982. "*Computer simulation experiments to assess the performance of measures of quantity of pottery*", World Archaeology, 14: 1-20.

PERLINGIERI C., n.d. *La ceramica aksumita da Bieta Giyorgis, Aksum (Tigray, Etiopia). Tipologia ed implicazioni storico-culturali e socio-economiche*, Ph.D. Dissertation, Istituto Universitario Orientale, Naples.

PHILLIPSON D.W., 1998. *Ancient Ethiopia. Aksum: Its Antecedents and Successors*, London: British Museum Press.

RICCI L., 1974. "*Scavi archeologici in Etiopia*," Africa (Roma), 29: 435-441.

RICCI L., 1990. "*Appunti archeologici*," Rassegna di Studi Etiopici, 32: 129-165.

RICCI L. & R. FATTOVICH , 1988. "*Scavi archeologici nella zona di Aksum. B. Bieta Giyorgis*," Rassegna di Studi Etiopici, 31 (1987): 123-197.

STUIVER M. & P.J. REIMER, 1993. "*University of Washington Quaternary Isotope Lab Radiocarbon Calibration Program Rev. 3.0*," Radiocarbon, 35: 215-230.

TOMBER R., 1993. "*Quantitative approaches to the investigation of long-distance exchange*", Journal of Roman Studies, 6, 142-166

USAI D., 1997. "*Early Aksumite Lithic Workshop Evidences from Bieta Giyorgis, Axum, Ethiopia,*" Rivista di Archeologia, 21: 7-12.

Landscape Archaeology in Tuscany: Cultural resource management, remotely sensed techniques, GIS based data integration and interpretation

Stefano Campana[1], Riccardo Francovich[2]

[1] Department of Archaeology and History of Arts, Section of Medieval Archaeology - University of Siena
Via Roma 56, 53100 Siena, Italy
campana@unisi.it
[2] Department of Archaeology and History of Arts, Section of Medieval Archaeology - University of Siena
Via Roma 56, 53100 Siena, Italy
francovich@unisi.it

Abstract. The Department of Medieval Archaeology at the University of Siena has long been engaged in the study of the ancient landscape, essentially through intensive field survey and surface collection from agricultural surfaces, the analysis of vertical aerial photographs and the stratigraphical interpretation of monuments. In the last ten years the Laboratory of Information Technology Applied to Medieval Archaeology (LIAAM) has directed its work towards the development and application of global management systems for archaeological documentation. This work has involved not only archaeological excavation but also the preparation and publication of archaeological cartography through the development of GIS dedicated systems. This type of solution allows the realization of three main objectives: acceleration of investigative times, sophisticated elaboration of "intra" and "inter" site data, and real possibilities for tutelage. In particular, we can now for the first time produce efficient instruments for integrating archaeology into the politics of teaching and for achieving a fuller appreciation of the landscape around us. In the past, there have been in essence two fundamental problems: insufficient precision in describing the location of discoveries and the absence of efficient instruments for processing the resulting data. Now, with the georeferenced definition of site boundaries, based on differential GPS technology managed using Geographic Information System, this gap has been conclusively filled. In recent years we have been working towards the integration into the GIS base of a remote sensing module able to process, integrate and maximise the data received from various sources: vertical and oblique aerial photography, high-resolution satellite imagery, geophysical surveying and micro-digital terrain modelling using differential GPS. Basically, this work is aimed at the creation of an interconnected series of databases composed of alphanumeric archives (features, images, the results of ground-truthing exercises and dGPS surveys) along with the essential tools for processing the images in georeferential environments in which it is possible to define surfaces and to correctly calculate dimensions, distances and possible relationships between anomalies. This organization of the data greatly assists the potentially difficult task of surface ground-truthing, allowing anomalies to be located quickly and easily on the ground through a dGPS system and facilitating the subsequent management of the site documentation. The prototype system has been developed in two sample areas, chosen for their differing geo-morphological characteristics, landscape and cultural complexity. These lie in the southern and southeastern part of the province of Siena and along the coastal strip in the province of Livorno between Populonia and San Vincenzo. They have a combined extent of about 470 sq. km. In recent years both areas have been subject to intensive surveys and the multi-temporal analysis of vertical aerial photographs. They therefore present an excellent opportunity for testing the potential of multistage remote sensing in combination with conventional methodologies for landscape analysis.

1 Introduction

The Department of Medieval Archaeology at the University of Siena has been actively engaged in programmes of landscape archaeology for over twenty-five years. The current research projects fall into two categories: the creation of diachronic archaeological maps, and the thematic cataloguing of individual sites. The archaeological maps now cover the provinces of Siena, Grosseto, and various communities around Livorno, taking in a total area of about 9000 sq km. Lately we have carried out systematic field survey within sample areas totaling 800 sq km and have located a total of about 6000 new archaeological sites. The processes of archaeological cartography are for the most part based on three methods of investigation: systematic field survey in sample areas (20-30% of the total landscape); analysis of vertical air photos combined with selective ground-truthing; and field examination to assess the significance of individual monuments[1]. The second category of projects concentrates on thematic research such as that leading to the *Atlas of Tuscany Hill Forts,* which includes more than 4500 anomalies identified on vertical aerial photographs, 1500 of them attributable to medieval castles[2].

We take the view that the creation of archaeological maps has two main objectives, one scientific, the other politico-administrative. The combination of these two objectives facilitates on the one hand the study of the socio-economic and settlement patterns of the region, and on the other gives us access to public administration processes concerned with development and conservation.

To achieve this last objective it is necessary to raise our instrumentation and functionality to the level demanded by the responsible public authorities. New digital technology, in addition to offering powerful tools for the cataloguing, storage, and analysis of archaeological data, has become the new common language for the communication of information. For example, in Tuscany one only has to look at the widespread use of GIS, now present in every local authority office. In addition to traditional methods, archaeologists must use the power of technology to facilitate the dialogue with public authorities that is vital for truly co-operative work. The revolution in technology has opened up new prospects for the public application of archaeology[3].

Our chosen solution is directed towards two fundamental requirements: firstly, the education of professionals who are able to respond to, improve and promote new standards of production; and secondly, communication with different kinds of users through the multi-media presentation of archaeological information.

For at least the last ten years we have been working on the training of archaeologists in computer sciences. This puts a group of people in the position to be educated working, experimenting, and sharing their most valuable experiences for activate new teaching. Presently, we are working with various technologies and applications, including image processing of remotely-sensed data, GIS management of excavations and landscapes, multimedia cataloguing of archaeological resources, 3D modeling, photo-quality rendering and animation, morphing of excavated structures, electronic video documentation, organization and CAD management of sites, QuickTime Virtual Reality setting, video/film editing and management, and the creation of Internet sites and pages concerning research and archaeological parks. Communication with the general public, as well as with private individuals and institutions, is enhanced by multimedia support and educational Internet sites, which are quickly becoming the gateway to Medieval Archaeology in Italy. By including archaeological sites and information in digitally-available cartography we are increasing the value and impact of our research while at the same time satisfying the needs of public authorities for the protection and management of archaeological sites, and the desire of the general public for cultural enrichment[4]. The principal scientific objective has been achieved through the construction of a management system for all of the information gathered in the research programmes of the Department of Medieval Archaeology. Archiving this information in the most comprehensive way allows us to bring together a vast amount of digital data, alphanumeric, graphical and moving images. It also speeds the processing time and significantly reduces the loss of data. Ultimately, it means that the research results become available to the whole scientific community. Digital archiving, along with the Internet availability of research data, opens up our laboratories and secures the public use of archaeological information. Our chosen solution combines the use of data-management packages (ArcView GIS, MacMap, FileMaker Pro, Canto Cumulus etc) in combination with the design and implementation of an application which we have called Open-Archeo. This application, through a simple user interface, compares and creates multi-directional dialogue between the basic data and the many different applications in use at any one time. Essentially, it permits optimal access to the archaeological data[5].

Riccardo Francovich

[1] We note here only the last few in a long series of papers on the methodology used in the Siena Province Archaeological Map Project: VALENTI 1999, pp.10-14; FRANCOVICH-VALENTI 2001, pp.83-116; CAMPANA 2001, pp. 47-71. A detailed introduction on this subject can also be found on the Siena University Internet Site (Webmaster: L. Isabella, A. Mignani):<archeologiamedievale.unisi.it>.

[2] FRANCOVICH-GINATEMPO 2000.

[3] FRANCOVICH-VALENTI 2001, pp.83-116.

[4] FRANCOVICH-ZIFFERERO 1999.

[5] FRONZA *et al.* 2001, pp.173-177; ISABELLA *et al.* 2001, pp.31-64.

2 Research methods

In the last few years we have turned our attention to the development of remotely sensed techniques[6]. Within the Department, the Laboratory of Aerial Photographic Interpretation has been active since 1984. In addition to teaching, the Laboratory has carried out numerous research projects, leading to the identification of over 5000 air-photo anomalies in Tuscany alone[7]. The laboratory is dedicated to the stereoscopic examination of vertical aerial photographs. The photographs at our disposal belong to Regional Mapping Office or to Military Institute for Cartography. They are taken as a matter of routine, every ten years between August and September, to generate maps. Despite good archaeological results, we have been conscious throughout of the inherent limitations of this method of survey. The main problem is the cartographic nature of the data and the impossibility of planning the flights to coincide with times when conditions for the detection of archaeological features are at their best. In addition, there are other problems with vertical photographs, such as the inherent inflexibility of paper documents, the difficulties of magnifying details, and the limited capacity to distinguish between tones of grey, etc.
To try and overcome these limitations in pursuit of our own objectives we have changed our focus to the experimental application and evaluation of new techniques in the study of the Tuscan landscape. After looking at experience elsewhere, we realized that there was not an ideal technique capable to exclude all the others. Each data source has its imperfections. In short we started from known concepts summarized by Lillesand and Kiefer as "more information is obtained by analysing multiple views of the terrain than by analysis of any single view" and that "successful application of remote sensing is premised on the integration of multiple, interrelated data sources and analysis procedures"[8]. This is the reason why we turned to oblique aerial photography, to the latest generation of multi-spectral high-resolution satellite imagery, to geophysical survey and to micro-digital terrain modelling using differential GPS.

Our progress in developing this approach can be highlighted by looking at two sample areas, representative of the landscape complexities and settlement patterns of Tuscany.

The first, consisting of primarily flat land, is situated in the province of Livorno and includes the coastal strip between Populonia, Campiglia Marittima and Donoratico (Fig.1). The second area is situated in the province of Siena. This geo-morphologically hilly countryside, as well as its known archaeology, is representative of considerable stretches of Siena province (Fig.2)

[6] Campana-Forte 2001.
[7] Cosci 2001, pp.55-64.
[8] Lillesand-Kiefer 1994.

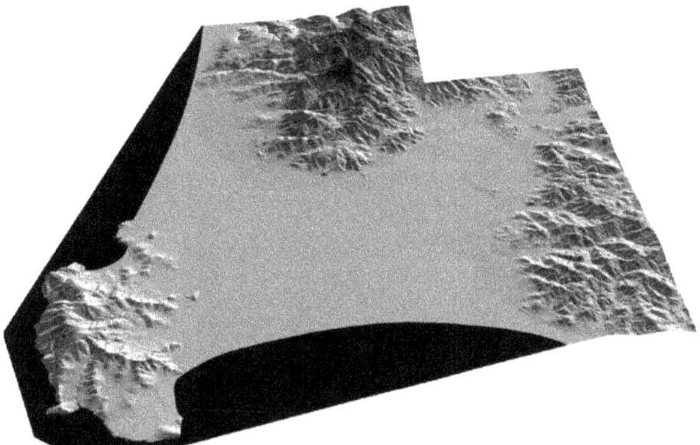

Fig.1 – DEM of coastal strip between Populonia, Campiglia Marittima and Donoratico (sample area 1)

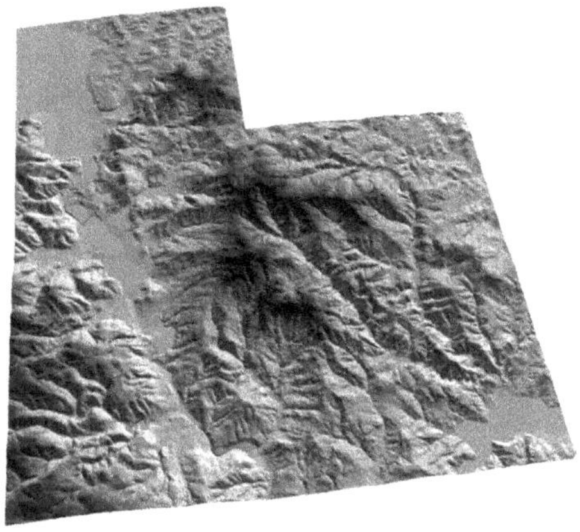

Fig.2 – DEM of the typical hilly countryside of Siena Province (sample area 2)

The total extent of these sample areas is around 470 sq km. Both areas have recently been the subject of numerous socio-archaeological studies, field-walking surveys, excavations, vertical air-photo interpretation and geological and geomorphological analyses. Presently there are in our DBMS more than 1800 archaeological sites, from Palaeolithic to Late Medieval[9]. When setting up the research project we paid particularly close attention to the systematic collection of data. The first objective of the operation was to acquire as many individual pieces of information as possible for comparison between oblique aerial photography and satellite imagery. The second stage will aim to integrate and reinterpret the whole body of information using GIS based technology, and from this to postulate new settlement patterns. To manage all of the related documentation we have designed an archaeological GIS system using a data model which combines alphanumeric and multimedia databases, basemaps (topographic raster and vector data), remotely-sensed data, Triangulated Irregular Network (TIN) objects, differential GPS data and a variety of derived datasets.

It is possible to divide the graphical information into three categories: technical, thematic and historical. The maps can be presented in raster and/or vector format with Gauss-Boaga geographical coordinates. The availability of source

[9] Cortemiglia *et al.* 1983, pp. 148-173; Fedeli *et al.* 1993; Mazzanti 1995; Ceccarelli Lemut-Garzella 1996; Bianchi *et al.* 1997; Cosci 2001, pp.55-64.

maps has increased during the project, the material now at our disposal being shown in Table-1. At the same time, we have designed an ArcView GIS extension for the better management of the raster and vector data[10]. The basic concept of the extension is to use a friendly wizard to select the source-material that we want to present visually - Ikonos-2 MS, vertical photographs from 1954, vector data, TIN data, or any other kind of data - and then to add automatically the data corresponding to the area that we are looking at on-screen. This utility allows us to organise and manage - without the need for grids or other reference tools – very large numbers of datasets, thereby reducing dead time and giving other researchers easy access to the data[11]. The DBMS was developed by the Laboratory of Information Technology Applied to Medieval Archaeology (Siena University). We have extended it with the addition of new modules for remotely sensed data, GPS data and ground-truth check data[12].

2.1 Vertical aerial photographs

Despite their inherent limitations, vertical aerial photographs, with their wide temporal range, represent an irreplaceable source for the analysis of the Tuscan landscape. Anyone interpreting the photographs used in our own study will see a 3D replica of the whole landscape as it was in 1938, 1954, 1976 and 1994. In addition to their historical content, vertical photographs are of course an important source for the conduct of "aerial reconnaissance". Our way of working can be defined as hybrid, based on analog and digital analysis of every frame. All the photographs at our disposal have an overlap of 65%, providing stereoscopic views. Firstly the whole of each photograph is examined thoroughly using a stereoscope. In a second phase each frame is scanned and orthorectified. Because this process is very time-consuming it has been applied only to recent research projects (Fig.3). In the case of the 5000 anomalies detected in earlier studies we have not orthorectified the entire frame but only the areas of the already-identified archaeological features[13]. It would have been more interesting, of course, to make GIS-ready mosaics of all of the photographs, suitable for image processing, interpretation and reinterpretation, and for comparison with other archaeological layers within the GIS. But by mapping only the anomalies we were able, in a relatively short time, to obtain data for "intra" and "inter" site analyses. At the moment we are working in collaboration with the Remote Sensing Laboratory of the University of Florence to achieve desk-top management of the whole process through the use of ERDAS Imagine® software[14].

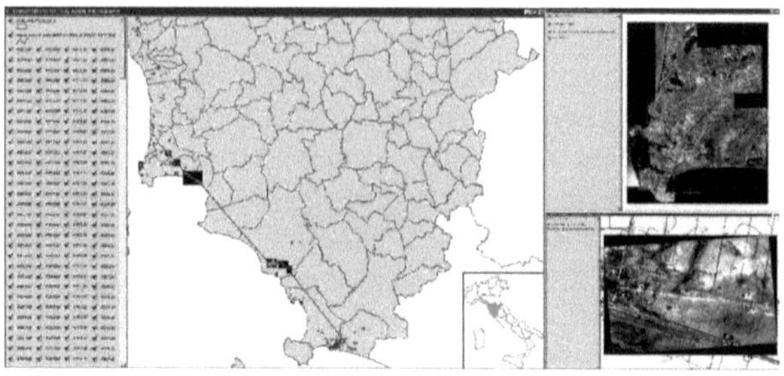

Fig.3 – Example of GIS management of vertical aerial photographs and mosaic

2.2 Oblique aerial photography

The opportunity to develop a programme of oblique air-photography for landscape investigation first presented itself in 1999 during the University of Siena's XIth International School in Archaeology. In anticipation of imminent changes in Italian law on aerial photography, the possibility of organizing an air survey training course in Siena was discussed with two colleagues from Britain, firstly Chris Musson, from Wales, and then Robert Bewley, Head of Aerial Survey at English Heritage[15]. Depending on the weather conditions during the spring, our reconnaissance season usually runs from the middle of May to the middle of June. In the spring of 2000 we carried out preliminary aerial reconnaissance to check at first hand the potential and limitations of oblique air-photography in the landscape of Tuscany. Then, at the end of May 2001, twenty-one post-graduate students engaged in the areas of research and heritage conservation took part in the first Aerial Archaeology Research School to be held in Italy. In addition to instruction, the objective of the school was to start building up an archive of oblique air photographs for the recording of the archaeological resource throughout Tuscany. Over a week of aerial photography we completed 127 hours of flying, took 5500 air photographs and made three hours of video film.

At the moment all the images have been scanned and, along with the video-film, have been imported into a multimedia database. The chosen software, Canto Cumulus, is of some interest[16]. It permits the creation of extensive catalogues of images and other media data, apparently without record limits. This software allows the building of a system of hierarchical and thematic categories that allows us to order the various items according to their intrinsic archaeological content. The pictures can be presented on-screen at various scales, from thumbnail to screen-size or greater, with related space for descriptive text (Fig.4). Following generation of the catalogue the next stage was to map all the sites that had been photographed.

[10] Lorenzo Bianchini, of the Department of Engineering at the University of Florence, wrote the programme for the extension in ESRI AVENUE language.

[11] We are now developing the extension for use on an intranet network.

[12] For the characteristics of the DBMS see FRONZA 2000, pp. 125-137; for the new modules see CAMPANA 2001, pp.61-63.

[13] After pre-processing, the images are geo-referenced, generally using at least four ground control points on the regional technical map at a scale of 1:10.000.

[14] The Remote Sensing Laboratory of Florence University is directed by Prof. Enzo Pranzini.

[15] DRIVER-MUSSON 2000, pp. 51-54.

[16] See <http://www.canto.com>

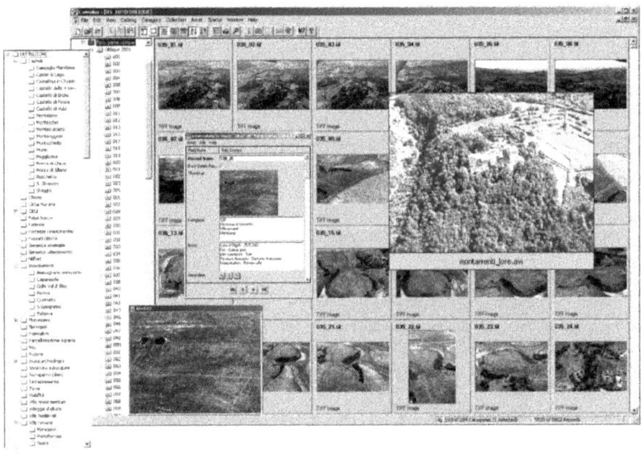

Fig.4 – Canto Cumulus media catalogue: pictures, movies and thematic categories

At this stage of the work each sequence of frames defining a site, tied to the GPS data-points acquired during flight, was placed into the GIS environment. In this phase the availability of detailed topographical maps represented the main pre-requisite for the correct siting of each sequence of frames and thence of the sites themselves[17]. Altogether, we acquired 616 GPS data-points, corresponding to approximately 281 air photo anomalies within Tuscany and a still-unidentified number in various parts of Lazio, Abruzzo and Umbria. For working in these latter regions there is a typically Italian problem - the bureaucratic difficulty of obtaining technical cartography. Moreover, the absence of comprehensive archaeological maps for these areas prevents a rapid comparison between known sites and those represented in the photographs.
Within Tuscany, by contrast, the situation is more favourable. The comparison between the known archaeological data and the oblique air photographs was carried out within the archaeological GIS, by checking the correspondence between the images and the known and mapped sites. To do this, we first concentrated on locating the images on a points layer within the GIS system. At the moment we are still working on the rectification and mapping of the photographs[18]. Considering the high number of photograph, our strategy for mapping has been to use only the most representative images, limiting ourselves in most cases to a single photo per site. This was the only way that we could hope within a reasonable time to integrate the aerial information into the GIS system and to compare it with other layers of information at our disposal. In addition, we will create a thematic vector layer showing the anomalies or the perimeters of the monuments which they represent.

During the Research School we recorded approximately 281 sites in Tuscany, the great majority being castles, medieval settlements, monasteries, old field boundaries, archaeological excavations, Roman villas and ancient roads (Fig. 5). About 37% of the sites, identified simply as 'crop marks', 'soil marks' or 'unidentified anomalies', await allocation where possible to a more precise site type. Upon first examination the comparison with the archaeological GIS of the Department of Archaeology suggests that of the 281 sites catalogued 132 are new to the record. This preliminary interpretation of the sites has made a significant contribution to our search for understanding of settlement patterns and road systems in the region. We can perhaps see this as a growing trend, even though 66% of the new sites are for the moment only generically defined, as crop marks, soil marks and unidentified anomalies.

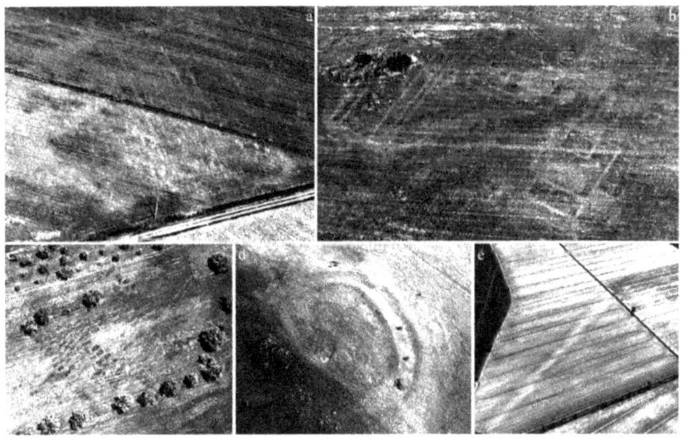

Fig.5 – Examples of anomalies recorded during the Aerial Archaeology Research School; a-b) roman villas; c) necropolis; d) hill fort; e) road

Granted that our work with oblique photography has only just begun, and that individual anomalies still need to be confirmed in the field (at least on a sample basis), our first experience of active aerial survey has been on the whole positive. Among the objectives that we would like to achieve in the rest of 2002 and the early part of 2003 is the publication of the whole photographic archive on the Internet. Altogether, the results of the Research School are of considerable significance, permitting the identification of a large number of new sites, the monitoring of well-known sites and, last but not least, the acquisition by the Department of the knowledge and experience to continue the work independently. In this last respect another goal that we are working towards is the development of a systematic programme of aerial photography and air photo mapping in Tuscany.

2.3 High-resolution satellite imagery (HRSI)

It was only in February 2001 that we begun to build up our experience in the use of HRSI. The aim is the evaluation of multispectral data for archaeological research, firstly using Ikonos-2 imagery and then from June 2002 working on QuikBird-2 imagery.

The central question was to see whether Ikonos-2 imagery could be useful in the identification of archaeological sites and heritage resources in an area like Tuscany. In the past the successes achieved through the use of satellite imagery have usually been obtained in desert landscapes or in areas where such imagery represents the only available source of

[17] DOLATOWSKA-GOLIASZ 1999, pp.41-42.
[18] For the mapping techniques used see DONEUS 2001, pp.17-27.

remotely sensed data[19]. In Italy, for instance, there have been relatively few studies using satellite imagery, and these have been mainly restricted to Roman centuriation and geomorphological or paleoenvironmental analyses[20]. Two factors have been in operation here: on the one side the low resolution of Landsat and Spot imagery, on the other the principal advantages of satellite imagery, its capacity to capture large section of the landscape and to record these at a number of different wavelengths in the visible, reflected or emitted infrared parts of the electromagnetic spectrum. In addition, computer enhancement of the digital data places less dependence on the time of year for revealing archaeological features. It is possible to identify the main differences between Landsat and Ikonos-2 as being that Ikonos-2 imagery has a higher ground resolution but a lower number of bands, including the complete absence of medium and thermal infrared.

The continuing improvement in the resolving power of the new generation of satellites is changing the possible uses of satellite imagery so that in the right circumstances the information drawn from Ikonos-2 imagery is beginning to stand comparison with that of vertical air photography[21]. Theoretically the level of detail visible in Ikonos-2 imagery allows the identification of line features 3-4 meters in width and of area features within the range of 1000-2500 sq m. In archaeological terms, Etruscan-Hellenistic *oppida*, Roman villas, churches, monasteries, medieval castles and villages are all types of remains that would potentially be visible on the new generation of satellite imagery.

The two images used in our study were captured on 10th July 2000 at 10.05 in the morning by the Ikonos-2 satellite. The first, on the Livorno coastline, is characterised by excellent quality, very good visibility and a total absence of cloud and haze (Fig.6). The second, in the province of Siena, is of low quality, showing evidence of clouds and of degradation by haze (Fig.7).

Our methodological approach to Ikonos-2 imagery has been focused on 2D visual interpretation and the exploration of 3D representations[22].

The procedure followed in processing the Ikonos-2 imagery falls into two main phases.

Fig.6 – Ikonos multispectral imagery of sample area 1

Fig.7 – Ikonos multispectral imagery of sample area 2

[19] We are thinking for instance of the pioneering work of the NASA Space Centre in south America, see <http://www.ghcc.msfc.nasa.gov/archeology/archeology.html>.

[20] In particular we would point to the experience of MARCOLONGO - MASCELLARI 1978, pp. 131 ss.; BARISANO *et al.* 1984 ; PIERI-PRANZINI 1989, pp.1385-1388; ALESSANDRO *et al.* 1992, pp.547-551; COSTI *et al.* 1992; DICEGLIE 1992, pp.421-439; COLOSI *et al.* 1996 ; BAGGIO *et al.* 1998; FORTE *et al.* 1998, pp. 291-304; CREMASCHI-FORTE 1999, pp.207-226; PRANZINI-SANTINI 1999, pp. 283-291; MARCHISIO et *et al.* 2000; PASQUNUCCI-TRÉMENT 2000.

[21] Even if their geometric resolution is still somewhat lower than that of vertical aerial photographs, the high quality of Ikonos images, due to their digital nature and their eleven-bit radiometric resolution, overcomes this problem, making it possible to achieve a real comparison between the two different sources of information.

[22] For software we used mainly ERMAPPER, ERDAS Imagine, ArcView 3D analyst (ESRI) and ENVI.

Fig.8 – Examples of anomalies identified analysing Ikonos-2 data; a) Mound (RGB colour composite of bands 4-3-2); b) road and ditch (true colour composite); c) ancient riverbed (RGB Colour composite of bands 4-3-2); d) Etruscan settlement(RGB colour composite of BGW derived from TCT); e) medieval village (Second Principal Component); f) medieval village (RGB colour composite of bands 3-4-2); g) hillfort (NDVI); h) not identified (RGB colour composite of bands 4-3-2); i) medieval castle (true colour composite)

The first consists of a series of standard transformations of the whole image. In this stage of the processing some of the most commonly used techniques have been contrast stretching, density slicing, RGB colour composites of the original bands (3-2-1; 4-3-2; 4-2-1; 3-4-1) and arithmetic manipulation, in particular averaging (to reduce the noise component) and rationing (NDVI). This phase plays a central role in the identification of archaeological features.

Between the first and second phase of processing and visual interpretation we use the GIS system to make comparisons with the features identified in the various GIS layers that mentioned above. This step has been useful in preventing other kinds of misidentification, for instance of non-archaeological features. In all, we have recognized in our satellite imagery 104 anomalies, of which 45 had already been identified through the analysis of air photography. Comparison in the GIS reduced the number of probable archaeological features to eighty-four.

At a later stage (second phase of image processing), the focus of view was narrowed in order to isolate homogeneous textures around individual anomalies. The processing was carried out using Principle Component Analysis (PCA), Tasselled Cap Transformation (TCT), Decorrelation Stretch (DS) and RGB colour composites of the results of the various transformations. The filters, when applied - whether in the first or second phase - were primarily constituted of 3 by 3 matrices, for the most part confined to sharpening, smoothing, and edge enhancement[23].

On completion of the image processing we are able to recognize some trends. As we expected, all of the best results come from transformations in which the near infrared band plays a primary role, especially in NDVI, first Principal Component, brightness and Wetness Transformation and relative colour composites[24]. Certainly there is no single ideal technique, but rather a

[23] CAMPANA-PRANZINI 2001, pp.17-62.

[24] On account of the very similar spectral ranges of bands 1-2-3-4 of Ikonos with Landsat TM in the first approximation we used the algorithm developed by Crist-Cicone, 1984.

spectrum of techniques producing variable results[25].

Altogether, our present processing of the Ikonos-2 imagery has allowed us to identify 84 archaeological features. Firstly, we may note that 82% of these are in the coastal strip (sample area 1). In the first stage of analysis we interpreted features as settlements, hill forts, mounds, roads, ancient riverbeds and some not-identified (Fig.8). During the winter of 2001/2002 we checked on the ground, mainly through traditional field survey, a sample of 40% of the features. The results confirmed the presence of archaeological finds or features in 59% of the cases. In the 18% anomalies were the consequence of modern activities. In the 23% where we found no archaeological artefacts or structural remains there was also no modern activity that might have revealed such evidence.

Although this work is still in progress, with 60% of the features still to be confirmed in the field, our experience working with Ikonos-2 imagery has been, as a whole, positive. We think that multispectral imagery has characteristics which are entirely compatible with the needs of archaeological landscape investigation. The resolving power of the images allows us to identify a large range of archaeological sites. At the best, archaeological features are distinguishable at a size of 20 to 30 m across, and more commonly in the order of 50 to 60 m across.

In summary, the particular contribution of Ikonos-2 imagery should be recognized as lying in its multispectral properties, in the near infrared band and in the possibility of recording the whole of the landscape at times when crop marks or soil marks are at their best. Unfavourable aspects of recent satellite imagery remain substantially the same as for the preceding generation of satellites, in particular the impact of unfavourable morphology, the need for excellent atmospheric conditions, and the relatively high cost.

3 Linking real landscape and GIS environment: the GPS technology

Looking at poor archaeological literature produced in Italy about use of GPS for archaeological research, the general idea that emerged it is that GPS is rarely used and the real potential of this tool has been underestimated and in a certain way misunderstood. Moreover paradoxical is the widespread use in landscape studies of GIS technology in comparison with the low diffusion of handheld GPS data collection system.

In the 1999 we bought our first handheld GPS, a Trimble GeoExplorer2. At the beginning the main use we did of GPS was strictly related with mapping operations of artefact scatters and monumental structures during field survey. Soon we realized that GPS technology could be applied on the field or during aerial survey for numerous tasks as to map particular finds, photographs (ground and aerial) or Quick Time Virtual Reality movie, shovel test and ground control point for photogrammetry and geophysics prospection; to monitor the movement of artefacts scatters; to record aerial tracks and the true surveyed areas; to generate micro-digital terrain modelling, etc. Furthermore GPS not only may map point, line and area but allow us to navigate on site exporting geographic coordinates from GIS. Navigating with the support of the GPS is not only extremely useful in many situations, but also easy, quickly and accurate. For instance in relationship with anomalies a methodological problem we had before the introduction of GPS technology was to identify with accuracy features on the ground previously identified on the imagery. This is true above all in situation of scarce visibility (wood areas) or absence of artefacts or structures. In short, GPS is fast, easy to use and very accurate for land survey mapping and navigation[26]. That's why we improved this sector buying two others handheld Trimble GeoExplorer3 and a Trimble 5700 reference station for the differential correction of rover data. Using these GPS system very intensively we collected autonomously a wide quantity of topographic data in the main with an error of ±1 m.

In our experience we believe that a significant peculiarity of GPS system it the reduction of the distance between real world and his digital representation. GPS data collection system let us move from an approximate representation of reality (our archaeological GIS and more generally digital data) to the real world and vice versa making easier to link and tie drastically the two environments: reality and its GIS abstraction. On this topic it is already possible to do more using handhelds Pocket PC. During 2002-beguinning of 2003 we will flank to Geoexplorer3 an iPAQ Pocket PC with a Trimble GPS module (the mapping precision is ±2 m). The difference between the devices is easy to understand. Geoexplore3 is "only" a handheld GPS data collection system, iPAQ is a GPS and a small computer too. We believe that this device will allows us to bring an important part of our laboratories directly on the field or better in the landscape in order to build the documentation - GIS, DBMS, media catalogues, GPS and most of digital data - directly on the field. We trust applying this technology to increase the typology and improve the quality of data avoiding any kind of intermediation. Furthermore the information will be available and manageable during every phase of the research reducing work time and lose of data[27].

---

[25] An approach based on visual detection is affected by subjectivity, and the perception of anomalies varies from individual to individual.

[26] For an exhaustive description of our experience using GPS see CAMPANA 2001, pp. 47-71.

[27] In general on handhelds Pocket PC see, WILSON, 2001, pp.40-45; COOPER, 2001, 29-32. For archaeological application POWLESLAND in the CAA Conference proceedings of 2002 (in press).

4 GIS based interpretation of remotely sensed data and their integration with other data sources

It could be argued that it is possible to interpret remotely sensed data quite effectively without mapping. This may sometimes be true but to understand landscapes fully, and to combine multi-source data, there is no alternative to mapping and to management of the related documentation through GIS technology.

As has been recently noted, the archaeological interpretation of remotely sensed data is not simply the identification of anomalous features but the non-destructive recovery of archaeological information by producing accurate maps and completing entries for a descriptive DBMS. Only an experienced archaeologist can carry out this process in its entirety[28]. Moreover, archaeological interpretation in terms of detecting, accurate mapping and writing archaeological descriptions of relevant features is a dynamic process based on mental comparison of the detected features with known structures or landforms, a process driven by archaeological feedback. The interpretation of archaeological survey data, as of any other source material, represents the first stage of in the research process, but in contrast to excavation it is in this case a repeatable and non-destructive process.

As in every form of landscape analysis it is possible to distinguish two different levels of analysis: intra-site and inter-site analysis. For both kinds of analysis GIS represents the appropriate environment for the complex processes of interpretation and integration between different sources. Our approach has been based on 2D visual interpretation and the exploration of 3D representations.

In the first stage of analysis (intra-site) the remotely sensed data has to be combined and reinterpreted together if we are to retrieve the maximum information. The archaeological interpretation of the combined prospection data is done directly on-screen, through the overlaying of different forms of imagery, along with relevant geographical information (technical, thematic and historical maps) and known archaeological data. The archaeological features have been drawn onto different thematic layers (every individual element as a polyline and polygon)[29] and the alphanumeric information stored in a relational database linked to the graphical interpretation.

The quantitative results of our archaeological interpretations are summarised in Table-2. This shows clearly how every remotely sensed technique that we adopted increased the archaeological record. As expected, the vertical aerial photographs from 1938 and 1954 made a profound contribution to the study of the landscape. In Tuscany, and more generally in Italy, after the Second World War the landscape underwent great transformations. Mechanized agriculture, building operations, provincial road schemes, highways, aqueducts, pipeline, power lines and the like destroyed numerous archaeological sites. Only the appropriate use of historical air photo sorties can now show us these sites. Moreover, this source allows us to visualize the whole landscape as it was before such these destructive human interventions, giving a reasonably realistic idea of how it appeared in 'pre-industrial' times.

The differing returns, in our area of study, from vertical aerial photographs taken in 1976, 1994 and 1996, compared with those from the historical flights, has to be measured against the sometimes inappropriate "time-window" in which they were taken, as well as the loss of many sites because of recent changes in the landscape.

As regards the utility of oblique aerial photographs in the sample areas it is difficult to draw up a balance at this stage. The fourteen sites identified in these areas could seem a low number in comparison with the returns from other sources. It should be remembered, however, that the amount of flying undertaken in the sample areas during the Aerial Archaeology School provided only very partial coverage of the area. On the basis of the results achieved, and in the light of evidence from elsewhere, we feel that the high value of the oblique technique is undeniable. We recognise that it is too early to form a judgement for our own area and that we need to fly again and again, year after year in the right "time-windows". However, oblique aerial photography has already permitted us to recognize fourteen sites within our sample areas, three of them never identified before, while in four other cases we added to the existing archaeological information.

Positive results have been obtained using Ikonos-2 multispectral imagery, with the recording of 84 features in the sample areas, of which 39 are new sites. In fourteen cases where anomalies were identified previously through vertical or oblique photographs it was possible to add to the existing information. We should perhaps note one peculiarity of Ikonos-2 imagery. Through Ikonos-2 we can recognise many features that were visible in the early air photographs but which are no longer identifiable in those taken between 1976 and 1996. This situation perhaps derives from the inappropriate "time-window" in which the later photographs were taken, or alternatively from the higher sensitivity and computer enhancement capabilities of the Ikonos-2 data. If confirmed, however, this trend will indicate HRSI as an important tool for monitoring and exploration of the archaeological heritage.

In conclusion, we believe that most of the results obtained from analysis of the Ikonos-2 imagery depend very much upon the multispectral properties of the sensor. The Ikonos-2 scanner has four bands that range from 450 to 880 nanometers, although not all are equally useful. In our study we concluded that bands 2 (green), but above

[28] Doneus *et al.* 2001, p.31.

[29] It is possible to distinguish at least two groups of individual elements detected by the processing of imagery, area features and line features, the latter having length but no "shape" or breadth.

all 3 (red) and 4 (near infrared), show the most potential for the identification of archaeological features. Band 1, blue, suffers from atmospheric attenuation and scattering that degrades its definition. Red and near infrared images are less affected by haze and provide good definition for soil marks and crop marks. Above all the near infrared represents the most powerful band. This band is particularly sensitive to plant health and can often detect water stress in vegetation before it can be seen by the naked eye. Despite these promising early results the true potential of this type of imagery is still not fully clear and needs to be further evaluated to test its responsiveness under a broad range of environmental conditions.

Once we have reached an understanding of each site individually it is possible to proceed to the analysis of inter-site relationships. The GIS environment allows us to interpret sites in the landscapes not simply as a distribution of points but rather as real objects possessing shape, surface modulation, length and orientation. For instance using statistical tools and spatial analysis it is possible to classify orientations and compare the results for settlements, cemeteries, roads etc of the same or of different chronological periods.

It has to be admitted, however, that we have still a lot of work to do in terms of developing archaeological interpretation tools based on GIS technology for exploiting the full possibilities presented by this form of prospection.

5 Conclusions

In any approach to applying remote sensing to landscapes or to an individual site, it is necessary not only to choose the right mix of systems for data acquisition and data interpretation, but also to identify the right combination of remote sensing and "traditional" methods.

The evaluation and use of Ikonos-2 imagery, oblique aerial photography and geophysics forms part of a wider strategy aimed at understanding the peculiarity of every single source so that we can on each occasion employ the appropriate combination of remote sensing techniques to maximize our understanding of the ancient landscape.

By applying multistage sensing techniques to our landscape projects we are beginning to develop a system of modular prospecting. Starting from a broad over-view of the survey area, we move through a series of steps, ultimately to a level of detailed definition. The use of different data-sources allows us to work at a variety of spatial resolutions. But it also introduces two other key factors: spectral and temporal resolution. "Spectral resolution" refers to those parts of the electromagnetic spectrum that are employed in each technique: black-and-white panchromatic for vertical photography, colour or near infrared for oblique photography, and blue/green/red/near-infrared for Ikonos-2 satellite imagery. "Temporal resolution" refers to the frequency and length of time over which the images have been collected for the same area. The vertical photographs at our disposal have a wide temporal range, between 1938 and 1998. The oblique photographs and Ikonos-2 satellite imagery have been collected only since 2000.

Even at this early stage we can say that the introduction of this approach to landscape analysis, running hand in hand with continuing programmes of field-walking, has transformed both our way of working and our understanding of the ancient landscape.

Stefano Campana

Acknowledgments

Dr. Stefano Campana work was funded by the Swiss National Science Foundation (Research Commission, University of Italian-speaking part of Switzerland) at the Remote Sensing Laboratory of Florence University directed by prof. Enzo Pranzini.
The authors thank Chris Musson for reading the translation of the text and for the related comments.

References

ALESSANDRO, V. and PRANZINI, E., 1992. Il telerilevamento nella gestione dell'ambiente tropicale. In *Studi Geologici Camerati*, Special Issue, pp.9-27.

BAGGIO, P., BIROCCO, C.A., COLOSI, F., PIRO, S., VERSINO, L. E. and ZAMBONI, C., 1998. Telerilevamento, ricognizioni archeologiche e prospezioni geofisiche sulla Collina Orientale di Selinunte. In *Science and technology for the safeguard of cultural heritage in the Mediterranean basin*, 1st International Conference, Catania-Siracusa, November 27-December 2 1995.

BARISANO, E., BARTHOLOMES, E. and MARCOLONGO, B., 1984. Interprétation integrée de données télédetectées. HCMM, Landsat, photos aériennes, corrélée avec des aspect physiographiques et archéologiques dans la pleine vénitienne occidentale, CNRS, Paris.

BIANCHI, G., BOLDRINI, E., CASINI, A., CICALI, C., GUIDERI, S., and ZEFFIRERO, A., 1997. San Silvestro. Guida al Parco Archeominerario, Firenze.

CAMPANA, S., 2001. Carta Archeologica della Provincia di Siena. Murlo. Vol. 5, Siena.

CAMPANA, S. and FORTE, M., (eds.), 2001. Remote Sensing in Archaeology. XI International School in Archaeology, Certosa di Pontignano, 6-11 December 1999, Florence, pp.17-62.

CAMPANA, S. and PRANZINI, E., 2001. Telerilevamento in Archeologia. In Campana, S. and Forte, M. (eds.), *Remote Sensing in Archaeology*, pp.17-62.

CAMPANA, S., MUSSON, C., PALMER, R.,. In Volo nel Passato,

University of Siena, in press.

CECCARELLI LEMUT M.L., and GARZELLA G., 1996. Populonia e Piombino in età medievale e moderna. Pisa.

COLOSI, F., POMPEO, L., SANGIORGIO, D. and ZAMBONI, C., 1996. Elaborazione ed interpretazione di immagini telerilevate per lo studio sistematico delle testimonianze antropiche nel territorio. I casi di Palmyra e di Selinunte. In *Archeologia e Calcolatori*, 7.

COOPER J., 2001. Heading in the right direction? GeoEurope takes stock of the coming revolution in location-based services – and it promise of a bonanza for the GI industry. In *GeoEurope*, 10/1, pp. 29-32.

CORTEMIGLIA, G.C., MAZZANTI, R. and PAREA, G.C., 1983. Geomorfologia della baia di Baratti (Livorno, Toscana) e della sua spiaggia, Geografia Fisica e Dinamica Quaternaria, 6, pp. 148-173.

COSCI, M., 2001. Il contributo della fotografia aerea allo studio e alla valorizzazione del territorio: metodi adottati e risultati conseguiti. In Di Gangi, G., Lebole, C. M., *La gestione del territorio. Memoria, partecipazione, sviluppo della ricerca*, Proceedings of Saluzzo National Meeting, 11-12 November, 2000, Cuneo, pp.55-64.

COSTI, A., LAZZARO, L., MARCOLONGO, B. and VISENTIN, J., 1992. La centuriazione romana fra Sile E Piave nel suo contenuto fisiografico. Nuovi elementi di lettura. CNR, Padova.

CREMASCHI, M. and FORTE, M., 1999. Reconstructing a fossil landscape by Remote Sensing and GIS applications: sites, virtual models and territory during the Middle Bronze Age in the Po Plain (Northern Italy). In *Archeologia e Calcolatori*, 9, pp. 207-227.

CRIST E.P., CICONE R.C., 1984, *Application of the Tasselled Cap Concept to Simulated Thematic Mapper Data*, "Photogrammetric Engineering and Remote Sensing", vol.50, 3, pp. 343-352.

DICEGLIE, S., 1992. I "cerchi di pietra" di Lampedusa (Telerilevamento archeologico dell'Isola). In *Osservazione dello spazio dell'Italia e delle sue regioni: metodo, risultati e prospettive*, proceedings of 3th AIT National Conference, L'Aquila, 7-10 November 1990, Pisa, pp. 421-439.

DOLATOWSKA, A. and GOLIASZ, J., 2000. Aerial Archaeology Workshop Leszno 1998. The view from behind. In *AARG*, 20, pp.41-42.

DONEUS, M., 2001. Precision Mapping and interpretation of oblique aerial photographs. In *Archaeological Prospetion*, 8, pp.13-27.

DONEUS, M., EDER-HINTERLEITNER, A. and NEUBAUER, W., 2001. Archaeological Prospection. Proceedings of 4th International Conference on Archaeological Prospection (Vienna, 19-23 September 2001), Vienna.

DONOGHUE, D., 2001. Multispectral Remote Sensing for Archaeology. In Campana, S. and Forte, M. (eds.), *Remote Sensing in* Archaeology, pp.181-192.

DRIVER, T. and MUSSON, C., 2000. Italy from the ground. Remote Sensing in Archaeology, Summer School. Università degli Studi di Siena, Certosa di Pontignano, Siena, 6-11 novembre 1999. In *AARG*, 20, pp.51-54.

FEDELI, F., GALIBERTI, A. and ROMUALDI A., 1983. Populonia e il suo territorio. Profilo storico-archeologico, Florence.

FORTE, M., MONTEBELLI, M. and TUSA S., 1998. Il progetto valle del Belice: applicazioni GIS e di Remote Sensing su dati archeologici. In *Archeologia e Calcolatori*, 9, pp. 291-304.

FRANCOVICH, R. and ZIFFERERO, A., (eds.), 1999. Musei e Parchi Archeologici. IX International School in Archaeology, Certosa di Pontignano, 15-21 December 1997, Florence.

FRANCOVICH, R. and GINATEMPO, M., (eds.), 2000. Castelli. Storia e archeologia del potere nella Toscana medievale, Florence.

FRANCOVICH, R. and VALENTI, M., 2001. Cartografia archeologica, indagini sul campo ed informatizzazione. Il contributo senese alla conoscenza ed alla gestione della risorsa culturale del territorio. In Francovich R., Pellicanò A., Pasquinucci M. (eds.), *La carta archeologica. Fra ricerca e pianificazione territoriale*, Florence, 6-7 May 1999, Florence, pp. 83-116.

FRONZA, V., 2000. *Il sistema di gestione degli archivi nello scavo di Poggio Imperiale a Poggibonsi (Insegnamento di Archeologia Medievale dell'Università di Siena). Una soluzione all'interno della "soluzione GIS".* In D'Andrea A., Nicolucci F. (eds.), Archeologia e Calcolatori, XII, Proceedings of I National Workshop of Computer Archaeology, pp. 125-137.

FRONZA, V., NARDINI, A., SALZOTTI, F. and VALENTI, M., 2001. *The GIS solution for an excavation. Some considerations about the experience of the Medieval Archaeology Chair at the University of Siena.* In Stancic Z., Veljanovski T. (eds.), Computing Archaeology for understanding the past CAA 2000, proceedings of the 27th Conference, Ljubljana, April 2000, Oxford, , pp.173-178, (BAR International Series, 931).

ISABELLA, L., SALOTTI, F. and VALENTI, M., 2001. *L'esperienza dell'insegnamento di Archeologia Medievale nel campo dell'informatica applicata.* In De Marchi M., Scudellari M., Zavaglia A. (eds.), Lo spessore storico in urbanistica, pp.31-64.

LILLESAND, T.M. and KIEFER, R. W., 1994. *Remote Sensing and Image Interpretation*, New York.

MARCHISIO, E., PASCQUINUCCI, M., PRANZINI, E. and VIGNA GUIDI, G., 2000. The Pisa territory Project. In Pasquinucci, F. Trément M. (eds.), *Non-destructing techniques applied to landscape*, Oxford.

MARCOLONGO, B. and MASCELLARI, M., 1978. Immagini da satellite e loro elaborazioni applicate alla individuazione del reticolato romano nella pianura veneta. In *Archeologia Veneta*, I, pp.131-154.

MAZZANTI, R.,1995. La Scienza della Terra nell'area della Provincia di Livorno a sud del Fiume Cecina, Livorno.

PASQUINUCCI, M. and TRÉMENT F., (eds.), 2000. Non-destructing techniques applied to landscape archaeology. Oxford.

PAVLIDIS, L., C., FRASER, C.S. and OGLEBY, C., 2001. The application of High-resolution satellite imagery for the detection of ancient Minoan features on Crete. In Burenhult G. (eds.), *Archaeological Informatics: Pushing the Envelope CAA 2001, proceedings of the 29th Conference*, Gotland, April 2001, pp.393-400, (BAR International Series, 1016)

PIERI, M. and PRANZINI E., 1989. Geomorphological evolution of the Pisa plain in historic time deduced from Landsat TM data. In *Global Natural Resource Monitoring and Assessment: Preparing for the 21st Century*, proceedings, Venezia, Novembre 1989, pp. 1385-1388.

POWLESLAND D., in press. Managing Stratigraphy, the integration of stratigraphic, database and graphical information from excavation. The Digital Heritage of Archaeology CAA 2002, proceedings of the 29th Conference, Heraklion, April 2002, (BAR International Series).

PRANZINI E., SANTINI C., 1999, *Depositional environments mapping in alluvial plains based on wetness seasonal changes*, in *Remote sensing for Earth Science, Ocean, and Sea Ice Applications*, "SPIE Europto series", 3868, pp. 283-291.

VALENTI, M., 1999. Carta Archeologica della Provincia di Siena. La Val d'Elsa (Colle di Val d'Elsa e Poggibonsi). Florence,pp.10-14.

WILSON D.R., 2000. Air Photo Interpretation for Archaeologists. Tempus.

WILSON J.D., 2001. The Next Fronter. Discovers how GIS is empowering a new generation of mobile solutions. In *GeoEurope*, 10/7, pp.40-45.

Internet references

Canto Cumulus Software:
<www.canto.com>

Medieval Archaeology
– Department of Archaeology and Arts History
– University of Siena: <archeologiamedievale.unisi.it>
– Remote sensing pages:
<192.167.112.135/NewPages/REMOTESENS/REMOTE.html>

NASA archaeological section:
<www.ghcc.msfc.nasa.gov/archeology/archeology.htm

Table 1 *– GIS data entry and derived data*

		Raster	*Vector*	*GPS*	*Derived Datasets*
Basemaps	**Technical**	***regional maps 1:10.000***	***regional maps 1:10.000***	***micro-digital terrain modelling using differential GPS***	***TIN of sample areas***
		regional maps 1:5.000	***regional maps 1:2.000***		***TIN of Populonia area (1:2.000)***
		cadastral maps 1:5.000	***cadastral maps 1:5.000***		
	Thematic		published archaeological sites; administrative boundaries; technical reference grids, etc.; wetlands areas (1:10.000); vegetative surface cover (1:25.000); non-vegetative surface cover (1:25.000); regional geology, hydrography, transportation, climate, precipitation, mean annual temperature 1:100.000.		vector overlay of : land units and field-survey sample areas; texture mapping of vector overlays on the TIN.
	Historical	***various maps dating from between 1779 and 1830***	***1830 historical cadastral maps***		***1830 vegetative surface cover***
		1940 (Military Institute for Cartography - 1:25,000)	***1940 map of settlement (Military Institute for Cartography - 1:25.000)***		***1940 Thiessen poligon of settlement***
			old mineral-prospecting permissions (1900-1970)		
Remotely Sensed Data	**Satellite**	***landsatTM (summer-winter 1995); Ikonos MS (spring 2000)***	***line and area features***	***ground survey data: mapping and waypoints***	***texture mapping on TIN***
	Vertical photograph	***Fights of years: 1938, 1940, 1954, 1976, 1994, 1996, 1998***	***line and area features***	***ground survey data: mapping and waypoints***	***texture mapping on TIN***
	Oblique photograph	***Aerial survey of years: 2000 and 2001***	***line and area features***	***aerial tracks and positioning of photographs; ground survey data: mapping and waypoints***	***texture mapping on TIN***
	Geophisics	***magnetometry survey of years: 2001 and 2002***	***line and area features***	***ground survey data: mapping and waypoints***	***texture mapping on TIN***
Field survey		***geocoded photographs, QTVR and movies***	***point, line and area of: artefact scatters, archaeological sites divided up so as to show settlement patterns at different periods***	***Mapping of: artefact scatter areas, perimeter of monuments***	***texture mapping on TIN***

***Table 2** – Relationship between remotely sensed techniques*

	Features in the two sample areas	**Features visible only through a single source**	**Increased information for features visible from two or more source**
Vertical aerial photographs 1938-1954	92	36	32
Vertical aerial photographs 1970-1996	22	6	2
Oblique aerial photographs 2001	14	3	4
Ikonos multispectral imagery July 2000	84	39	14

Hyperspectral airborne remote sensing as an aid to a better understanding and characterization of buried elements in different archaeological sites

Cavalli R.M.[1], Marino C. M.[2] and Pignatti S[3]

[1] CNR IIA –LARA V. Mt. d'Oro 11, 00040 Pomezia, Italy
cavalli@lara.rm.cnr.it
[2] State University of Milano Bicocca, Department of Environmental Sciences
carlo.marino@unimib.it
[3] CNR IMAAA, C.da S. Loya, Tito, Potenza, Italy
pignatti@lara.rm.cnr.it

Abstract. In recent years the development of new technologies, such as Remote Sensing and GIS, allowed archaeologists to more easily gain insight into the relationships between buried remnants and subtle surface features also integrating the information derived from field work surveys. Due also to the advent of the imaging spectrometry, the research was focused on extracting information on subtle spectral features in different environments and landscape contexts. In particular, the analysis of airborne multi/hyperspectral images gave the opportunity to recognize peculiar networks of ancient settlements by investigating both spectral and spatial information. In this presentation the aim is to describe how the development of airborne hyperspectral scanner data sets, such as those recorded from the Daedalus AA5000 Multispectral Infrared and Visible Imaging Spectrometer (MIVIS) of the Italian National Research Council (CNR), has markedly increased and improved the information related to some ancient human witnesses in Italy.

1 Introduction

Archaeologists have been using air pictures since decades to detect buried structures by means of the study of effects brought to surface by the physical discontinuities present the first sub-soil layer.

In the last years the new technologies, in increasing development, like Remote Sensing and Informative Systems, have represented an additional investigation instrument, useful for the study of relationships existing between community and its territory by means of the synoptic land analysis, processing and integrating information of different kind collected in campaigns in situ.

Sensors generally used by satellite and airborne remote sensing have enabled to study natural phenomena by means of few spectral bands (AVHRR, Landsat MSS4, SPOT 4, TMS, TIMS, etc.), while the technological development of the space industry achieved in the last years have allowed to realize sensors which combine the advantages of spectroscopy to remote sensing. Imaging spectrometers and hyperspectral sensors are remote sensing instruments which join the analytical capacity of a spectrometer to the spatial representation of an imaging sensor. They can be characterized by hundred narrow bands with a spectral resolution of 10nm. The imaging spectrometers then produce a complete spectrum for each image pixel, and allow to identify materials, unlike low spectral resolution sensors which enable only their characterization.

The operative efficiency is therefore determined both by the quantity of contiguous bands and their width.

As a matter of fact, to characterize as better as possible materials' spectral signature, besides a high number of bands to sample a wider portion of the electromagnetic bands, bands' reduced spectral width is also fundamental to obtain more accurate measurements of spectral absorption.

The increase of spectral and spatial resolution, and then the possibility of observation given by Remote Sensing, allow to define quantitatively some surface parameters of ecosystems, thus facilitating the parameterization of natural processes useful for evolutionary models' development.

2. Airborne Laboratory Environmental Research

In 1991 the National Research Council of Italy (CNR) established LARA Project (Airborne Laboratory for Environmental Research) within the Strategic Project "Climate, Environment and Territory in Southern Italy" to give the national and international scientific community an advanced support activity helpful for environmental research by means of airborne remotely sensed hyperspectral data (Bianchi et al., 1994, Bianchi et al. 1996). Since 1994 LARA has been carrying out its research activities by means of MIVIS (Multispectral Infrared Visible Imaging Spectrometer - Daedalus AA5000) hyperspectral sensor (figure 1); LARA research group has been processing, assessing quality, distributing and analyzing MIVIS data in Pomezia ground station. Also archaeological investigations have being deployed through airborne remotely sensed hyperspectral data.

Fig. 1. MIVIS scanner on board

2.1 Multispectral Infrared Visible Imaging Spectrometer - MIVIS AA5000

MIVIS system's peculiarity and unicity depend on the characteristics of this modular hyperspectral scanner composed of 4 spectrometers which collect simultaneously the electromagnetic radiation coming from the earth surface by recording 102 spectral bands.

The simultaneous measurement of a high number of channels with a high spectral and spatial resolution has required an innovative technology for optics, sensors, and the electronic processing. The combination of a high resolution in the Middle Infrared spectral region with a good sensitivity in the thermal Infrared region has made necessary to solve several problem during system design phase (Bianchi et al. 1994, Bianchi et al. 1996)

2.2 Airborne platform

Remote sensing employs different kind of platforms to detect the earth surface: free and captive balloons, airplanes and spacecrafts. The functional aptitude of these platforms is determined by the altitude to which they can arrive and work, by the degree of "manovrability", by the greater or less difficulty to manage the platform. Aircrafts and spacecrafts have actually a grater practical utility.
The airborne platform aboard which MIVIS hyper-spectral sensor is installed is a CASA C212/200 of Compagnia Generale Ripreseaeree in Parma. This aircraft because of its operative characteristics and its load carrying capacity can be used as a real airborne laboratory.

The two-engined turbopropeller CASA C212/200 can reach the maximum flight altitude of 7500m, the maximum speed of 180 nods and has a stall velocity of 80 nods. During MIVIS hyperspectral data acquisition campaigns CASA airborne platform operates at a speed interval comprised between 180 and 110 nods, at a flight altitude between 1200m a.s.l. and 6000 a.s.l.

3.0 MIVIS Remote Sensing to study archaeological remnants

To test the investigation aptitude of the sensor and verify the reliability of MIVIS hyperspectral system the first data acquisition was carried out in July 1994. Many overflight sites were chosen for their geological importance (Etna, Vulcano island, Stromboli and Lipari Islands), while other areas presented an archaeological interest (Selinunte, Halaesa and Mozia) (Bianchi et al., 1997). Ever since numerous campaigns were conducted over archaeological sites to develop and improve MIVIS hyperspectral scanner's potentiality.

The most significant archaeological airborne campaigns were carried out by the University of Palermoo over the site of Halaesa, by LARA, Institute on Atmospheric Pollution of CNR in cooperation with the University of Rome, Ancient Topographic Section, over the site of Mozia (Bianchi et al. , 2000).

The Department of Hydraulic Engineering and the Institute of Archaeology of the University of Palermo have analyzed MIVIS data for archaeological researches over Halaesa to study relations existing between community and territory by means the realization of a GIS. The purpose of this GIS was to integrate and compare cartographic products, digital acquisition (MiVIS scenes and air pictures), direct surveys and epigraphic documentation. The integration of such data in the GIS enabled to produce the thematic map of settlement potentiality, to reconstruct the exploitation of water resources in the past and recognize buried structures by means of MIVIS data processing.

The research carried out on the Archeological Park of Selinunte and Mozia Island was instead based on the hypothesis that MIVIS sensor is capable to give through appropriate processing information which can represent superficial elements (evidences) of the presence of buried remnants.

As known, the presence of anthropic remnants in the first layer of subsoil can be detected on the basis of spatial discontinuity of mechanical characteristics of the soil. In particular grass vegetation is inclined to stress possible dishomogeneity of substratum both of physical (humidity) and thermal kind (day excursion of soil temperature, etc) by means of its different growing.

At the beginning to develop and improve the investigation methodology was necessary to identify a site diffusely detected, where results achieved can be compared with the archaeological information already acquired , to verify directly the method's soundness. For this reason the archaeological site of Selinunte has been chosen and studied since decades. The encouraging results obtained from the researches conducted over this site have driven to apply the same methodology on Mozia Island.

3.1 Archaeological Park of Selinunte

In the south-western coastline of Sicily (Italy) there are the ruins of the ancient Greek settlement of Selinunte founded in the middle of the 8th century B.C. by the colonists coming from the Greek town of Megara Iblea (Siracusa). The Selinunte Archaeological Park has been largely investigated with both excavation and geophysical surveys. To support these campaigns CNR LARA flew over Selinunte Archaeological Park, in July 1994 and May 1996, with the Multispectral Infrared and Visible Imaging Spectrometer (MIVIS) instrument.

These campaigns have been carried out to add more information to that acquired from field archaeological and geophysical campaigns, which have not yet answered all the archaeological questions related to the area.

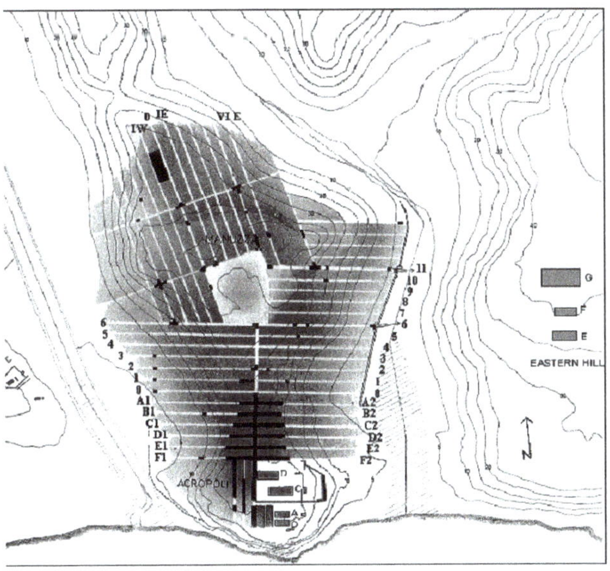

Fig. 2. Ancient Selinunte city plan, based on Mertens most recent field works (Di Vita 1996)

These remotely sensed data have been analysed in an attempt of identifying both unknown and hypothesized street axes of the urban system of the ancient Selinunte. A preliminary trial exploiting remotely sensed imagery for the detection of this ancient street network was performed on air photos by G. Schmiedt (1957), who focused on an hill (Manuzza), to the north of the "Acropolis", detecting an NNW-SSE trending urban network. Successive investigations on the area have partly confirmed the presence of these structures and relative junctions (Rallo 1977; Rallo 1984). The same archaeological features have been investigated by processing the hyperspectral MIVIS data, while aerial photos and radar images were useful for the visual interpretation of the processed MIVIS pictures.

They have been found to be in good agreement with an ancient buried street pattern, previously and only partially pointed out by archaeologists (fig.2).

3.2 Mozia

The encouraging results obtained in the study of Selinunte archaeological Park has pushed the research group to apply the same methodology, always using archaeological information, on the Phoenician settlement of Mozia, which occupies the tiny island (45 ha) of S. Pantaleo, in the "Stagnone of Marsala", Trapani, Sicily (CD-Rom Cavalli and Pignatti pg 45-46). Here an almost secular excavation activity has brought to light the remnants of a village, structured already at the end of VII cent. B.C, even though evidences of first Phoenician presence traced back to XII cent. b.C. and a prehistoric settlement is proved.

Fig. 3. Street network as derived from analysis of exclusive archaeological lineaments.

MIVIS data used for this study were collected on July 20, 1994 at 10.15 a.m. (local time) at an altitude of 2000 m a.s.l. at a ground resolution of 4 m. The comparison of the considerations derived from the analysis of the air photos and the remotely sensed images stresses some differences: most traces shown by the photograms dated between 1943 and 1968 are not present in MIVIS scenes that instead reveal numerous linear discontinuities in the South-western area of the Island.

On the basis of the information achieved by means of hyperspectral data analysis, which should be verified at ground in the future, it seems that the extension of the urban planning is wider than what assumed. This is also confirmed by the excavation activities, round the House of Amphorae and in the "zone B", which brought to light several housing units lying along road axes with NE-SW direction, according to what detected by air scenes (Fama. 1995)

The analysis of MIVIS hyperspectral data has proved to be an innovative mean to support the archaeological research, both as far as the Archeological Park of Selinunte and Mozia Island are concerned, as it has enabled to detect elements related to the urban structure which were not identified by the traditional photointerpretation of historical photos of the same areas (fig. 3).

References

AA. VV., 1992. Monograph on the geophysical exploration of the Selinunte Archaeological Park. BGTA, XXXIV, pp. 134-135.

BAGGIO P., BIROCCO C.A., COLOSI F., PIRO S., VERSINO L. E ZAMBONI C., 1998, Telerilevamento, ricognizioni archeologiche e prospezioni geofisiche sulla Collina Orientale di Selinunte. I Congress International on " Science and technology for the safeguard of cultural heritage in the Mediterranean basin". Novembre 27 - Dicembre 2 1995, Catania- Siracusa (Italia), vol. I.

BIANCHI R., MARINO C.M., PIGNATTI S., 1994, Airborne Hyperspectral remote sensing in Italy, Europto, Rome 27-29 September 1994, SPIE V. 2318, pp. 29-37.

BIANCHI R., CAVALLI R.M, FIUMI L., MARINO C. M., PIGNATTI S., 1996, Airborne imaging spectrometry: a new approach to environmental problems, Proceeding of the XVIII ISPRS Congress - Vienna 9-19 July 1996 -Vol. I - pp. 128-132.

BIANCHI R., BOGLIOLO M.P., BUONGIORNO F., CAVALLI R.M., MARINO C.M., PIGNATTI S., SALVI S., 1997, MIVIS per la Sicilia, Rivista Internazionale di Telerilevamento "Sistema Terra", dicembre 1997 anno 6, numero 1/3. pp. 24-27.

BIANCHI R., CAVALLI R.M., COLOSI F., MARINO C.M., PIGNATTI S., POSCOLIERI M., 1997, Airborne hyperspectral MIVIS data over Selinunte ancient town area (Sicily, Italy) as a support to classical archaeological investigation, Proceedings of the Third International Airborne Remote Sensing Conference and Exhibition Development, Integration, Application & Operations, Copenhagen, Denmark 07-10 July 1997. Vol. II, pp. 761-768.

BIANCHI R., CAVALLI R.M., CORSI C., MARINO C.M., PIGNATTI S., 2000, Ricerche topografiche in Sicilia: integrazione tra metodi tradizionali e dati iperspettrali da piattaforma aerea, Proceedings of the XVth International Congress of Classical Archaeology Amsterdam 1998, pp.15-16.

CAVALLI R.M., COLOSI F., PIGNATTI S., POSCOLIERI M., 1999, Il Telerilevamento aereo per lo studio dei beni archeologici. Applicazione dei dati iperspettrali sul Parco Archeologico di Selinunte, Bulzoni - Roma , pp. 339-361.

COLOSI F., POMPEO L., SANGIORGIO M. E ZAMBONI C., 1995, Elaborazione ed interpretazione di immagini telerilevate per lo studio sistematico delle testimonianze antropiche nel territorio. I casi di Palmyra e Selinunte. Archeologia e calcolatori, 6, pp. 35-60.

DI VITA A., 1996. L'urbanistica della Sicilia greca. in AA.VV., I greci in Occidente. Bompiani, Venezia (Italia), pp. 263-308.

FAMA M.L., 1995, Appunti per lo studio dell'urbanistica di Mozia, Actes du III Congres International des Etudes Phéniciennes et Puniques (Tunisi 1991), Tunisi, I, pp.442-434.

MERTENS D., 1989. Die Mauern von Selinunt, Vorbericht der Arbeiten des Deutschen Archeologischen Instituts Rom 1971-75 und 1985-87. RM, 96, pp. 87-154.

PIRO S. e VERSINO L., 1992. Primi risultati delle indagini geofisiche effettuate nell'area archeologica di Selinunte (Collina Orientale). Metodo magnetico, elettromagnetico induttivo ed impulsivo georadar. Monografia Selinunte 1 (Intesa di programma CNR-MISM), Bulzoni, Roma, pp. 9-32.

PIRO S. e VERSINO L., 1996. Metodi geofisici integrati nell'area archeologica di Selinunte. Monografia Selinunte 3 (Intesa di programma CNR-MISM), Bulzoni, Roma, pp. 95- 116.

RALLO A., 1976-77, Scavi e ricerche nella città antica di Selinunte. Relazione preliminare. Kokalos, 22-23, pp.720-723.

RALLO A., 1984, Nuovi aspetti dell'urbanistica seluntina. ASAtene, 46, pp.81-96.

ZOPPI C., 1996, Note selinuntine. 1. La necropoli di contrada Galera-Bagliazzo ed il limite nord dell'abitato di Manuzza. Monografia Selinunte 3 (Intesa di programma CNR-MISM), Bulzoni, Roma, pp.137- 152.

Archaeology at War

Armando De Guio

Università di Padova, Dipartimento di Scienze dell'Antichità, P.zza Capitaniato 7 - 35139 Padova (Italy)
tel +39 049/8274579 or 8274573; fax +39 049 8274613
E-Mail: deguio@interplanet.it

Abstract. CISAS (the International Centre for the Study of Surface Archaeology, based at the Department of the Sciences of Antiquity, University of Padua, Italy) has addressed a well-established domain of inquiry ('Landscape of Power' simulation and modeling) with special attention to the processes involved in the projection of social agency and complexity into the surface archaeological record. A number of lines of research (Object-Pattern-Scenery recognition, VR, GIS...) have been designed at the intrasite, but mainly at the 'near-site' and 'off-site' levels of spatial resolution. The main emphasis is on those 'fossil landscapes' (specific cells of the eco-mosaic that we find eg in both the lowland Po Plain and the pre-Alpine region) which preserve in a quite extraordinary way a wide range of power-related ('off-site power') network infrastructures, such as drainage/irrigation, terracing, land-divisions, connectivity. Moreover this analytical path led us recently to what seems to be the oldest known 'road' in the world (our 'dirt-road' at Brendola (VI): 4915-4735 cal BC (2□): Beta-107672). A fascinating perspective is related to the virtual exploration and enhancing of First World War images: this high-quality, easily geo-referenced package of aerial photographs (from Italian, Austro-Hungarian, French and British archives) offers a wealth of remote-sensing information, due to the unprecedented wartime deforestation: the critical circuit here is between an 'Archaeology of the War' (war = information vs archaeology = noise) and an 'Archaeology through the War' (archaeology = information vs war = noise), while a major series of later images give us the possibility to monitor a critical conjunctural evolution (over a century) of site/off-site formation processes in local landscapes.

1 Introduction

The core theme of this paper is a special approach to an 'Archaeology of War', as a source of both 'collateral damage' but also 'collateral benefits' for our research. Although in the present international situation this point might be seen as insensitive, little more than a frivolous intellectual *divertissement*, in our quite particular locally sensitive operational frameworks (see below) it does address a meaningful domain of interest and offers a stimulating frontier for archaeology[1].

We commenced this line of research quite recently, following a partially random critical path towards developing a theoretical and practical package for Surface Archaeology, which is the reason for and the overall goal of our ten-year-old CISAS (the International Centre for the Study of Surface Archaeology, based at the Department of the Sciences of Antiquity at Padua University: fig. 1).

We have in fact developed an ambitious programme for a thorough understanding of the complexity of the formation of what we call the 'Surfacies'. This is a hybrid English and Latin neologism (combining *Surface* and *Facies*), which considers the surface archaeological record not as a field for simple, blind sampling games, but as a complex, time, space, and functional palimpsest of 'objects' (ie operational containers of relevant information), starting from remote sensing objects and including, for instance, prospection objects, morphological objects, soil, core, stratigraphic window, density objects and expert (previous

[1] We are indebted to: a) for financial and/or logistical support: the Provincial Administration of Vicenza, Regione Veneto, Comunità Montana-Asiago, APT-Asiago, the Municipalities of Foza, Luserna, Lusiana, Rotzo, Camera di Commercio di Vicenza, Centro Documentazione Luserna. The Municipality of Luserna, in particular, financed (October 2001) the first campaign of a new collaborative project on the palaeometallurgy of the Vezzena-Luserna-Lavarone uplands. b) for crucial resources provided (in particular, aerial photographs, cartography and digital terrain models): the Regione Veneto and the Provincial Administrations of Vicenza and Trento, Museo Tridentino di Scienze Naturali; Dr Heniz von Lichen (photographs copyright: Sammlung Lichem/MediaDomain Verlag, D-82284 Grafrath) and Centro Documentazione Luserna; CGR-Parma, Museo Storico Italiano della Guerra - Rovereto; Museo Caproni – Trento; Museo del Risorgimento - Vicenza; the Municipalities of Asiago (Archivio Piatti) and Foza; Comunità Montana-Asiago, Fondazione Cassa di Risparmio di Verona; Air Photo Library - Dept of Geography – University of Keele - UK; Dr Luca Apolloni; Biblioteca Bertoliana – Vicenza; Archivio di Stato - Trento; dott. Giancarlo Bortoli. c) for information, and/or digital support and/or useful suggestions, in particular: dott. Luca Apolloni, dott. Mario Basso, Andrea Betto, dott. Gianni Bonaldo, dott. Giancarlo Bortoli, dott. Carlo Bressan, Michele Busato, dott. Gianni Ciurletti, Prof. Ermanno Finzi, Paolo Kirschner, dott. Gian Vittorio Martello, dott. Franco Marzatico, dott. Franco Nicolis, dott. Giuseppe Nicolussi, Cristiano Nicosia, Chiara Padovan, dott. Mauro Passarin, dott. Cristian Pezzi, Silvia Vicari, arch. Paolo Zammatteo. d) for sweating blood in revising the English of the text and offering very precious suggestions, my friend Dr Mark Pearce.

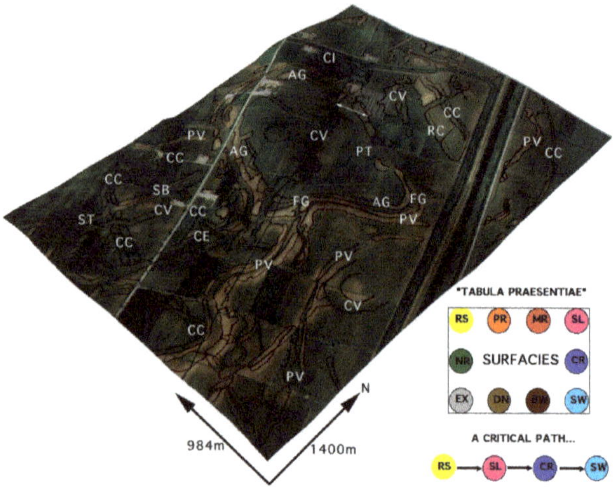

Fig. 1. CISAS (International Centre for the Study of Surface Archaeology) path to the surfacies': remote sensing, 3D-terrain modelling, first trial and preliminary photo interpretation (aerial photograph: CGR 26A, 20 1990) of the Fondo Paviani - Legnago (VR) major banked settlement in the lower Veronese Po Plain (Middle to Final Bronze Age: cf BALISTA et al. 1997). AG = main bank of the *terremara* (BA); AP = minor, embedded bank of the *terremara* (BA); CC = small, concentric channels (near-site: BA); CE = emissary-channel (BA); CI = tributary-channel (BA); CR = small rectilinear channels (off-site: BA); CT = Roman centuriation (RO); CV = "marsh-channel" (not BA); FG = main ditch of the *terremara* (BA); FP = ditch of the enclosure (BA); PT = "gate" (BA); PV = "palaeo-channel" (natural hydrography not synchronic with the BA landscape); RC = "enclosure" (BA); SB = causeway 'road' (BA); SR =causeway 'road' (RO). Lines with arrows = main stratigraphic sections. Lower-right: The *tabula praesentiae* approach, showing different objects to be recognised (in a ranked object to pattern to scenery recognition analytical climax) and a locally operational minimum spanning-three relational strategy (cf DE GUIO 1995c, 1996a).

knowledge) objects (DE GUIO 1995c). Our Surfacies thus becomes a metaphorical table (a *tabula praesentiae*) with many dishes, constituting, that is, food for thought: the objects (dishes) must be recognised and related via a dedicated, inter-object relational grammar using a number of detailed algorithms. This grammar is based on strict criteria of cognitive economy and minimum path and then projected into a behavioural formative scenario of past actors (physiogenetic, biogenetic and social) in an analytical climax consisting of an ideal object to pattern to scenery recognition. This was, avoiding procedural detail, the basic underlying philosophy: starting from ploughsoil complexity for a more ambitious 'deep ploughing into complexity' (both social complexity and complexity *tout court*) within the frame of a much wider 'Archaeology of Complexity' agenda (see DE GUIO 1992, 1995c, 1996a, 1996b, 1997a, 1997b, 1997c, 1998b, 2000a, 2001b, in press).

Following this course our critical path took us, via a sequence of three major projects in very different north Italian environments: the AMPBV project (in the lower Po Plain: fig. 1-4), the Berici Project (in an intermediate group of hills: fig. 5) and the Altopiano di Asiago Project (on a pre-alpine plateau: fig. 6-7).

Fig. 2. The three major C.I.S.A.S. field-projects (in the background: a Landsat 5TM image: EURIMAGE 1996) in collaboration with local CRM agencies (Soprintendenza Archeologica per il Veneto and Ufficio Beni Archeologici - Trento) and a variety of national and foreign institutions (including: the Universities of Padova, Napoli, London, Nottingham, Boston, and the ACCORDIA Research Centre, London): 'AMPBV Project' (Alto-Medio Polesine - Basso Veronese), 'Monti Berici Project', 'Altopiano dei Sette Comuni Project'.

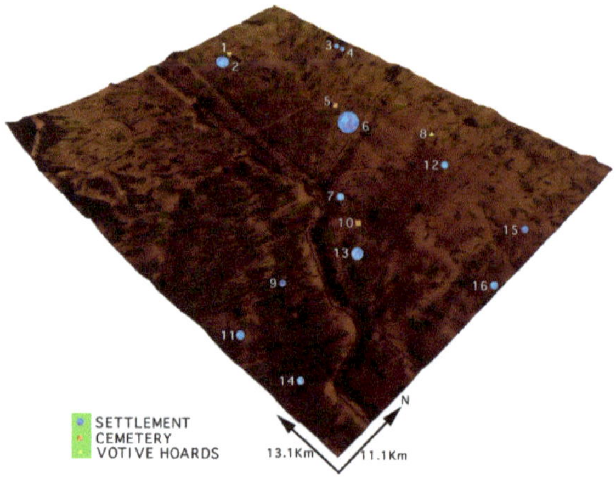

Fig. 3. The research area of the Alto-Medio Polesine - Basso Veronese Anglo-Italian pilot project: 3D model from a satellite image (Sojuz KFA 100, 28 May 1989), with location of sites: 1 = Castello del Tartaro - cemetery; 2 = Castello del Tartaro - settlement; 3 = Morandine di Cerea; 4 = Perteghelle; 5 = Scalvinetto; 6 = Fondo Paviani; 7 = Stanghelle; 8 = Corte Lazise; 9 = Canar di S. Pietro Polesine; 10 = Franzine; 11 = Canova; 12 = Venezia Nuova; 13 = Fabbrica dei Soci; 14 = Marola di S. Pietro Polesine; 15 = Massaua; 16 = Stanghelletti) (cf DE GUIO 1997a; BALISTA et al. 1997, 1998).

Our interest was progressively focused on the 'off-site' in order to detect crucial instances of 'fossil landscapes' (whether emerging or submerged) - that is specific instances of those cells of the eco-mosaic which preserve, in a quite extraordinary way, a wide range of infrastructures, such as drainage/irrigation, terracing, land-divisions, connectivity. They are essential network entities, often critically related to ecotones and overexposed, due to their spatial diffusion, to a number of sources of impact and deserving of a much stronger policy of monitoring, protection and *mise en valeur*.

These infrastructures are often power-related, thus giving us a critical opportunity to investigate our key domain of the 'Archaeology of power' from an off-site analytical perspective (which we call 'off-site power'). This analytical path led us, following the critical space/time progression of our key projects, to detect some extraordinary instances of 'off-site fossil landscapes', such as:

in the lower Po Plain: the first clear and detailed evidence of a very complex hydraulic and agrarian management system connected to the local central places - banked and ditched settlements (*Terremara* sites) - in the context of the first appearance of a formal landscape of power (a complex chiefdom polity: fig. 4).

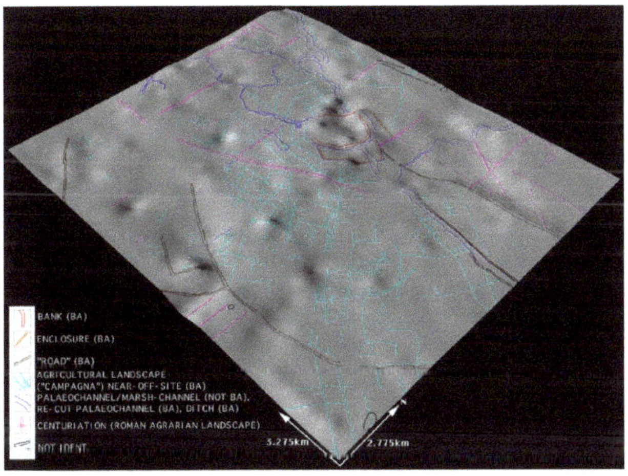

Fig. 4. AMPBV Project. Photo-interpretation of remotely-sensed traces from Castello del Tartaro (Cerea: VR) (cf DE GUIO 1997a; BALISTA et al. 1997, 1998).

in the Berici hills: what seems to be the oldest known 'road' in the world archaeological literature (DE GUIO & CATTANEO 1997; fig. 5): our 'dirt-road to Brendola' (VI), dated 4915-4735 cal BC (2 σ; 5930±40 BP - Beta-107672). It is in fact a proper 'road' – rather than a simple wooden track - and is a thousand years older than the much more famous 'Sweet Track to Glastonbury' in the English Somerset Levels (COLES & COLES 1986).

in the Altipiano of Asiago (fig. 6): extraordinary and growing evidence for a complex diachronic palimpsest of Bronze Age connectivity, terracing, land division and boundaries. At the end of the Bronze Age we also find exceptional evidence for large-scale copper smelting, with an extensive supporting infrastructure and a dramatic impact on the environment . It might be mentioned here that the metallurgical *chaîne operatoire* offered us a precious and rather unexpected thread connecting the mountains to the Po plain ecological and social environment in a relational network that is at the core of Late Bonze Age

Fig. 5. Monti Berici Project. The site of Soastene (Brendola - VI): the oldest known 'road' in the world? (4915-4735 cal BC) (2σ: Beta- 107672; SEE text and DE GUIO & CATTANEO 1997). Digital Terrain Model (DTM) of the 'road II' (Neolithic: VBQ III) with a combination of photo-realistic texture (ortho-rectified image of the excavation) and a 'virtual' one (simulating a post-abandonment scenario: Bryce 3D - MetaTools Inc.).

social complexity. This gives us an opportunity to combine, at a higher explanatory level, the different of evidence produced by our two major field projects. As a further step in the mountains we have now begun to implement a practical programme of eco-cultural resource management and Public Archaeology, involving the creation of a number of Archaeological and Ethno-archaeological theme-parks and trails.

In the Altipiano we have been conducting, since 1995, a variety of campaigns (using remote sensing, prospection and excavation) on different areas and sites, targeted at different sources of archaeological, ethno-archaeological and palaeo-envirommental evidence. For example we were able to implement further crucial steps in our Surface Archaeology Project agenda, by conducting a pilot experiment on a later Iron Age hillfort, the Bostel di Rotzo (fig. 7). This was a specific application of our revised inter-object relational grammar to the locally relevant sources of evidence (mainly remote sensing, prospection, micromorphologies, stratigraphic windows and expert previous knowledge) and gave us the opportunity to make many suggestions for a virtual reconstruction of the entire plan of the village and the construction of an on-site archaeological park.

Among the many research lines, I should like now to concentrate on an emerging, highly stimulating domain of inquiry, which is the core of this paper: it is a pilot approach that I call 'Archaeology of War', a multi-dimensional domain with a number of lines of interest that I shall try to sketch briefly.

Fig. 6. The Uplands of Asiago (VI) and Vezzena-Lavarone (TN). Rendering on DTM. a) Upper left: satellite image from Landsat 7 ETM: Telespazio. 8/10/1999; b) upper right: geology (Veneto Region: Geological Map at 1:250,000); c) lower left: geomorphology (Veneto Region: Geomorphological Map at 1:250,000); d) lower right: soil-use (Veneto Region: Soil-Use Map at 1:250,000).

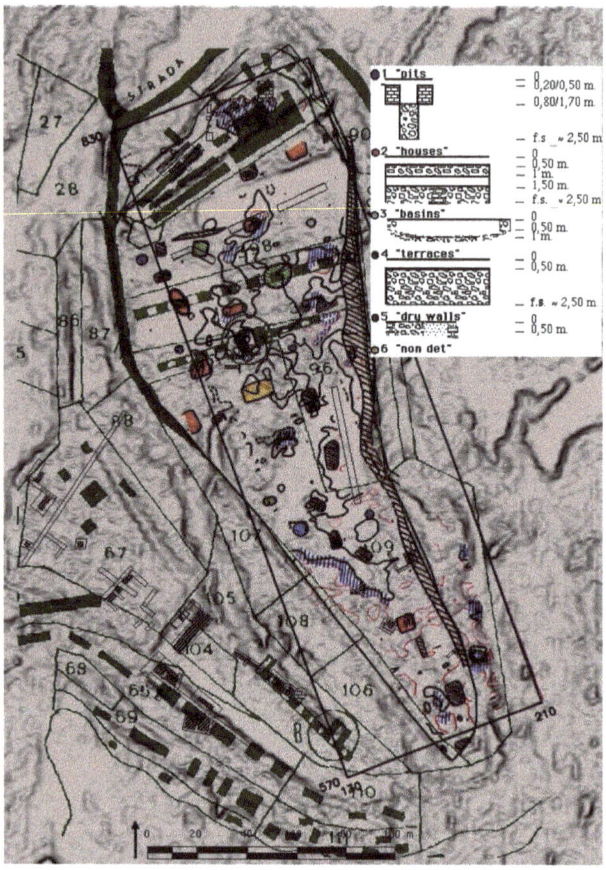

Fig. 7. 'Altopiano dei Sette Comuni Project'. The site of Bostel di Rotzo: overlay of different themes: previous knowledge (continuous trenches: 1912 excavations; hatched trenches: 1969 excavations), simplified photo-interpretations of major anomalies (white fill and various symbols), the 1994 magnetic prospection by E. Finzi (iso-lines); in the background: enhanced (Optilab- Graftek) aerial image (IGM n.446, 13/8/1986: SEE DE GUIO & FINZI 1995).

2 AN ARCHAEOLOGY OF WAR FOR THE ASIAGO PLATEAU

The Asiago plateau: the geophysical layout

Using the phrase 'Archaeology of War', in the local scenario, we intend first to evoke what is locally known as the Great War or simply 'The War': the First World War, in which the Altipiano of Asiago was one of the major and most bloody European theatres.

So let us first have a quick look to the physiographic setting of the plateau (see SAURO et al. 1991; SAURO 1993, SAURO & LANZINGER 1993; fig. 6).

The Asiago Plateau is the most central mountain group in the Venetian Pre-alpine belt and represents a transition between the Venetian Po plain and the Dolomitic Alps.
It covers an area of about 600 km^2, ranging between 600 and 2,300 m in altitude and consisting mainly of Mesozoic carbonatic rocks. The main rock units are (BARBIERI 1993):
- Dolomia Principale of Upper Triassic age
- Limestone units of Jurassic age (Calcari Grigi and Rosso Ammonitico);
- marly limestone units of Cretaceous age (Biancone and Scaglia).

The relief is strongly controlled by tectonic structure with fluvio-karstic landforms in its southern and central part and glacio-karstic landforms in its northern part.

The plateau has a roughly quadrangular shape and is delimited on the west side by the deep valley of the Astico river, on the north and east sides by the Brenta river valley and on the south side by a belt of hills.
Precipitation on the plateau is quite abundant (about 1600 mm per year on average).

This prealpine group is the most inhabited plateau of the Venetian Prealps. The main settlements are in the central basin, called the "Asiago basin", named after the major town of Asiago.

3 THE IMPACT ISSUE

The War impact

The First World War (see APOLLONI 1991-92, 1993a, 1993b; SAURO et al. 1991) had a massive impact in the plateau: for about 4 years there was fierce fighting. From 1915 to 1918 between two and four hundred thousand people were settled here and about one million and a half passed through (see LUSSU 1937; WEBER 1965; MATTALIA 1989, 1992; ACERBI et al. 1994; COVOLO 1996; MEREGALLI 1997; DI MARTINO 1998, 1999;

VON LICHEM & CENTRO DOCUMENTAZIONE DI LUSERNA 1998; BRANGIAN 2000; BRUNELLI 2000; ISNENGHI & ROCHAT 2000; RIGONI STERN 2000; FABI TODERO 2001).

Hundreds of kilometers of roads (nearly 400 km), trenches (about 200 km) and tunnels were built; large areas were under constant bombardment by shells and mortars: about 1500 cannon fired more than 200 tons of projectiles a day and many millions of them exploded. Most of the villages were abandoned and nearly completely destroyed. At the end of the war the plateau was a stony desert. Deforestation was, of course, at the same time the common denominator and major compound result.

A special issue of relevance is the ubiquitous bomb impact: analytically monitored from a number of relevant points of view, such as spatial layout, detailed dimensions of remnant morphologies, serious pollution effects due to heavy metal percolation in the karst subsoil and a frequent (roughly 50%) selective evolution as karstic dolinas, with a serious impact on surface and underground water flow (see CELI 1991).

4 IMPACTS ON IMPACTS

We need to consider other major sources of impact, which sometimes critically interfere with war impact, and were produced pre-war, during the war or post-war (see SAURO 1977, 1993; SAURO et al. 1991; APOLLONI 1991-92, 1993a, 1993b):

1) Deforestation due to a number of crucial activities such as the opening of grazing-land, different forms of woodland exploitation for construction, heating or charcoal and lime burning. A strict and careful woodland management policy was progressively implemented by the Republic of Venice, in consideration of a number of sometimes conflicting issues such as its critical need for wood supply and the environmental disasters (including the silting of the Lagoon) caused by massive, unplanned deforestation in the mountainous hinterland, not to mention the political and economic privileges that were granted to frontier mountain populations for defensive reasons and so on (see CACCIAVILLANI 1984, 1991, 1992; CASTI et al. 1988; APOLLONI 1991-92, 1993a, 1993b; SAURO et al. 1991).

2) Pastoralism and seasonal herding with overgrazing and cross trampling, causing soil coverage reduction, accelerated erosion and selective pressures on vegetal biodiversity (after a first prehistoric climax phase in the Middle to Late Bronze Age and a decline in Iron Age and Roman times, an impressive recovery took place from Late Medieval times: in the seventeenth century 300,000 transumant sheep are mentioned for the area, with a progressive decline since then as a result of the increasing economic marginalisation of this activity and newly introduced legal restrictions on mobility (see SAURO 1977, 1993; MIGLIAVACCA 1985, 1993; MIGLIAVACCA & VANZETTI 1988; APOLLONI 1991-92, 1993a, 1993b; SAURO et al. 1991; Storia dell'Altipiano dei Sette Comuni 1994; DE GUIO 1994; LEONARDI & RUTA SERAFINI 1994; BONETTO 1997)

3) Prehistoric metallurgic activity (as we will see later) with an impressive local concentration of a specific segment of the metallurgical chain (roasting and smelting of copper metals).

These formation processes acted at different time and space resolutions: the compound palimpsestic result, critically affected by post-depositional phenomena, mainly erosion, is sometimes difficult to dissect and de-stratify. This is because, for example, bomb impact (CELI 1991), deforestation (cf especially the 'trous avec monticules' features: SAURO 1977; BERNI et al. 1991) and paleometallurgy (SEBESTA 1992, MARZATICO 1997) may sometimes, very surprisingly, produce quite equi-final or similar micro-morphologies, while cross-trampling can coincide or partially obliterate old tracks, and even war related chemical pollution is crucially preceded by a metalwork-related pollution (as we shall see later; fig. 8).

Fig. 8. Micromorphologies related to human impact (for equifinality and interference phenomena, see text): sequential eradication (trous avec monticules. Left: see SAURO 1997, photo 1-3: Alti Lessini), animal cross-trampling (sentiérage. Centre: see SAURO 1997, photo 4-6: Alti Lessini) and bomb impact (right: cf APOLLONI 1993a, fig.6: Monte di Valbella; VON LICHEM & CENTRO DOCUMENTAZIONE DI LUSERNA 1998: Belvedere fort - Lavarone).

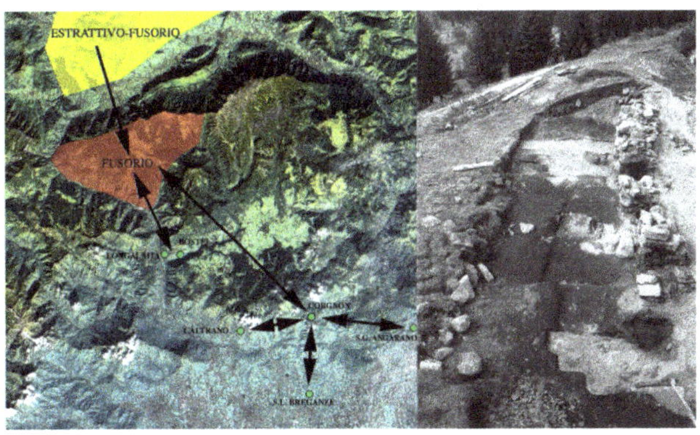

Fig. 9. Palaeo-metallurgy in the Uplands. Left: flow-chart model (see text) connecting the extractive zone, the smelting zone and the 'logistic' sites. Right: reference case-study for smelting apparatus on slope below the Redebus pass (Acquafredda - TN: cfr. MARZATICO 1977).

Post-war land reclamation, the long-lasting practice of recovering metal war relics by specialist operators (so-called 'recoverers': see RIGONI STERN et al. 1985), and more recently urban development and mass tourism, have progressively affected - at different and sometimes dramatic rates - the complex and hyper-fragile evolving scenario (see SAURO et al. 1991).

5 ARCHAEOLOGY of THE WAR vs ARCHAEOLOGY through THE WAR

This is a quite basic dichotomy between two divergent approaches (DE GUIO 2001b, in press).

By 'Archaeology *of* the War' we intend research properly aimed at the 'warscape', with the explicit intention of a virtual or even physical reconstruction of the war scenario by means of any available analytical procedures and sources of information (from the archaeological record itself to documentary or iconographic evidence). Our experiments on VR panoramics of GIS intervisibility can be included in this definition (fig. 14-15).

This can be understood as a proper archaeological exercise (Historical Archaeology) and is indeed increasingly required of archaeologists as a matter of specialist expertise to be used for a variety of purposes, from historical research to more practical issue such as Archaeo-War tourism (ACERBI et al. 1994; COVOLO 1996; BRUNELLI 2000; FABI & TODERO 2001). This is a quite popular local practice, now in need of more qualified support - both on site, with guided trails and restoration work, archaeo-reconstructions, Museums and so on, or as remote support, for instance, for detailed interactive information and possibly virtual reconstructions, panoramas and tours (see DE GUIO 1998a, 2001b). In this frame of reference the warscape is the legitimate target, a direct source of information, while anything else from previous archaeological traces to postwar impact is noise.
With a slightly different meaning, still under the instrumental label of 'Archaeology *of* the War', we might examine the warscape archaeological record, as an enormous reservoir, 'one century deep', for developing depositional and post-depositional theory (see SCHIFFER 1987, 1995, LEONARDI 1992b, 2000). Special attention should be devoted not to the more visible monumental warscape (such as a fort) but to the more vanishing and mundane evidence (like barracks and enclosures), of a 'submerged warscape' along the relevant pre-, sin- and post-abandonment formation processes. This is quite close to a 'then and now' perspective which is increasingly popular in the archaeo-war tourism literature (BRANGIAN 2000).
Here the challenge is to substitute the rough 'then and now' dichotomy with a sounder, continuum morphogenetic theory. This should be able cope with critically distinct time resolutions, the *evénémentielle* (like a single battle or a single-day land movement of a mole or post-depositional excavation by a war relics hunter), the *cononctural* (the war itself), or the *longue durée* such as pastoralism or frontier and boundary dynamics).

Using a Remote Sensing Metaphor, this would be an excellent exercise to recognize the signature of complexity, as the compound result of a number of actors (physio-, bio-, anthropogenetic, and even mechanical), each one performing its morphogenetic script with various degree of freedom and along quite different space, time, functional frequencies of the formative scenario.

The opposite approach 'Archaeology through the War' uses instead any relevant source of information (starting from wartime photographs and movies and military maps) as a medium to access other-target sources of archaeological or ethno-archaeological or palaeo-environmental evidence. Here the warscape itself, with its epidemic spatial diffusion, is a source of noise which since a century ago has progressively entered the archaeological record, increasingly obliterated by a variety of post-depositional processes.
It is now very easy to confuse the resulting remote sensing evidence with a variety of older proper 'archaeological' traces. Using wartime photographs (especially aerial) and maps and comparing them with both pre- and post-war evidence in a time series analysis it is much easier to extrapolate the source of war-noise and, at the same time, to address quite profitably traces and related source of evidence pertaining to our target archaeological scenarios.

This constitutes an extraordinary opportunity, particularly for remote sensing. Indeed the war, while causing substantial collateral damage to the archaeological and environmental heritage, offers archaeology a number of crucial collateral benefits.

Firstly the impact itself has largely acted as a colossal sampling device at different levels of spatial resolution (from intrasite to intersite) with a purposive and only partially random distribution of test pit transects (for instance trenches and bomb pits), triggering subsequent erosion and frequent opening of 'stratigraphic windows' which are sometimes still of use today.

The major opportunity, however, is related to the massive deforestation both before the war began (to assure for instance visibility to forts, or to build the massive communication network and other extensive infrastructure) or during the war itself (for instance due to bombardments). This was certainly the most extensive clearance in the whole history of human impact on this environment and offers an exceptional, unprecedented surface of information to the remote sensing specialist. Since the war, other sources of impact (such as urban development) or massive re-forestation of previously open space has limited the usable surface by about 30% (see DE GUIO 2001b, in press).

6 WAR: FROM A MEDIA EVENT TO AN HYPERMEDIA HEURISTIC OPPORTUNITY.

The First World War is commonly reputed to have been the first true media war with a variety of media covering the events (radio, photography, both aerial and on the ground, sometimes using special apparatus for making panoramic pictures or telephotography along with the first movies: see DI MARTINO 1998, 1999; fig. 10).

The most useful resource for our research is possibly offered by a massive number of aerial photographs (both vertical and oblique) from a variety of official (such as military, museum or other institutional archives from the Austro-Hungarian and Italian/French/British sides) and even private sources.

Fig. 10. Above: a Farman aircraft (12rh squadron) at Asiago airport, November 1915 (DI MARTINO 1999, p. 51). Below: the Italian 1° Army tele-photographic team (DI MARTINO 1998, p.16).

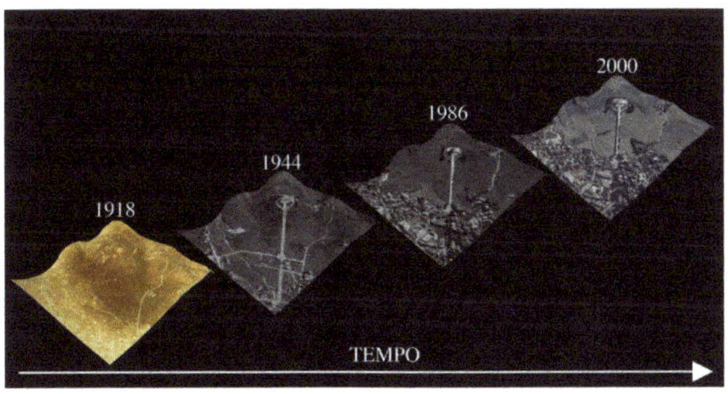

Fig. 11. Asiago. Time series of images in a photo-realistic solid model: Aerial photographs: a) "Archivio Piatti" 9/5/1918, 8h30, Biblioteca Civica di Asiago; b) Air Photo Library - Dept of Geography – Univ. of Keele - UK, 3144; 3145; 4142 (1943-1944); c) IGM , 412, 3/8/1986, 7h48; d) digital orthophoto, sez. CTR 082110: CGR-Parma, 2000 (Comunità Montana - Asiago).

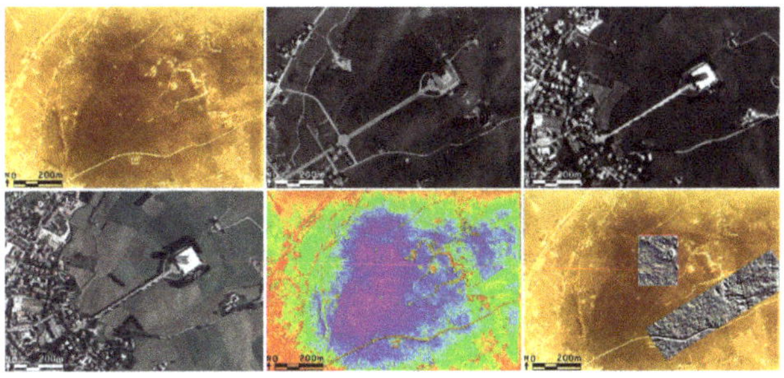

Fig. 12. Asiago. From upper left: a-d) time series of the four 1918- 2000 aerial photographs; e-f) global and local image enhancing, evidencing anomalies particularly in the two target areas of f (traces of 'slab road', land divisions and drainage: cf. text).

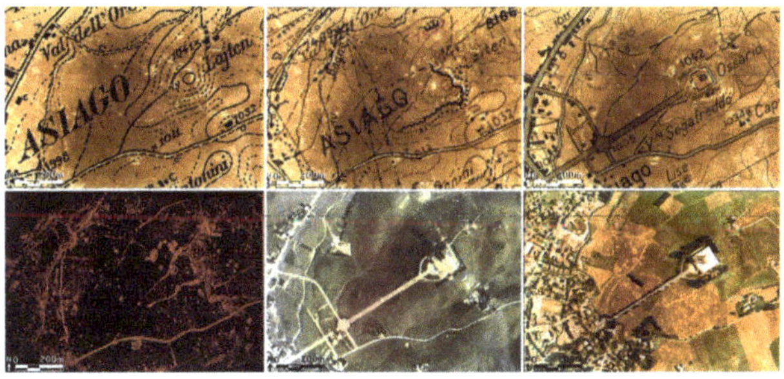

Fig. 13. Asiago. Above: historical cartography overlaying the 1918 aerial photograph: a) IGM Carta d'Italia alla scala di 1:25.000, F.37 IV SO Asiago (1886); b) French military topographic map: Biblioteca Bertoliana di Vicenza ('Album topographique offert à la Bibliothèque de Vicence'), 1918; c) IGM Carta d'Italia alla scala di 1:25.000 F.37 IV SO. Asiago. Below: d) image enhancement of the 1918 aerial photograph; e) 'difference image' between the 1943 and the 1918 aerial photographs (brighter areas = impact between 1918 and 1943); f) digital overlay of the 2000 orthophoto and the 1918 photo (with contour enhancing)

We therefore have the opportunity to use this material to implement a hypermedia line of research, by using the image processing, GIS, simulation and modeling, virtual reality toolkits not just for amusing computer games but with a real heuristic strategy in mind. We shall consider two examples Asiago (VI) (fig 11-13) and the Vezzena-Luserna uplands (TN) (fig 14-22).

7 ASIAGO: A GRANDFATHER'S ARCHAEOLOGY

For the first case-study (fig. 11-13) we can diachronically examine a time series of a limited number of layers of information from aerial photographs and maps, dwelling a little on the wartime image, to highlight some useful archaeological traces.

We find, for example, the traces of an abandoned traditional track paved with slabs and of an obsolete agrarian land division. The date of both is still uncertain (they are possibly late-medieval or early-modern in origin).

This case study points to a basically historical, ethno-archaeological and ethno-historical archaeology. Within this major domain and following this line of research we were able to accomplish substantial advances in five specific sub-domains of inquiry, relating to traditional activities which have now fully or nearly fully disappeared (see DE GUIO 1995b, 1996b, 1997b, 1997c, 1997d, 1998a, 2001b; BRESSAN 1996-97; KIRSCHNER 1997-98; ZILLI 1996-97; DE GUIO & BRESSAN 2000; DE GUIO & ZANUSSO 2000)

1) pastoralism and seasonal herding
2) charcoal burning
3) lime burning
4) wild bird snaring (*roccoli con le reti*)
5) toponyms (to which we devoted a special "TopoGis"[2] project, using GIS and other spatial and cognitive disciplines for a pilot formal approach.

8 VEZZENA-LUSERNA: AD METALLA

The most relevant results are perhaps related to the second case-study (fig 14-22), where we meet with a proper archaeological phenomenon: palaeo-metallurgy (see FRANCOVICH 1991, SEBESTA 1992; GIANNICHEDDA 1996; CIERNY 1998; GIARDINO 1998).

At the core of this key domain of enquiry there is an apparent paradox: an impressive and unprecedented space/time concentration of copper smelting in an area completely devoid of mineral resources.

The time window seems to be a conjuncture of perhaps no more than a few generations (probably a century) during the Late Bronze Age (more precisely at the end of the local Recent Bronze Age and beginning of the Final Bronze Age: PERONI,1996; BIETTI SESTIERI,1997).

This is the climax of the development of social complexity in the Po Plain (mainly related to the Terremare society: see BERNABÒ BREA et al. 1977) and comes just before a punctuated, dramatic and widespread collapse, whose special 'topology' we have been able to investigate in an original approach to seriation techniques (DE GUIO 2000). Among the various explanations for the collapse, covering a wide range from endogenous to exogeneous origin and from eco-determinism to social engineering, some scholars have suggested the disruption of an inter-regional long-distance metals trade network centred on Mycenaean and post-Mycenaean state formations (see DE GUIO 2000).

Fig. 14. Vezzena-Luserna uplands. Above: a) Digital Terrain Model (DTM) of digital ortho-photo, sez. CTR 081040, 081080, 082010: CGR - Parma, 2000 (Comunità Montana - Asiago); b) DTM of digital ortho-photo at 1:10.000 scale: Provincia Autonoma - Trento, 1996; c) stratigraphic palimpsest ('inverse stratigraphy': the oldest layer - 1915-18 - is above, the most recent – 2000 - below) of the 1915-18, 1944, 1986, 2000 aerial photographs analysed in this case-study. In the middle: d) viewshed of the Busa Verle fort on a DTM draped with the 2000 digital ortho-photo; e) viewshed of the small 'Fortino Basson' fort draped with the 2000 digital ortho-photo; f) viewshed of the Campo Luserna fort draped with the 2000 digital ortho-photo. Below: g) copper furnaces on a 'chorographic' map of the Vicentino territory - by Giovanni Molino: Biblioteca Bertoliana di Vicenza, map collection n.106, 1608; h) copper furnaces on a composite image: topographic map of the Lombardo-Veneto Kingdom, 1833 and topographic map from the Vienna State Archive (c/o dott. Giancarlo Bortoli), 1805; i) copper furnaces overlaying the 2000 digital ortho-photo

Fig.15. Vezzena-Luserna uplands. Above: a) DTM draped with the 2000 digital ortho-photo; b) DTM draped with a Landsat 7 ETM image: Telespazio, 8/10/1999. Below: c) DTM draped with the viewshed of the three forts overlaying the 2000 digital ortho-photo. Areas of exclusive visibility: yellow for the Busa Verle Fort, blue for the Basson small fort, red for the Campo Luserna fort; d) DTM draped with a map of the copper furnaces overlaying the topographic map of the Lombardo-Veneto Kingdom, 1833 (left) and topographic map from the Vienna State Archive (c/o dott. Giancarlo Bortoli), 1805 (right).

[2] In Italian *topo* means mouse, so our logo is dedicated to a famous cartoon character, Topogigio.

The spatial activity area is restricted to the Lavarone-Vezzena plateau (southern Trentino-Alto Adige) and, to a lesser extent, the nearby Altipiano di Asiago. The nearest copper mineral source, recently confirmed by slag provenance analysis, is the Valsugana district, especially its northern, opposite slopes.

The mines are situated at about 1500 m above sea level, and the ore (which we should imagine as having first been 'enriched' by pounding or grinding and sorting, in a form known as 'mineral flour') would have had to be transported down into the Valsugana - at this point about 500 m above sea level - and then up the steep wall of the southern, opposite side of the valley to the Lavarone-Vezzena plateau, at over 1200 m above sea level. The difficulties of such transportation cannot be overemphasized, and there are few steep paths up the sheer slopes. On this crucial area and domain of enquiry we have recently implemented a collaborative research project involving the Local CRM Authorities, the Universities of Padua and Nottingham (with a possible near-future involvement of Boston University itself especially for Remote Sensing and for ECRM and Public Archaeology). On the basis of previous research (see PREUSCHEN 1973; PERINI 1989; VIDALE et al. 1989; SEBESTA 1992; DE GUIO 1994a; MARZATICO 1997; CIERNY 1998; CIERNY et al. 1998) and some lines of evidence emerging from our initial fieldwork we have proposed a model (PEARCE & DE GUIO 1999; fig. 9) to explain the local anomaly and to recover its overall meaning within a complex critical flow of the metallurgical chain at its relevant inter-regional scale of resolution.
The main reasons seem to be:

1) need for fuel: copper exploitation had caused deforestation in the area close to the mines (with a further need to conserve wood supplies for fuel for fire-setting and mining infrastructure), and so the copper smelting took place where there was availability of wood and water. In this respect our plateaux certainly provided an excellent opportunity for the complex series of roasting and smelting operations, requiring a large labour force and a massive amounts of wood (for roasting) and charcoal (for smelting). The mixed forest and especially beech was highly suitable (MARZATICO 1997, 575-576; SEBESTA 1992, 9-10). A still unpublished pollen profile seems to offer an extraordinary confirmation of woodland exploitation at that time (end of Sub-Boreal) with a dramatic and highly diagnostic decrease in arboreal vegetation and a parallel increase of herbaceous species (V. Martello, pers. comm.).
2) the smelting chain: the local calcareous Biancone, lacking in the ore areas, would have provided, with its flint interstrata, another crucial ingredient for the smelting process (see SEBESTA 1992);
3) the supporting alpine economy (cf ZANETTI et al. 1988), with the practice of alpine grazing in the summer: the production of cheese would have provided a useful source of protein (PEARCE & DE GUIO 1999) for the metallurgical workforce (possibly amounting to a hundred for any production unit: SEBESTA 1992);
4) connectivity: the area is located at the heart of the network of major natural (and traditional) communication lines connecting the alpine region to the Po plain (see REICH 1910; CACCIAVILLANI 1991; BRIDA 1989, 2000; BRIDA et al. 1981; CAROTTA 1997; ZAMMATTEO in press).
5) a critical proximity to an ethnic and political boundary (or frontier) between communities (DE GUIO 1994, 1997, 2000; PEARCE & DE GUIO 1999). The smelting, like the mining, was firmly in the hands of the Luco / Laugen cultural group, but they may have chosen to smelt on the Lavarone-Vezzena plateau because it was ideally placed for trading the smelted metal with the lowland polities of the plain. This trade was probably located at the boundary between the two culture groups - the Luco / Laugen A group of the Alpine areas and the Po plain group - and it is striking how this boundary (perhaps we may speak of a 'frontier') between the metallurgical / pastoral Lavarone-Vezzena plateau and the neighboring pastoral Altipiano di Asiago still survives - indicating an extraordinary *longue durée* - in the administrative boundary between the Trentino and Veneto regions. It is not unreasonable to expect that as well as differences in material culture, there may also be territorial markers as in other similar situations (for example, Fleming's (1988, 1992) work on the Dartmoor *reaves*). In fact we are now detecting complex palimpsests of land divisions and markers, some of them very probably prehistoric in origin and crucially clustered around smelting furnaces or even bordering them.

The overall model was that of a dendritic central place system, with a number of key gateway communities located in the Po plain which drained raw material (ingots) from the Alps as well from other regions (such as Tuscany) and controlled the long-distance network of metals reaching the east Mediterranean palatial system.

So, along our critical path to the archaeology of complexity, we uncovered an unexpected, crucial thread connecting the social and organisational complexity of two critical stages of the metallurgical chains: they are situated at two extremes of the local ecological range of variation (the lower Po plain and the mountains) with a very rewarding coincidence with our above mentioned key-project areas. It is probably the first instance of a very impressive labour division (across ecological, economic and political boundaries) for mass production (we may call it 'proto-industrial') within an encompassing dynamic and meta-stable world system involving a nested level of 'cores', peripheries and 'margins'(according to Sherratt's taxonomy: SHERRATT 1994).

With reference to our present research programme we are developing an increasingly elaborate strategy based on our preceding surface archaeology package and a critical mass of specific previous and recently acquired knowledge in the metallurgical domain: basically we are looking for metallurgical sites (roasting and smelting) with a number of possible markers and expectations that can be detected and tested at various stage of spatial resolution from remote sensing to ground survey (fig 14-22).

The most fundamental are 1) vegetation marks (slags are a critical limitation of plant growth, with the exception of marker species like *Silene inflata*) and related snow marks; 2) specific positive/negative post-depositional micro-morphological features for furnace sites (batteries of up to a dozen of furnaces on a terraced platform with a nearby spoil heap: see SEBESTA 1992; MARZATICO 1997; fig. 9) sometimes evidenced by shadow marks; 3) proximity to a source of flowing or still water (streams and natural or artificial ponds, both fossil/abandoned and active); 4) proximity to suitable pastures; 5) proximity to the natural or traditional connectivity network; 6) correspondence with slag-related local place-names (eg 'slacche' in the local Germanic dialect means slags: see SEBESTA 1992).

We are at the beginning of a long process of hypothesis testing: our 'ground-truth' impact, following our 'remote' approaches and detections, now seems very encouraging. The expected markers (above: 1-6) constitute a polythetic but very robust table of expectations, showing a very high degree of spatial redundancy and clustering. We are building a substantial expert-knowledge base for our inferential engine by interfacing through our special 'relational grammar' (DE GUIO 1995c) the different sources of remote and field-evidence.

Fig. 16. Vezzena-Luserna uplands. DTM draped with a composite multi-layered image made up of: 1) IGM Carta d'Italia alla scala di 1:25.000 F.36 I NE Monte Verena (1886, published 1918) and F.36 I NO Caldonazzo (1886, published 1908); 2) French military topographic map, at 1:20.000 scale: Biblioteca Bertoliana di Vicenza ('Album topographique offert à la Bibliothèque de Vicence'), 1918, overlaying a Landsat 7 ETM image: Telespazio, 8/10/1999.

A wide range of field procedures, from surface sampling, stratigraphic window inspections, prospecting, micromorphology, coring, test-pits ... (see fig 14 - 22) is providing what are still preliminary but very promising results: for instance we have accurately geo-referenced dozens of already known but poorly localised sites and features and discovered about 20 new sites (see fig 14 - 15). We are identifying an impressive number of metallurgical, connectivity and agrarian management (mountain) infrastructures, as well as territorial markers that seem to be at least in a substantial part synchronous: our horizon of expectation is a very complex, palimpsestic scenario, whose 'emergence/persistence' (physical and/or socio-cognitive) or simply optimal allocation of relevant entities often act as a trap (or - better- a 'web') for capturing a number of post-abandonment human behaviors: the extraordinary long-term 'frontiers and boundaries' issue and the 80% coincidence of metallurgical infrastructures with present-day 'malghe' and 'pozze d'alpeggio' (seasonal herding huts and artificial ponds: see fig. 14-15, 22) are a striking evidence in that direction.

It is a challenge, it is an '(archaeology of) war' that seems, so far, to be of great 'collateral benefit' ...

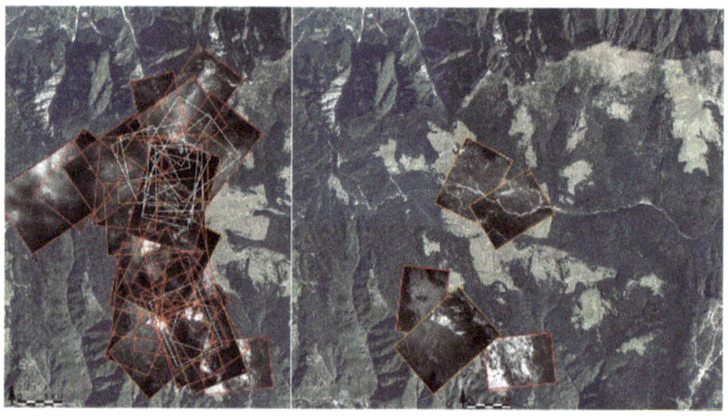

Fig. 17. Vezzena-Luserna uplands. Left: palimpsest of two dated photographic series: 1) Museo Storico Italiano della Guerra - Rovereto, 1915-1918 (red number, red border, white border where the photo is overlain by others); 2) Museo Caproni – Trento: 'Fondo Costantino Cattoi', 1915-1918 (orange number, orange border, white border where the photo is overlain by the others), overlaying the 2000 digital ortho-photo. Right: target areas: T1: 'Malga Fratte' (photo n.33: Museo Caproni – Trento: 'Fondo Costantino Cattoi'; legend: absent; data: absent; shot elevation: absent; size: 16,9 x 12,4; calculated scale: 1: 9662.94); T2: 'Vezzena' (crossing: photo n.31: Museo Caproni – Trento: 'Fondo Costantino Cattoi'; captions: absent; date: absent; shot elevation: absent; size: 16,99 x 12,38; calculated scale: 1: 9902.94); T3: 'Malga Millegrobe' (photo n.50: Museo Storico Italiano della Guerra - Rovereto; captions: 'malga Millegrobe di sopra e di sotto'; date: 04/11/1915; shot elevation: 1900; size: 17,22 x 23,46; calculated scale: 1: 6399.03); T4: 'Malga Campo - Malga Croiere' (photo n.34: Museo Storico Italiano della Guerra - Rovereto; captions: 'zona a sudest del forte Luserna' date: 18/11/1915; shot elevation: 1800; size: 17,14 x 23,31; calculated scale: 1: 5562.61); T5: 'Luserna' (photo n.19: Museo Caproni – Trento: 'Fondo Costantino Cattoi'; captions: '7° Gruppo Aeroplani (per artiglieria)/ 49° Squadriglia/ fotografia n°:215/ quadretti: 4759/ pilota: S.Ten. Oppi, Osservatore: S. Ten. Cattoi/Forte Luserna e Paese di Luserna'; date: 10/5/1916, 8h00; shot elevation: 3500m; size: 16,98 x 12,2; calculated scale: 1:12206.65), overlaying the 2000 digital ortho-photo.

Fig. 18. Vezzena-Luserna uplands. 'Stratigraphic palimpsest' of dated aerial photographs: 1915 (blue), 1916 (green), 1917 (red), overlaying the 2000 digital ortho-photo.

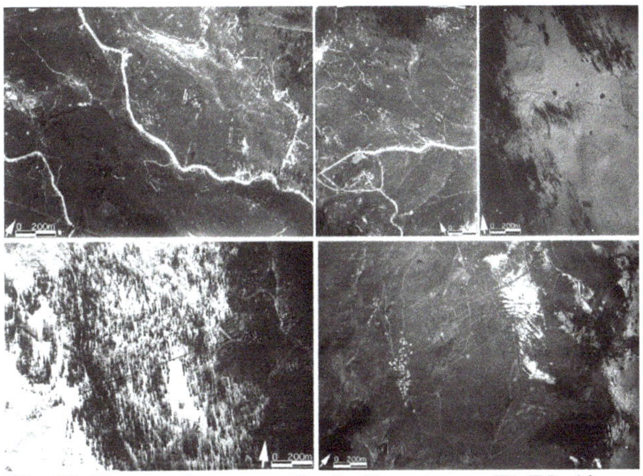

Fig. 19. Vezzena-Luserna uplands. Target-areas T1-T5. Above: a) target-area T1: 'Malga Fratte'; b) target-area T2: 'Vezzena'; c) target-area T3: 'Malga Millegrobe'. Below: d) target-area T4: 'Malga Campo - Malga Croiere'; e) target-area T5: 'Luserna'.

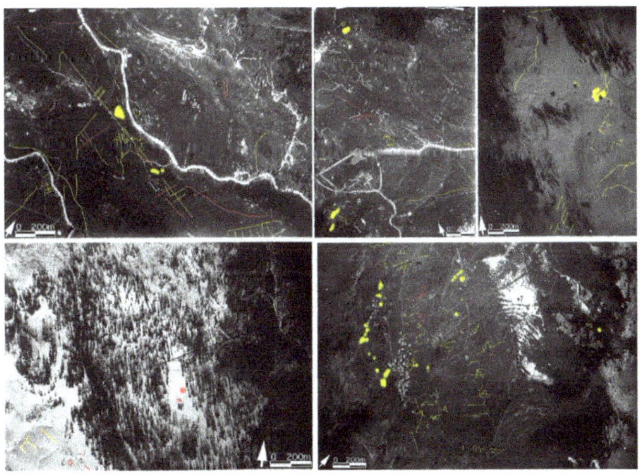

Fig. 20. Vezzena-Luserna uplands. Photo-interpretations of the T1-T5 target-areas (by Andrea Betto: bright traces in yellow, dark traces in red). Above: a) target-area T1: 'Malga Fratte'; b) target-area T2: 'Vezzena'; c) target-area T3: 'Malga Millegrobe'. Below: d) target-area T4: 'Malga Campo - Malga Croiere'; e) target-area T5: 'Luserna'.

Fig. 21. Vezzena-Luserna uplands: target-area T1 – 'Malga Fratte'. DTMs draped with various images (textures). Above: a) the 1915-18 aerial photograph (see above); b) the1943 aerial photograph (see above). Below: c) the 1986 photo (see above): d) the 2000 aerial ortho-photograph (see above); e) the photo-interpretation (by Andrea Betto) of the 1915-18 aerial photograph (see above); f) the 2000 aerial ortho-photo with digital bumping 'inherited' (derived from) the enhanced 1915-18 image.

Fig. 22. Vezzena-Luserna uplands: target-area T1 – 'Malga Fratte', 2001 field campaign. Above: a) sub-area A (copper furnace facility conforming to the 'Redebus model' (on slope, terraced with multiple furnaces: see SEBESTA 1992; MARZATICO 1997), close to the malga (seasonal farm building); b) zoom-in to the same sub-area; c) sub-area B (close to the monument, to the west of the malga). In the middle: d-f) other shots of the same sub-area B (in f: residual traces of some boundary features with fitted slabs: photos by dott. Gianni Bonaldo). Below: g-h) two 360° digital panoramic views centred on sub-area A (see the above-named copper furnace facility morphology) and sub-area B (see a war-impact morphology: shots and QTVR processing by dr. P. Kirschner).

References

ACERBI E., POVOLO E., GATTERA C., MALTAURO M 1994, *Guida ai forti italiani e austriaci degli altipiani. Itinerari e storia*, Novale, Gino Rossato Editore.

ALDENDERFER M., MASCHNER H.D.G. 1996, *Anthropology, space and geographic information systems*, New York and Oxford, Oxford University Press.

ALLEN K.M.S., GREEN S.W., ZUBROW E.B.W. (eds) 1990, *Inter-*

preting space: GIS and archaeology, Taylor & Francis, London.

APOLLONI L. 1991-92, *Modificazioni antropiche e impatto ambientale legato alla pastorizia ovina nella parte alta dell'Altopiano dei Sette Comuni*, tesi di laurea, Università di Padova - Dipartimento di Geografia.

APOLLONI L. 1993a, *Aspetti dell'impatto antropico nella parte alta dell'Altopiano dei Sette Comuni dal tempo della Serenissima ai giorni nostri*, "Acta Geologica", Museo Tridentino di Scienze Naturali, 70, pp. 153-164.

APOLLONI L. 1993b, *La pastorizia nell'Altopiano dei Sette Comuni dal tempo della Serenissima ai giorni nostri*, "Acta Geologica", Museo Tridentino di Scienze Naturali, 70, pp. 143-152.

BALISTA C., CAFIERO F., DE GUIO A. 1997, *Castello de Tartaro, Fondo Paviani, Fabbrica dei Soci*, in BERNABO' BREA M., CARDARELLI A., CREMASCHI M (eds), *Le terramare. La più antica civiltà padana*, Milano, Electa, pp. 240-249.

BALISTA C., BAGOLAN M., CAFIERO F., DE GUIO A., LEVI S., WHITEHOUSE R.D., WILKINS J. 1998, *Bronze-Age "Fossil Landscapes" in the Po Plain*, in B. HÄNSEL (ed.), *Mensch und Umwelt in der Bronzezeit Europas*, Kiel, Oecter-Voges Verlag, pp.493-499.

BARAGIOLA A. 1908 (1980), *La casa villereccia delle colonie tedesche veneto-trentine,* Asiago, Comunità Montana dei Sette Comuni, ristampa anastatica.

BERNI P., SAURO U., VARANINI G.M. (eds) 1991, *Gli alti pascoli dei Lessini Veronesi. Storia, Natura, Cultura*, Vago di Lavagno, La grafica Editrice.

BARBIERI G.1993, *Le rocce dell'Altopiano dei Sette Comuni*, "Acta Geologica", Museo Tridentino di Scienze Naturali, 70, pp. 9-20.

BERNABÒ BREA M., CARDARELLI A., CREMASCHI M. (eds) 1997, *Le terramare. La più antica civiltà padana*, Milano, Electa.

BERNARDI M. (ed.) 1992, *Archeologia del paesaggio*, Firenze, All'Insegna Del Giglio.

BEVILACQUA E. 1994, *L'Altipiano nella cartografia Storica*, in AA.VV. 1994, pp. 371-382.

BIETTI SESTIERI A.M. 1997, *Italy in Europe in the Early Iron Age*, in "Proceedings of the Prehistoric Society", 63, pp. 371-402.

BLANKHOM H.P.1991, *Intrasite spatial analysis in theory and practice*, Aarhus, Aarhus University Press.

BONATO M. 1857-1905 (1978), *Storia dei Sette Comuni e Contrade Annesse dalla loro origine alla caduta della Veneta Repubblica*, Milano, Studio Editoriale Insubria, ristampa anastatica.

BONETTO J. 1997, *Le vie armentarie fra Patavium e la montagna*, Padova, Provincia di Padova, Assessorato ai Beni Culturali, Dipartimento di Scienze dell'Antichità dell'Università di Padova.

BRANGIAN G. 2000, *Altopiano dei Sette Comuni "Zona Alta". Quasi un secolo dopo. Percorsi storico-fotografici sui luoghi di guerra, Asiago*, Bonomo.

BRENTARI O. 1884 (1973), *Storia di Bassano e del suo territorio*, Bologna, Arnaldo Forni, ristampa anastatica.

BRENTARI O. 1885 (1974), *Guida storico-alpina di Bassano-Sette Comuni vicentini,* Bologna, Arnaldo Forni, ristampa anastatica.

BRESSAN C 1996-97, *Archeologia ed Etnoarcheologia delle calcare: dal survey, allo scavo, all'Eco-Cultural Resource Management*, unpublished dissertation, Università di Padova-Dipartimento di Scienze dell'Antichità.

BRIDA L. 1989, *Alle antiche radici. Rilevazioni storico-economiche nella zona di competenza della Cassa Rurale di Caldonazzo,* , Gardolo di Trento, Luigi Reverdito Editore.

BRIDA L. 2000, *Caldonazzo. Contributi storici*, Pergine, Associazione 'Amici della Storia'.

BRIDA L., MASTRELLI ANZILOTTI G. 1981, *I nomi locali del comune di Caldonazzo*, Trento, Società Trentina di Studi Storici.

BRUNELLI F. 2000, *Itinerari di guerra sull'Altipiano di Asiago*, Colognola ai Colli, Demetra.

CACCIAVILLANI I. 1984, *I privilegi della Reggenza dei Sette Comuni*, 1339-1806, Limena/Padova, Signum Verde.

CACCIAVILLANI I. 1991, *La confinazione veneziana con gli imperiali*, 1339-1806, Limena/Padova, Signum Verde

CACCIAVILLANI I. 1992, Le autonomie locali nella "Serenissima", Limena/Padova, Signum Edizioni.

CAMASSA G., DE GUIO A., VERONESE F. (eds) 2000, *Paesaggi di potere: problemi e prospettive*, Atti del Convegno di Udine,16-17 Maggio 1996, Roma, Quasar.

CAMPANA S., FORTE M. (eds) 2001, *Remote sensing in archaeology*, Firenze, All'Insegna del Giglio.

CANTELE G. 1991-92, *Impatto agrario e archeologia di superficie* , tesi di laurea, Università di Padova- Dipartimento di Scienze dell'Antichità.

CAPUIS L. 1993, *I Veneti. Società e cultura di un popolo dell'Italia preromana*, Milano, Longanesi.

CAROTTA A.1997, *Le nostre radici. Brancafora. Notizie storiche sull'ospizio, la parrocchia e le comunità che ne facevano parte*, Vicenza, La Serenissima.

CASTI MORESCHI E., ZOLLI E. 1988, *Boschi della Serenissima. Storia di un rapporto uomo-ambiente*, Venezia, Arsenale Editrice.

CELI M.. 1991, *The impact of bombs of World War I on the limestone slopes of Monte Grappa*, in SAURO, BONDESAN, MENEGHEL M.1991, pp. 279-288.

CIERNY J. 1998, *Lo sfruttamento dei giacimenti di rame nell'area alpina dalle origini fino agli esordi dell'età moderna*, in RAFFAELLI U. (ed.), *Rame d'arte. Dalla preistoria al XX secolo nelle Alpi centro-orientali*, Trento,

Provincia Autonoma di Trento, Servizio Beni Culturali, Castello del Buonconsiglio 1998, pp. 75-93.

CIERNY J., MARZATICO F., PERINI R., WEISGERBER G. 1988, *Prehistoric copper metallurgy in the southern alpine region*, in MORDANT C., PERNOT M. RYCHNER V. 1998, *L'atelier du bronzier en Eurpope du XXe au VIIIe siècle avant notre ère*, Paris, CTHS, pp. 25-34.

CLARK A. 1996, *Seeing beneath the soil. Prospecting methods in archaeology*, London, B.T. Batsford.

DAL POZZO A.1820 (1976), *Memorie istoriche dei Sette Comuni Vicentini*, Bologna, Arnaldo Forni, ristampa anastatica.

COLES B., COLES J. 1986, *Sweet track to Glastonbury. The Somerset Levels in prehistory*, London.

COVOLO R.1996, *Altopiano dei Sette Comuni. Itinerari della Grande Guerra. Museo di Canove – Sacrario Militare*, Bassano, Ghedina & Tassotti.

CONSIGLIERE E. 1998-99, *Progetto Brendola: approccio analitico per una carta archeologica ipermediale*, tesi di laurea, Università di Padova - Dipartimento di Scienze dell'Antichità

DAL POZZO A. 1993 (ed. by G. Bortoli), *Memorie istoriche dei Sette Comuni Vicentini. Libro secondo, che contiene la storia particolare dei Sette Comuni e delle loro chiese. Libro terzo, che contiene memorie istoriche intorno alle contrade annesse e luoghi contigui*, Vicenza, Banca Popolare Vicentina.

DE GUIO A. 1985a, *Archaeological applications of survival analysis*, in VOORIPS A., LOVING S.H. (eds), *To pattern the past*, PACT, 11, Souvain, pp.361-381.

DE GUIO A. 1985b, *Archeologia di superficie ed archeologia superficiale*, "Quaderni di Archeologia del Veneto" I, pp.176-184.

DE GUIO A. 1988-89, *Analisi funzionale dei paesaggi di potere*, "Origini", XIV, pp.447-478.

DE GUIO A. 1992, *Archeologia della complessità" e calcolatori: un percorso di sopravvivenza fra teorie del caos, "attrattori strani", frattali e...frattaglie del postmoderno*, in BERNARDI M. (ed.), *Archeologia del paesaggio*, Firenze, All'Insegna Del Giglio, 1992, pp. 305-389.

DE GUIO A. 1994a, *L'Altipiano di Asiago dal Bronzo Medio agli inizi dell'Età del Ferro*, in *Storia dell'Altipiano dei Sette Comuni* 1994, pp. 157-178.

DE GUIO A. 1994b, *Sulle orme del Dal Pozzo: proposte per un <<ritorno al futuro>> per l'archeologia di montagna*, "Quaderni di Archeologia del Veneto," X, pp.206-209.

DE GUIO A. 1995a, *Alto-Medio Polesine - Basso Veronese: from a "landscape archaeology " to an "archaeology of the mind"*, in CHRISTIE N., *Settlement and economy in Italy 1500 bC to AD 1500*, Papers of the Fifth Conference on Italian archaeology, Oxford, Oxbow, 1955, pp. 13-24.

DE GUIO A. 1995b (ed.) , *"Ethnoarchaeology of us": calcare e carbonare a Lusiana*, "Quaderni di Archeologia del Veneto," XI, pp. 166-178.

DE GUIO A. 1995c, *Surface and subsurface: deep ploughing into complexity*, in HENSEL W., TABACZYNSKI S., URBANCZYK P. (eds), *Theory and practice of archaeological research* , II, Warszawa, Institute of Archaeology and Ethnology, Commitee of Pre- and Protohistoric Sciences, Polish Academy of Sciences, pp. 329-414.

DE GUIO A. 1996a, *Archeologia della complessità e "pattern recognition di superficie"*, in MARAGNO E. (ed.), *La ricerca archeologica di superficie in area padana*, Stanghella, Linea AGS. pp. 275-317.

DE GUIO A. 1996b, *Dall'archeologia "raccogliona" all'eco-cultural resource management: storie ordinarie di predatori della superficie perduta...*, "Quaderni di Archeologia del Veneto," XII, 212-215.

DE GUIO A. 1997a, *Alla periferia del mondo terramaricolo: "archeologia della complessità" nelle Valli Grandi Veronesi*, in BERNABO' BREA, CARDARELLI, CREMASCHI 1997, pp. pp. 147-165.

DE GUIO A. 1997b, *Archeostratigrafia di superficie ed Eco-Cultural Resource Management*, BORIANI M. (ed.), *Patrimonio archeologico*, progetto architettonico e urbano, Firenze, Alinea Editrice, pp. 53-59.

DE GUIO A. 1997c, *"Landscape Archaeology "e impatto archeologico: una rivoluzione annunciata*, in QUAGLIOLO M. (ed.), *La gestione del Patrimonio Culturale. Cultural Heritage Management. Lo stato dell'arte*, Roma, DRI - Ente Interregionale, pp. 50-67.

DE GUIO A.(ed.) 1997d, *Ricerche a Montecchio Maggiore (VI) 1994-1996: un percorso critico dalla "archeologia delle masiere" alla "archeologia cognitiva"*, Quaderni di Archeologia del Veneto," XIII, pp.87-95.

DE GUIO A (ed.). 1998a, *Archaeology of us, Archaeology for us...: sistemi di Eco-Cultural & Human Resource Management nel "Progetto Altopiano-Risorse" (Asiago-VI)*, in QUAGLIOLO M. (ed.), *La gestione del Patrimonio Culturale. Cultural Heritage Management. Sistemi di Beni Culturali e Ambientali*, Roma, DRI - Ente Interregionale, 1998, pp. 222-241.

DE GUIO A. 1998b, *"Off-site powerscape": il potere "fuori porta". Nuovi orizzonti di attesa per l'Età del Bronzo padana*, in PEARCE M., TOSI M. (eds), *Papers from the EAA Third Annual Meeting at Ravenna 1997, Vol I: Pre- and Protohistory*, Oxford, British Archaeological Reports, International Series 717, pp. 165-171

DE GUIO A. 2000, *Ex Occidente lux: linee di un percorso critico di rivisitazione del Bronzo Finale nel Veneto*, in HARARI M., PEARCE M. (eds), *Il Protovillanoviano al di qua e al di là dell'Appennino*, Como, New Press, pp. 259-357.

DE GUIO A. 2001a, *Lusiana: le testimonianze archeologiche di Lusiana*, in BOSCARDIN M., (ed.), Lusiana, G.A.L. n. 6, 2001, pp. 125-135.

DE GUIO A., (ed.) 2001 b, *"Superfici di Rischio" e C.I.S.A.S. Se lo conosci, non lo eviti*, in GUERMANDI M.P. (ed.), *Rischio Archeologico: se lo conosci lo eviti*, Firenze, All'Insegna del Giglio, pp. 265-306.

DE GUIO A. in press, *Archeologi in guerra*.

DE GUIO A., BRESSAN C., 2000, *Tra archéologie évenementielle" e "longue durée": il caso di studio delle calcare dell'Altopiano di Asiago*, "Archeologia Postmedievale", 4, pp. 73-86.

DE GUIO, A., CATTANEO P. 1997. *"Dirt roads to Brendola": le strade preistoriche di Soastene-Brendola (VI)*, *"Quaderni di Archeologia del Veneto"*, XIII, pp. 168-182.

DE GUIO A., FINZI E. 1995, *Il "Progetto Rotzo" (Altopiano di Asiago): un caso di studio di pattern recognition incrociata*, in D'AMICO C., FINOTTI F. (eds), *Le scienze della terra e l'archeometria*, Rovereto, Museo Civico di Rovereto, Museo di Mineralogia "L. Bombicci", Università di Bologna, pp.33-41.

DE GUIO A., SECCO G. 1984, *A new computer seriation algorithm*, "Computer Applications in Archaeology", pp.199-209.

DE GUIO A., SECCO G. 1988, *Archaeological applications of the "Percolation Method" for data analysis and pattern recognition*, in RAHTZ S.P.Q. (ed.), *Computer and Quantitative Methods in Archaeology 1988*, Oxford, British Archaeological Reports, International Series, 446(i), pp. 63-93.

DE GUIO A., ZANUSSO S. 2000, *Attorialità sociale ed etnoarcheologia: micro-storie carbonare dalla Lessinia (VR)*, "Archeologia Postmedievale", 4, pp. 53-72.

DI MARTINO B. 1998, *Spie Italiane contro forti austriaci. Lo studio della Linea fortificata austriaca sugli altopiani trentini*, Novale, Gino Rossato EdItore.

DI MARTINO B. 1999, *Guerra aerea. Vicende e immagini dell'aviazione italiana sugli altopiani veneto-trentini*, Novale, Gino Rossato Editore.

DEMERS M.N. 1997, *Fundamentals of Geographic Information Systems*, John Wiley & sons, Inc., New York.

EASTMAN J.R. (ed.) 1997, *IDRISI for Windows, User's guide-Version 2.0*, Clarke University, Worcester (Ma).

EPSTEIN J.M. 1999, *Agent-based computational models and generative social science*, "Complexity", 4 (5), pp. 41-60.

EPSTEIN J.M., AXTEL R. 1996, *Growing artificial societies. Social science from the bottom up*, Washington, Brookings Institutio Press.

FABI L., TODERO R. 2001, *Tre giorni sugli altopiani. itinerari trentini della Grande Guerra*, Lavarone, Guide Gasperi.

FRANCOVICH R. (ed.) 1991, *Archeologia delle attività estrattive e metallurgiche*, Firenze, All'Insegna del Giglio.

GIANNICHEDDA E. (ed.) 1996, *Antichi mestieri. Archeologia della produzione*, Genova, Sagep.

GIARDINO C. 1998, *I metalli nel mondo antico. Introduzione all'Archeometallurgia*, Roma-Bari, Laterza.

ISNENGHI M., ROCHAT G. 2000, *La Grande Guerra 1914-18*, Roma, La Nuova Italia.

LEONARDI G., RUTA SERAFINI A. 1981, *L'abitato protostorico di Rotzo (Altipiano* di Asiago), in "Preistoria Alpina",17, pp. 7-75.

FAGGIANI L., *Le malghe del trentino. Viaggio nel tempo sulle tracce di un mondo antico*, Torino, Vivalda Editori.

FLEMING A. 1988, *The Dartmoor Reaves. Investigating prehistoric land divisions*, London, Batsford.

FLEMING A. 1992, *Landscape archaeology in the British uplands: opportunities and problems*, in BERNARDI 1992, pp. 67-88.

FORMAN R.T.T. 1995, *Land mosaics. The ecology of landscapes and regions*, Cambridge, Cambridge University Press.

FORTE M. (ed.) , *Archeologia. Percorsi virtuali nelle civiltà scomparse*, Milano, Mondadori.

FRANCO T. s.d., *Lavarone. Incontri con l'antico. Antropologia, arte e indizi di preistoria*, Vicenza.

FRANCOVICH R., ZIFFERERO A. (eds) 1999, *Musei e parchi archeologici*, Firenze, All'Insegna del Giglio.

JAMERSON J.H. 1997, *Presenting archaeology to the public. Digging for truths*, Walnut Creek, Altamira Press.

KIRSCHNER P. 1997-98, *Archeologia Pubblica e Gestione delle Risorse Culturali: Applicazioni multimediali Relative ai casi di studio del "Progetto Rotzo" e del "Progetto Brendola"*, unpublished dissertation, Università di Padova - Dipartimento di Scienze dell'Antichità.

KVAMME K.L., KOHLER T.A. 1988, *Geographic Information Systems: technical aids for data collection, analysis, and display*, in JUDGE W.J., SEBASTIAN L. 1988, *Quantifying the present and predicting the past: theory, method and application of archaeological predictive modeling*, U.S. Department of the Interior Bureau of Land Management, Denver (Colorado).

LEONARDI G. 1992a, *Le prealpi venete tra Adige e Brenta tra XII e VI secolo a.C*, in METZGHER, GLEIRSCHER 1992, pp. 135-144.

LEONARDI G. (ed.) 1992b, *Processi formativi della stratificazione archeologica*, Padova, Dipartimento di Scienze dell'Antichità, Saltuarie Del Piovego.

LEONARDI G. 2000, *Processi formativi*, in MANACORDA D., FRANCOVICH R. (eds), *Dizionario di Archeologia. Temi, concetti e metodi*, Roma-Bari, 2000, Laterza, pp. 228-231.

LEONARDI G., RUTA SERAFINI A. 1994, *L'Altipiano dei 7 Comuni dall'Ottavo secolo alla romanizzazione*, in *Storia dell'Altipiano dei Sette Comuni. I. Territorio e istituzioni, Vicenza, Neri Pozza*, 1994, pp. 179-191.

LUSSU E. 1937, *Un anno sull'Altopiano*, Torino, Einaudi.

LOCK G., STANCIC Z. (eds) 1995, *Archaeology and Geographical Information Systems*, London.

MACINNES L., WICKHAM JONES C. 1992, *Archaeology and the green debate*, Oxford, Oxbow.

MALCOM A., COOPER M.A., FIRTH A., CARMAN J., WHEATLEY G. (eds), 1995, *Managing archaeology*, London, Routledge.

MANTOVANI F., MARCOLONGO B. 1992, *Il telerilevamento nelle scienze della terra*, Roma, La nuova Italia Scientifica.

MARZATICO F. 1997, *L'industria metallurgica del Trentino durante l'età del Bronzo*, in BERNABO' BREA, CARDARELLI, CREMASCHI (eds) 1997, pp. 570-594.

MARTELLO G.V. 1993, *Gli ambienti dell'Altopiano dei Sette Comuni: evoluzione recente e correlazione con le attività antropiche*, , "Acta Geologica", Museo Tridentino di Scienze Naturali, 70, pp. 135-142.

MATTALIA U. 1989, *La guerra dei forti sugli Altopiani. Vezzena – Lavarone – Folgaria - Altopiano dei Sette Comuni*, Novale, Gino Rossato Editore.

MATTALIA U. 1992, *Cronache della Grande Guerra 1915-1918. Altipiani – Valsugana – Pasubio – Isonzo – Piave*, Novale, Gino Rossato Editore.

MEREGALLI C. 1997, *Grande Guerra Sull'Altopiano di Asiago. Il Cannone sconvolse la quiete dei monti*, Bassano, Tassotti.

METZGER I.R., GLEIRSCHER P. 1992, *I Reti*, Bolzano, Athesia.

MIGLIAVACCA M. 1985, *Pastorizia e uso del territorio nel Vicentino e nel Veronese nelle età del Bronzo e del Ferro*, «Archeologia Veneta», VIII, pp.27-62.

MIGLIAVACCA M. 1993, *Lo spazio domestico nell'Età del Ferro. Tecnologia edilizia e aree di attività fra VII e I secolo a.C. in una porzione dell'arco Alpino orientale*, "Preistoria Alpina", 29, pp. 5-161.

MIGLIAVACCA M., VANZETTI 1988, *Progetto Rotzo-Sette Comuni. Campo Manderiolo 1 Nord: un deposito archeologico in formazione da una struttura di malga in blocknau*, "Quaderni di Archeologia del Veneto", IV, pp? 407-412.

MUIR R. 1999, *Approaches to landscape*, Houndsmill, Macmillian Press.

NALLI G. 1895 (1974), *Epitome di nozioni storico economiche dei Sette Comuni Vicentini*, Bologna, Arnaldo Forni, ristampa anastatica.

NELSON T., 1981, Literary Machines, Pennsylvania.

OROMBELLI G.1977, *Le condizioni climatiche durante il II millennio a. C.*, in BERNABO' BREA, CARDARELLI, CREMASCHI (eds) 1997, pp. 51-55.

PARISI D. 1989, Intervista sulle reti neurali. cervello e macchine intelligenti, Bologna, Il Mulino.

PASSARIN M., VIAZZI L. 1998, *Panorami della Grande Guerra*, Vicenza, Fondazione della Cassa di Risparmio di VR VI BL e AN.

PEARCE M. 1998, *Reconstructing prehistoric metallurgical knowledge: the northern Italian Copper and Bronze Ages*, in "European Journal of Archaeology", 1, pp. 51-70.

PEARCE M., DE GUIO A.1999, *Between the mountains and the plain: an integrated metal production and circulation system in later Bronze Age north-eastern Italy*, DELLA CASA P. (ed.), *Prehistoric Environment, society, and economy*, Papers of the international colloquium PAESE '97 in Zurich, Bonn, Rudolf Habelt, pp. 289-293.

PEISER B.J., PALMER T., BAILEY M.E. (eds) 1998, *Natural catastrophes during Bronze Age civilisations. Archaeological, geological, astronomical and cultural perspectives*, Oxford, British Archaeological Reports International Series 728.

PELLEGRINI E. 1998, *Aspetti della metallurgia protovillanoviana in Etruria*, in NEGRONI CATACCHIO1998b, pp. 23-34

PERINI R 1989, *Testimonianze dell'attività metallurgica dall'Eneolotico alle fasi finali dell'Età del Bronzo nel Trentino*, in *Per Giuseppe Sebesta*, Trento 1989.

PERONI R. 1996, *L'Italia alle soglie della storia*, Roma-Bari, Laterza.

PREUSCHEN E. 1973, *Estrazione mineraria dell'età del Bronzo nel Trentino*, "Preistoria Alpina", IX, pp. 113-1150 (trad. di L. Dal Ri).

PEZZI C. 2001, *Partir bisogna. Economia e storia di Luserna tra Ottocento e Novecento*, Luserna, Centro Documentazione Luserna.

RANDSBORG K. (ed.) 1996, *Absolute chronology. Archaeological Europe 2500-500 BC*, Kobenhavn, Munksgaard.

REICH D. 1910, *Notizie e documenti di Lavarone e dintorni*, Trento, Società Tip. Ed. Trentina.

RIGONI STERN M. (ed.) 2000, *La guerra sugli Altipiani. Testimonianze di soldati al fronte*, Vicenza, Neri Pozza.

RIGONI STERN M, OLMI E., KEZICH T. 1985, *I recuperanti*, Asiago, Tipografia Moderna.

RIZZOLO D. 1996, *Asiago e le sue contrade nei nomi di luogo di origine cimbra e veneto-italiana. Toponomastica storica del comune di Asiago*, Roana, Istituto di Cultura Cimbra.

ROSSIGNOL J., VANDSNIDER L. (eds) 1992, *Space, time and archaeological landscape*, New York, Plenum Publishing Corporation.

SAURO U. 1977, *Aspects de la morphogenèse anthropique dans le milieu karstique Alti Lessini*, "Revue Norois", 95, pp. 149-163.

SAURO U. 1993, *Human impact on the Karst of the Venetian Fore-Alps*, Italy, "Environmental Geology", 21, PP. 115-121.

SAURO U., BONDESAN A., MENEGHEL M. (eds) 1991, *Proceedings of the International Conference on Environmental Changes in Karst Areas*, Quaderni del Dipartimento di Geografia", Università di Padova, 1991.

SAURO U., LANZINGER M *1993, Le ricerche sull'unità morfocarsica dell'Altopiano dei Sette Comuni (Prealpi Venete*, "Acta Geologica", Museo Tridentino di Scienze Naturali, 70, pp. 5-6.

SAURO U., MARTELLO G.V., FRIGO G. 1991, *Karst Environment and human impact in the Sette Comuni Plateau (Venetian PreAlps)*, in SAURO, BONDESAN, MENEGHEL 1991, pp. 269-278.

SCHIFFER M.B. 1987, *Formation processes of the archaeological record*, Albuquerque, University of New Mexico Press.

SCHIFFER M.B.1995, *Behavioral Archaeology. First Principles*, Salt Lake City, University of Utah Press.

SEBESTA G. 1992, *La via del rame*, supplemento a "Economia Trentina", 3, Calliano.

SHERRAT A. 1994, *Core, periphery and margin: perspectives on the Bronze Age*, in MATHERS, STODDART 1994, pp.335-345.

Storia dell'Altipiano dei Sette Comuni. I. Territorio e istituzioni, Vicenza, Neri Pozza 1994.

TOMASI G. 1997, *Il territorio trentino-tirolese nell'antica cartografia*, Ivrea, Priuli & Verlucca.

VAN DER LEEUW S., MCGLADE L.J. (eds) 1997, *Time, process and structured transformation in archaeology*, London, Routledge.

VANZETTI A. 1997, *Lo «off-site»: transetto di ricognizione fra Fondo Paviani e Fabbrica dei Soci*, in BERNABO' BREA , CARDARELLI , CREMASCHI (eds) 1997, pp. 161-162.

VIDALE M., EHRENREICH R.M.R., MICHIELII M., VANZETTI A. 1988, *Nuovi indizi sulle attività di trasformazione dei metalli nell'insediamento protostorico di Rotzo*, in "Archeologia Veneta", 1988, XI, pp. 15-43.

VON LICHEM H., CENTRO DOCUMENTAZIONE DI LUSERNA 1998, *Per non dimenticare. Luserna e gli Altipiani nella Prima Guerra Mondiale*, München, MediaDom Verlag.

WEBER F.1965, *Tappe della disfatta*, Milano, Mursia.

ZANETTI L. BERNI P., LIGUORI G. 1988, *Formaggi e cultura della malga*, Verona. Giuseppe Liguori

ZAMMATTEO P. in press, *Cimbri-Walser. Le case di Luserna.*

ZAMMATTEO P. in press, *Luserna. Introduzione storica.*

ZAMMATTEO P. in press, *Il metodo e la ricerca. Maestranze minerarie sugli Altipiani.*

ZAMMATTEO P. in press, *Le miniere medioevali sugli Altipiani e nel settore sudorientale del Trentino.*

ZAMMATTEO P. in press, Santa Margherita e la porta naturale. Il margine dell'insediamento silvo-pastorale e il passo del Cost a Lavarone.

ZAMMATTEO P. in press, Uno sguardo dalla montagna. La "pianura" vista dagli Altipiani.

ZILLI S 1996-7, I Sistemi Informativi Geografici (GIS) in Archeologia: problemi e casi di studio, unpublished dissertation, Università di Padova - Dipartimento di Scienze dell'Antichità.

The Power of GIS and Remote Sensing: Multi-Scalar Spatial Analysis of settlement data in SE Pacific Coastal Guatemala and the Southern Maya Lowlands

Francisco Estrada-Belli

Vanderbilt University, Department of Anthropology,
Box 306050 Station B, Nashville, TN 37235.
Francisco.Estrada-Belli@vanderbilt.edu

Abstract. Two quite different environmental regions and settlement datasets provide excellent case studies to illustrate recent fruitful applications of Geographic Information Systems (GIS) and Remote Sensing in archaeological analysis at the regional and community scale. The potentialities of GIS and other digital technologies are fully exploited when a wide array of data types from different sources are integrated in a single contextual database. The coastal studies focus on regional land-use, settlement history, environmental change, demography, political organization, and inter-regional interaction. Analysis of the La Milpa and Holmul cities, on the other hand, focus on finer scale analysis of resource allocation within Lowland Maya cities as well as their community-level political organization, and ideational signatures imposed by the Classic Period Maya on their landscapes.

1 Introduction

Geographic Information Systems (GIS) are generally and simply defined as computer programs for recording, storing, retrieving and analyzing spatial data (Burrough and McDonnel 1998). On the other hand, under the rubric of Remote Sensing fall a variety of data from a number of electromagnetic sensors used in near surface, airborne or space borne surveys. These technologies were introduced in north American archaeology in the late seventies and today are often used as aids to archaeological survey and excavation. However, compared to disciplines in the natural sciences, and in spite of a long time of experimentation use since the elaboration of these technologies, few Mesoamerican archaeologists utilize these methods to the fullest of their potential.

Geographic Information Systems (GIS) were once considered a set of expensive and complicated tools rather exclusive to large-budget research institutions capable of supporting supercomputing laboratories and highly trained personnel, however today are implemented by large and small research projects throughout the world in order to analyze regional datasets or nation-wide inventories or to solve problems relating the integration of multiple data sources, such as digital imagery, traditional aerial photography and topographic maps, all in the same analytical coordinate-based environment.
Just in the last 5 years, GIS programs have reached a high level of sophistication. New algorithms and new graphical interfaces are constantly produced as software packages are updated in a biannual or quarterly basis. The technological evolution is accompanied by decreasing costs in hardware and software resulting in wider accessibility and favored by a steadily flattening learning curve such that today we can say that GIS analysis is accessible to the widest ever population of researches and end-users working from any Personal Computer or laptop computer.

In the Maya region the transition to fully digital and automated spatial analysis has been slower than in other areas, primarily due to the difficulties of achieving large area coverage under the forest canopy, and due to a dearth of extensive regional surveys which have been considered traditionally the natural testing ground for GIS.

Today, it has become readily apparent to archaeologists working with site-specific data that a GIS database can be beneficial in every step in the analysis of spatial patterns, from the inventories, exploratory data analysis to model-testing and publishing. Likewise, the archaeological scale of analysis can range from small household layouts to regional systems and most GIS applications are applicable to regional scale problems as they are to micro-scale phenomena.

In this paper, I would like to present an overview of two types of research in very different contexts in which GIS analysis has provided a new way of approaching spatial analysis through visual and quantitative methods allowing us to explore new patterns and pose new questions on human behavior and landscape cognition that had been nearly impossible to frame with analysis on traditional map data.

The first case study involves a simple application of GIS and remote sensing on a regional scale in a non-forested area of the southeastern coast of Guatemala. This is a region in which most landforms and archaeological features can be seen from the air and human-environment and intersite relationships can be analyzed in a large contiguous area with extensive survey coverage.

The second type of applications involves two separate sites in the Maya Lowlands, La Milpa, located in Northwestern Belize, and Holmul, located in Northeastern Guatemala, 45 km east of Tikal. Existing data from La Milpa were used to prepare the first-full fledged GIS database of a Maya city which readily gave its fruits in the identification of patterns of spatial organization both regionally and locally with some ramifications in current discussions of social organization, economy and landscape cognition within the Maya state. Some of the hypotheses stimulated by patterns identified at La Milpa were tested against new data gathered from the site of Holmul, which was reconsidered after 90 years of neglect, with much potential to reveal the dynamic structure of Maya urban landscapes due to its long-spanning sequence. In all cases, the research goals could be met thanks to the new technologies of GIS, remote sensing, and Global Positioning Systems (GPS) with greater efficiency and accuracy than with traditional methods.

2 PACIFIC COAST

The first case study is a systematic regional survey on the Southeastern Coast of Guatemala which was led in collaboration with a number of field archaeologists and specialists between 1995 to 1997 (Estrada Belli, 1997a, 1997b, 1999a, Estrada Belli et al. 1996, 1997, 1998, Estrada Belli and Kosakowsky 1996, 1997, Kosakowsky and Estrada Belli, 1997, Kosakowsky et al. 1999, 2000, Neff et al. 1999). We surveyed in "extensive coverage" about 1300 km2 of coastal plain and about 10% of the region in intensive coverage through 1km-wide transects. The choice of such a large contiguous study area was intended to possibly include more than one political entity, or settlement system and more than one river basin.

Fig. 1. Map of Central America showing the Guatemalan Southeastern coast and the Lowland Maya city of La Milpa, Belize study areas.

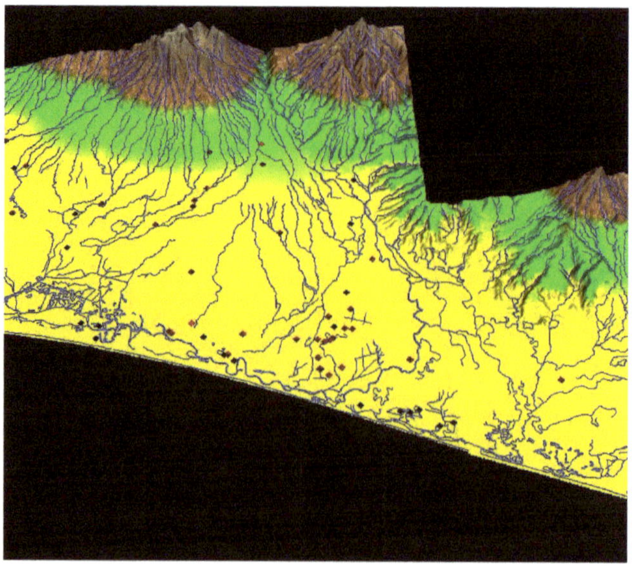

Fig. 2. View of the Southeastern Coast survey area from the south. The Río Esclavos forms the main drainage in this region flowing between the Tecuamburro volcano and the Cerro Consulta (center).

It eventually included four major drainages, a total of 156 sites grouped in four main polities, and one major settlement system, with three levels of settlement rank-size during Late Preclassic to Late Classic times. This is a riverine environment, with a tropical wet-dry climate, and a well-defined 6 month-long rainy season. Throughout prehistory, proximity to rivers was apparently very important for the growth of settlements. A linear arrangement of sites along the major rivers is apparent with the major centers located along the largest watercourses. This settlement pattern recalls the linear patterns of settlement in Mesopotamia and other alluvial plains (Fig.2).

As in Mesopotamia, sites of lower rank-size are located along the river routes both upstream and downstream of the main center. As in Mesopotamia, river morphology is somewhat complex with many rivers disappearing into sinkholes and only a few ancient sites are currently located away from modern river courses.

The central area of the region in the flood plain of the Esclavos river presented a complex sequence of occupation and abandonment at two specific intervals, the first about 400-600 A.D. and the second post 800 A.D. The Nueve Cerros site was abandoned about A.D. 400 and a new ceremonial core was in place 1 km to the south (La Maquina site) by about A.D. 600 (Fig.3). The new site consisted of a main rectangular acropolis platform with dense residential architecture in the surrounding area stretched along a new "major" river channel (Middle Classic, see Fig.4). The new site itself would be abandoned finally in the period between 800 and 950 A.D.

Close scrutiny of the Late Preclassic map the Nueve Cerros site on high-resolution aerial photographs showed a ceremonial core consisting of temple-pyramids and range structures aligned in narrow N-S oriented plazas on a knoll surrounded by two minor water channels flowing N-S detectable through soil and crop marks (Fig.3).

Fig.3. Geo-referenced aerial photo (taken in 1991) of the alluvial area surrounding the Nueve Cerros and La Maquina. Image ground resolution after geo-referencing: 1 m. Crop and soil marks reveal major artificial mounds and former river courses. Detected Late Preclassic structures and water courses are shown as overlays.

Fig.4. Geo-referenced aerial photo (taken in 1991) of the alluvial area surrounding the Nueve Cerros and La Maquina. Image ground resolution after geo-referencing: 1 m. Detected Middle Classic structures and water courses are shown as overlays.

While the importance of the river channels to the major settlement seemed obvious, the processes that caused the river channels to shift and what wider impact any geo-morphological process might have had on human settlement was difficult to ascertain from traditional aerial photography and topographic maps.

Observation of false color Landsat imagery using a combination of near-infrared, infrared and visible green bands (7-4-2) covering a much larger area, readily revealed a major geological feature of the area, a large alluvial fan of probable Pleistocene age immediately to the west of the Esclavos main river channel (Fig.5). This suggested the existence of a very long-term process of possible progression of the Esclavos river from an original river channel flowing away from the alluvial fan and resulting in a general eastward trend.

Fig. 5. Landsat 5 TM image of the Southeastern coast of Guatemala taken on 12/1989 showing bands 7,4,2 as RGB. A large alluvial fan west of the Esclavos river is visible in the center of the image.

Only two existing river channels were visible in the field, and the low resolution of the imagery prevented further detailed analysis to link the channels spotted on aerial topography to regional river morphology.

The landsat imagery was then sharpened using the Intensity/Saturation/Hue procedure to combine the high-resolution low-greyscale contrast aerial photos with the low-resolution high-spectral-contrast imagery. The byproduct of this operation allowed the identification of numerous new river channels on a W-E progression confirming the relationship of the early Nueve Cerros and late La Maquina occupation with two meandering episodes of the Esclavos River (Fig. 6). Following this phase, at about 900 A. D. the next river channel to the east appears to have taken a much more braided course. The new courses may have been caused by faster and greater sediment transport than ever before and more frequent and destructive floods may have impaired the agricultural activities and residential stability of the local populations, and therefore may have lead to the long-term abandonment of this area until modern times, as the settlement chronology suggests.

Among the few additional and equally potential factors for the hypothesized change in nature of the Esclavos River we shall include tectonic plate movements, climatic

Fig.6. Landsat 5 TM geo-referenced and sharpened with aerial photo at 5 m ground resolution showing buried river courses undetectable from aerial photo and satellite image alone.

change, earthquakes, volcanic eruptions, deforestation and other causes still under evaluation. According to our survey, the first shift from the Nueve Cerros to the La Maquina channel and the abandoned and re-building of the ceremonial core was a fairly well contained episode within the 400-600 A. D. period. Such period also marks a general dip in the population history of the region. Could this population decline be possibly linked to an environmental disaster of major proportions? Newly reworked dates for the Ilopango volcanic eruption in El Salvador, 90 km to the east, to about A.D. 450 may provide support an hypothesis linking this most destructive volcanic episode with the depopulation of coastal sites (Dull et al. 2001). The Ilopango eruption may have coincided with powerful earthquakes on the Guatemalan coast, and is known to have left 5-10 cm of ash in SE Guatemala. In addition, larger amounts of ashfall documented in the Chalchuapa and nearby highland areas would have caused widespread deforestation in the headwaters valleys of the La Paz and Esclavos river, the two main watercourses in SE Guatemala. The deforestation in turn could have caused greater erosion coupled with ashfall being added to the water flow to result in unprecedented level of water flow and sediment transport in the drainage systems which would have consequently flowed at much greater force opening new courses. While this scenario at the time of writing remains highly hypothetical, in the future, a sampling of coastal levees and sedimentation levels in lakes, might produce data with which to test the possible effects of the Ilopango eruption on changes in the river morphology in the coastal plain.

In conclusion, the final abandonment of the area appears to be connected to a longer-term shift in the nature of the river from meandering to braided finally reaching its current location 8 km east of the Middle Classic course. While the general timing for the progression is at the moment suggested only by archaeological remains, we are awaiting publication of new paleoclimatological data from the nearby lakes of SE Guatemala by Neff, Persall and associates in order to evaluate possible climatic changes ensuing at about 850-950. The enhancement of satellite imagery provided crucial data beyond the visible spectrum and at a greater spatial scale which was beneficial in identifying the long-term process linking river channel to settlement abandonment.

3 SOUTHERN MAYA LOWLANDS

La Milpa, like many other Maya cities in the Southern Maya Lowlands, is situated at the intersection of large expanses of low-lying areas and an upland hilly region (Fig. 7). Six seasons of survey and excavation at La Milpa led by Gair Tourtellot and Norman Hammond (Tourtellot et al. 2000, 2002) have provided finely detailed maps in which the spatial arrangement of ritual and residential structures is interspersed with land modification features creating a nested heterarchy of landscapes: constructed, conceptualized and ideational. Economic and demographic forces certainly played a major role in determining the siting of landscape features and social order associated with them. These features can be easily identified and quantified using Geographic Information Systems (GIS) computerized capabilities (Kvamme 1992). Their relationships are often visually clear when plotted against cost-distances, population sizes, and available resource locations. Especially important is the ability to remove from the landscape the forest cover which is the biggest obstacle today for everyone attempting to get a sense of the distribution of structures within a Maya city from a vintage point or even from the air. However, the La Milpa GIS has the additional potential to explore the symbolism associated with landscape features and to attempt a reconstruction of the conceptualized structure of a Maya city.

The La Milpa GIS (Estrada Belli 1999b) offered an opportunity to test hypothesis on the nested structure of its landscape without much initial guidance from site-specific epigraphic and iconographic evidence, but with the ability to create a virtual experience to the La Milpa landscape through 3D graphics and spatial measurements (Estrada Belli and Tourtellot 2002, Tourtellot et al. 2000, 2002). For example, La Milpa, and its environment are more clearly visible on artificially created surface models than in any real conditions, either from the air or from the ground. The 3d model of the La Milpa region up to a 14 km distance showed the large extent of Classic Period settlement cross-cutting upland and lowland eco-tones (Fig.7). The source for the DEM used in this study were elevation data obtained from archaeological maps at 1:1000 scale and 1 m vertical resolution, national topographic maps at 1:50,000 scale and at 20 m vertical resolution, and remote sensing imagery provided by NASA JPL to Norman Hammond.

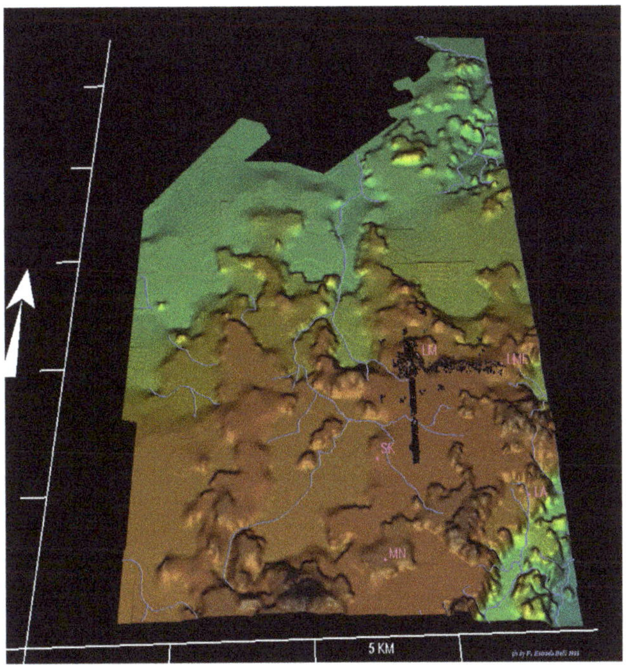

Fig.7. View of the La Milpa regional topography (DEM at 20 m vertical resolution). La Milpa residential structures mapped by Boston University and neighboring sites are also shown. MN= Ma'ax Na, LA= Las Abejas, LME=La Milpa East.

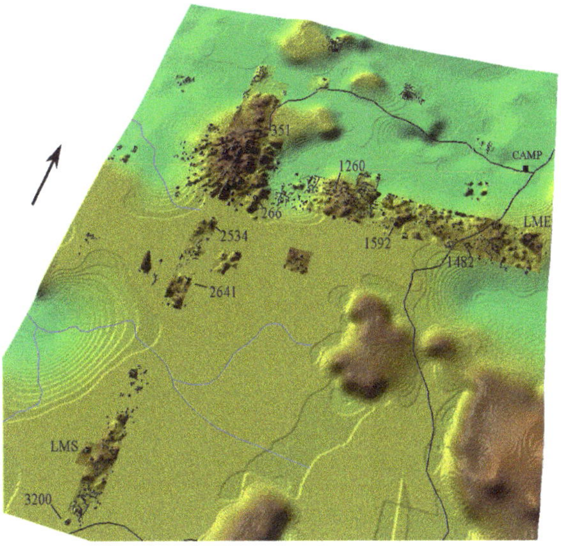

Fig. 8. View of the DEM and settlement data from the La Milpa settlement mapping project by Gair Tourtellot, John Rose and others.

Settlement Analysis. Once the data entry was completed, simple GIS procedures were used to measure and rank structures and features within the settlement areas mapped (ca. 20,000 structures), and identify top-ranking elite groups at regular intervals. Visualization of top-ranking group location suggested a spatial ordering (at 1 km intervals) and other possible regularities which may be linked to a variety of administrative or ritual functions (Fig. 8, 9).

In addition, the relationship of settlement and agricultural features with hilltop groups revealed a pattern of hilltop-centered neighborhoods with clusters of residences with terracing in the intervening areas as documented in all survey transects . Interpretation: barrios administrated by hilltop elite to control agricultural production, labor, and ritual services to the suburban population. The pattern observed supported the model of Maya center as garden cities, with widely spaced sub-urban settlement with interspersed orchards and fields.

The regular spacing of nodes, on the other hand, suggested a high degree of integration of the population through a lattice of large and small community nuclei of elite ritual and administrative function. In addition, the top ranking groups were found to be similar in size and form, with long palace-like buildings and eastern pyramid temples. Several groups were found in an inner ring at 1 km and other groups in an outer ring at 3.5 km to the east and south. La Milpa East is the largest found of all peripheral groups. The distribution of larger and more formal groups at the outer ring from the center further suggested integration of the population though the de-centralization of services at those outer centers.

Further analysis of the La Milpa settlement, made use of viewsheds. Working under the assumption of no-tree

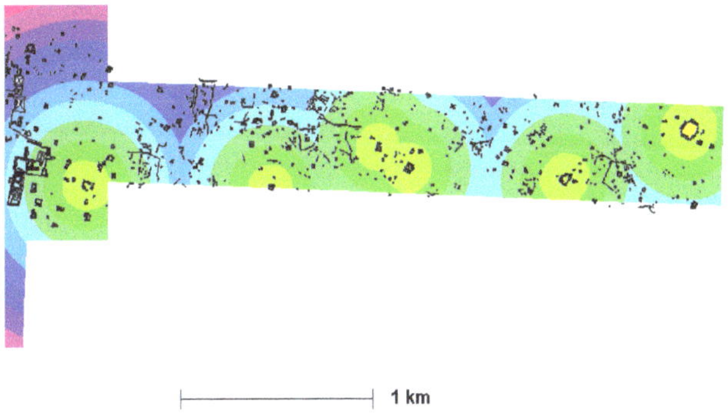

Fig. 9. View of topography surface including a combined DEM resulting from addition of high-resolution DEM at 1 m elevation intervals to low-resolution DEM at 20 m elevation intervals. Notable hills and drainages are obscured in the low-resolution DEM. Large groups along the east and south transects are identified.

coverage during the maximum population density in this area, a viewshed map was created to show all areas on the landscape that had an unobstructed view from Pyramid 1. The question to be answered was whether the view to and from this major feature of the La Milpa landscape affected household location. A simple chi-square statistic of the total set of La Milpa Late Classic houses against the viewshed areas indicated a deliberate attempt to place houses with an obstructed view of the main sacred mountain at the site (Fig.10). All the top-ranking elite groups have a clear line of sight with the main temple.

The large elite group La Milpa East shows deliberate attempt to link eastern shrines with Pyramid 1 (Fig.11). Was intervisibility a metaphor for ritual and political interconnectivity of the rural people vis-à-vis the elite temple? A second viewshed along the south identified the group named La Milpa South at 3.5 km within view of

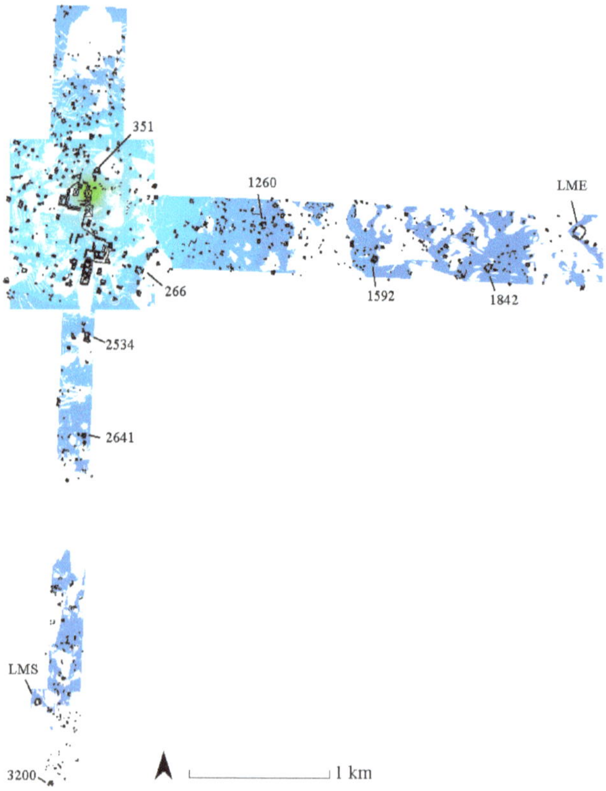

Fig. 10. Overlay of structures of east, south, north and central maps of La Milpa on viewshed area from Pyramid One in the main plaza. Large elite groups with eastern shrine are also identified.

Pyr. 1, directly south of it. This raised the question of whether there is a wider pattern of intervisible nodes around the main La Milpa hill.

At the predicted distance, 3.4 km to the west of Pyramid 1, a new site was located with a large plaza and western pyramid shrine; a mirror image of the La Milpa East group. La Milpa East and La Milpa West had eastern and western shrines tilted to look at sunrise and sun set during the solstices.

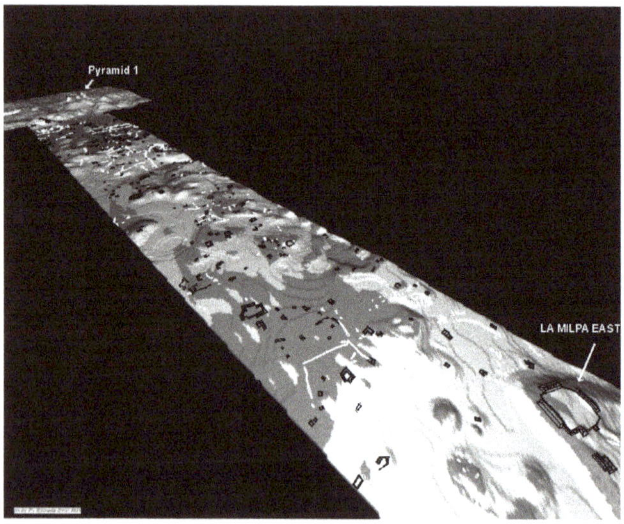

Fig. 11. Viewshed on 3d model of eastern transect looking towards the plaza. Structures and berms are displayed on viewshed areas (shaded). La Milpa East group is in the foreground.

The 7 km-long line connecting them across the La Milpa landscape is oriented to east-west or roughly in line with the equinoxes (Fig.12). The equinoxes and solstices form four stations in the path of the sun in the tropical sky's horizon and mark four divisions in the solar year. Furthermore, the line that connects the two distant groups passes over the site's main plaza and intersects with a north south line connecting a northern group (La Milpa North) to La Milpa South thus forming a quincunx or cross on the La Milpa landscape. A cosmogram for the Late Classic Milpeños. The Maya cosmos is made of four corners and the center and this is replicated in the layout of a large plaza, a chultun/cave, and pyramid at the center of La Milpa (Fig.13).

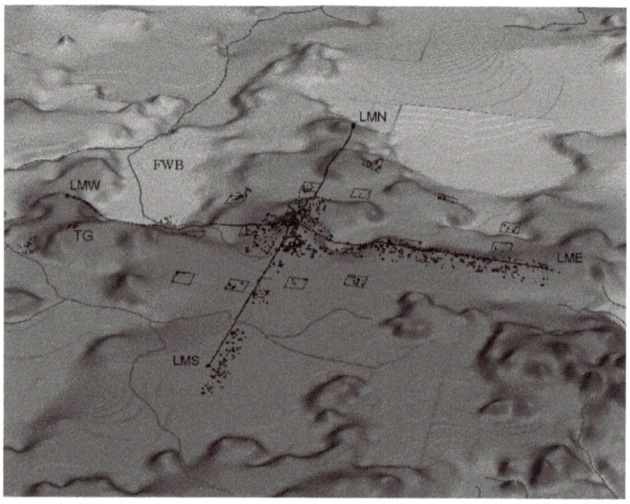

Fig. 12. View of the La Milpa settlement area showing mapped structures and line-of-sight connection La Milpa Eeast, La Milpa South, La Milpa West and La Milpa North groups passing over the main plaza area.

Fig. 13 View of the La Milpa plaza showing line-of-sights from four cardinal groups intersecting in coincidence with Str. 10 and chultún in plaza center.

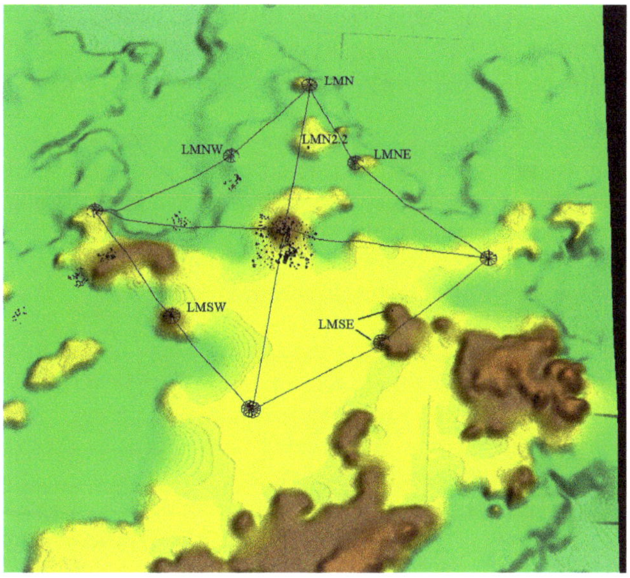

Fig. 14. View of La Milpa region showing cosmograms four cardinal hill-top groups discovered and hypothetical intercardinal hill-top groups at regular distance intervals.

The cosmos is frequently represented as a quincunx in Maya art and it is also represented as a circle of towns on the horizon of a major center in colonial period Maps of Yucatan (see map of Mani, Stephens 1843, plate XXV). Is it possible that other, less important centers might have existed at intercardinal positions on the La Milpa horizon? On our landscape model we were able to identify four hills with unobstructed view of LM and at fairly regular distances from LM and from one another (Fig. 14). Or instead, could the Maya elite have filled the landscape with sacred and secular nodes at regular intervals to serve the souls and land of their kingdom with efficiency in mind? A new season of survey in 2002 by Dr. Tourtellot's team of surveyors will help evaluate the hypothesis that intermediate hills were included in the La Milpa cosmogram forming an eight-point cosmogram. This grandiose image of the cosmos might have been created by the very last La Milpa rulers by linking intervisible cardinal and intercardinal nodes on the suburban La Milpa landscape as a way of marking the unity of the territory around the ruler's center.

Furthermore, when considered in a wider regional context, the viewshed from La Milpa's Pyramid 1 over the surrounding region shows the central and dominant position of the La Milpa hill in relation to a very large region (Fig. 15). Within this region, the models shows possible intervisibility among centers 5-9 km apart which were located on prominent hills, as well. As noted elsewhere, the pattern of links among the most visible spots on the hilly ground within a 10 km radius maybe significantly and adversely affected by poor resolution of our topographic data as well as potential remnants of forest in the intervening areas between major populated areas (Madry and Rakos 1996). However, it is interesting to note that when the viewshed from the La Milpa hill is compared to the views from a few neighboring centers the difference in viewshed areas is remarkable. Could it be that importance of the location on the landscape and the viewshed associated with a certain location affected the growth of centers and therefore a relationship might exist between the viewsheds of a center, its population and architectural size, and the territory controlled exclusively by each polity?

Following this observation of GIS viewshed and available site data we could tentatively surmise that viewshed area or position on the landscape and such practical effects of long-term Maya history as population size and position in the regional hierarchy were indeed directly correlated. Therefore, if ritual power and political aspirations within and among Maya centers are symbolically encoded by their relative position within ideational and political landscapes at the regional scale, are physical and political hierarchies of landscapes relevant at the macro regional scale as well? Were Tikal and Calakmul, both physically and politically prominent among Maya cities, as possible centers of Maya superstates (Martin and Grube 1995)? In order to shed more light on these intriguing hypothesis a thorough analysis of a sample of viewsheds of sites across the Maya lowlands should be undertaken. In order to achieve accurate viewshed models, we will likely have to wait for higher-resolution terrain information than is currently not available from published maps (1:250,000 or 1:50,000 scale). Fortunately, high-resolution (30 m) global coverage elevations data from interferometry survey of a SIR-C/XSAR radar mission flown by the Shuttle Endeavor in February 2000, will be available to the public in late 2002 (URL: *http://www.jpl.nasa.gov/srtm/*).

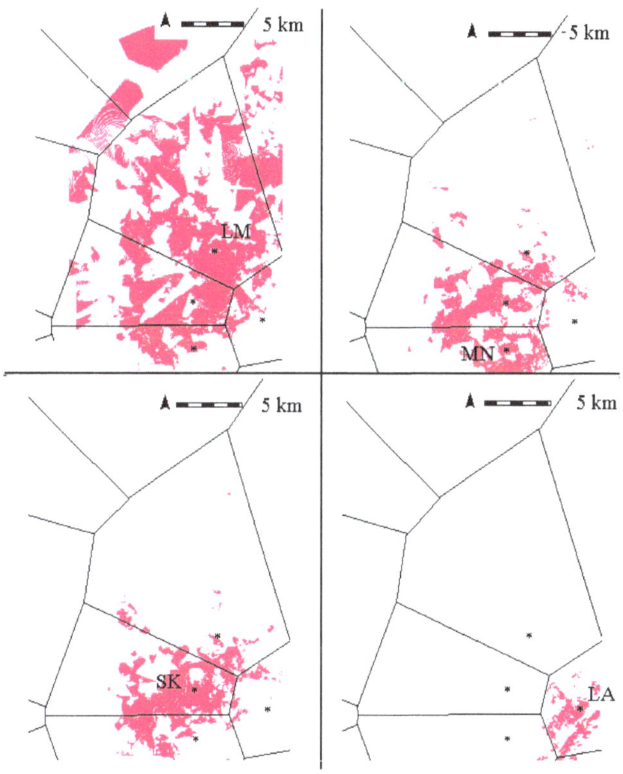

Fig. 15. Size of viewshed maps and hypothetical territories (identified by Thiessen polygons) of La Milpa, Ma'ax Na, Say Ka and Las Abejas compared.

4 HOLMUL

In order to explore further the above described models on the structure of Maya city landscapes we have initiated a new long-term field project at the site of Holmul situated 45 km east of Tikal and 15 km from the Guatemala Belize border, in densely forested NE Petén (Estrada Belli 2001). This site is located the typical position that cross-cuts upland and swamp ecotones as for most of the major Lowland Maya cities. A major feature of its landscape are the limestone escarpment ridges that run SW-NE west and east of the site. The site itself sits on a limestone upland peninsula surrounded by seasonal swamps providing security and rich flora and fauna to the upland community. Classification of environmental and geo-morphological zones around Holmul is a relatively simple matter using Landsat imagery which otherwise would be extremely difficult and time-consuming if relying exclusively on field observations. For the Holmul survey, navigation in this difficult terrain is greatly aided by targeting locations on satellite imagery (i.e. Cival site) and use of GPS hand-held receivers to reach the target destination (Fig. 16).

Although the data collection and analysis are still at a very initial stage, at a micro-regional scale, several patterns have become evident from GIS, remote sensing and foot survey data:

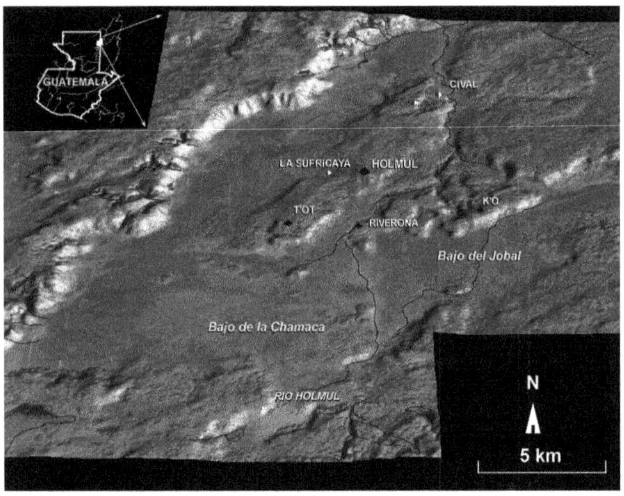

Fig. 16 View of 3d model of topography of Holmul area showing the location of sites mentioned in the text.

1) outside Holmul, three minor centers are located at intercardinal cardinal directions and at a regular distance of ca. 4 km from the center in the southwest, south and southeast quadrants.

2) At two of these minor centers (T'ot and Riverona [Wolf et al. 2002]) plaza layouts closely replicate layout of Holmul main plazas and groups.

At these sites, the plaza with its main northern group platform recalls Holmul's main plaza and Group 1. On the other hand Holmul's Group 1 and adjacent east plaza (but see also Late Classic Complejo Ritual Publico in SE Petén, La Porte and Mejia 2001) recall Naranjo,s and Calakmul's main plaza's layouts (see a parallel situation with Xunantunich/Naranjo/Calakmul; Ashmore 1998).

Also, are Holmul's minor centers part of a grander design on landscape as in the example from La Milpa discussed above? Or is a ring of centers around Holmul with cardinal intercardinal directionality part of administrative as well as conceptual model (e.g. Marcus' (1993) central place theory model around Calakmul)?

3) Cival and La Sufricaya: Cival is a large center with a 37 m high triadic group pyramid and is abandoned in the Late Preclassic (6 km from Holmul). La Sufricaya is a small elite groups with several Early Classic carved monuments with early inscriptions located only 1.2 km from Holmul (Estrada Belli 2001). Holmul is a large center with occupation from late Middle Preclassic and major architecture and monuments dating to Late-Terminal Classic. These centers all exhibit royal architecture and/or monuments which may not overlap chronologically. If this is considered in light of the wider context of the dynastic history of non-overlapping sites such as Naranjo and Xunantunich (Ashmore 1998), could there possibly be a broad-regional and long-term pattern of shifting capitals from the Late Preclassic to the Terminal Classic period in eastern Petén?

CONCLUSIONS

The examples presented in this chapter were intended to emphasize the analytical advantages of GIS and Remote Sensing over more traditional way or recording and analyzing archaeological settlement data in Mesoamerican archaeology. The case studies were designed as an experimental use of GIS and Remote sensing but in an attempt to solve essential archaeological questions. In these examples, I sought to emphasize the versatility of satellite data to answer environmental questions of regional scope in non-forested area as well as in heavily forested regions. Ehere the visibility of ancient structures is low or non-existent, GIS and Remote Sensing have much to offer to the archaeologist concerned with questions about settlement and environment. Moreover, the integration of data from disparate sources in a common coordinate spaces within the GIS environment, permits data exploratory analysis and hypothesis testing from a variety of perspectives, environmental, social, and conceptual not available with traditional analogical methods of analysis. An additional strength of this type of analytical platform is that all scales of analysis are possible without major reworking of the data. On the other hand, while traditionally GIS analysis has been carried out on a regional scale, the creation of high-resolution maps of transects and survey areas allows a fly-through approach to ancient Maya cityscapes at the site-specific scale of analysis. Thus, by virtually removing forest covers, intra and intersite relationships are revealed through powerful graphics for easier identification of patterns. New models created through a series of visual iterations can subsequently be tested with quantitative methods. Furthermore, proximity analysis in GIS introduces a new powerful method for exploring multiple links between central and rural populations in Maya cities. Viewshed and least-cost analysis are still relatively underused lines of investigation in Maya archaeology but will provide useful insights on the conceptual structure of "vistas" and movement paths within Maya cities once a diverse sample of Maya cityscapes is considered.

Finally, while extensive survey coverage of ancient settlement in the Maya Lowlands will probably never be accomplished, several adjacent sites have been or will be investigated using Remote Sensing and GIS archaeological datasets in NE Petén, at Yaxha, Naranjo, El Pilar, and BRASS's data from Xunantunich (Culbert et al. 1996, Fialko 1996, Ford and Fedick 1992, 1999, Ashmore 1998). These research projects will be producing a large continuous area coverage. As Maya settlement and GIS data are produced and made available over the world-wide web we hope that research collaborations across projects will quickly lead to exciting new discoveries of regional patterns on the structure, function and change of Maya landscapes.

Acknowledgements

I would like to thank the organizers of the 2001 USA-Italy Remote Sensing workshop and editors of the present volume for inviting me to participate. Work in Southeastern Guatemala was possible thanks to grants from the National Science Foundation (Dissertation Improvement grant# SBR9521471) and from National Geographic Society (5934-97, 5560-95, 5404-95), the Alice Brennan Humanities Foundation and additional funds from Dr. Norman Hammond. GIS work on the La Milpa data was possible thanks to a grant from National Geographic Society to Dr. Norman Hammond. I thank Dr. Gair Tourtellot and Dr. Norman Hammond for inviting me to create a GIS database for La Milpa and for giving me access to their datasets. GIS and survey work at Holmul was possible thanks to grants from National Geographic Society (6394-98 6979-01), Ahau Foundation (200104,0208, [9904 to Dr. Hammond]) and F.A.M.S.I. (90010). I thank IDAEH of Guatemala for granting permission to work in Southeastern Guatemala and at Holmul, Petén. Special thanks to Dr. Ken Kvamme and Dr. Magaly Koch for their crucial advice on GIS and remote sensing problems. Many thanks to Dr. Laura Kosakowsky for her assistance with the ceramic analyses and to Dr. Payson Sheets and Dr. Frederick Bove for their advice on SE Guatemalan archaeology. Finally, the field work associated with GIS work was made possible my many dedicated students and friends to whom goes my gratitude. A special thank you to the Boston University system administrator Tony Soares for his help in keeping the systems running.

References

ASHMORE, W. A., 1998. Monumentos politicos: sitio, asentiamiento y paisaje en el area de Xunantunich. In Anatomía de una civilización. Ciudad Ruiz, A. et al. (eds.) pp. 161-184. Madrid: Sociedad Española de estudios Mayas.

BURROUGH, PETER A. AND RACHEL A. MCDONNEL , 1998. Principles of Geographic Information Systems. Oxford: Oxford University Press.

CULBERT, T.P. L.J. LEVI, B.M. MCKEE, AND J. L. KUNEN, 1996, Investigaciones Arqueológicas en el Bajo La Justa entre Yaxhá y Nakum. En IX Simposio de Aqueología en Guatemala, J.P. Laporte y H.L. Escobedo (eds.), pp. 51-57. Museo Naciona de Arqueología e Historia, Guatemala.

DULL, R. A., J.R. SOUTHON, AND P. SHEETS, 2001. Volcanism, Ecology and Culture: a Reassessment of the Volcán Ilopango TBJ eruption in the Southern Maya Realm. Latin American Antiquity 12(1):24-44.

ESTRADA BELLI, F., 1997A. GPS And GIS as Aids For Mapping Archaeological Sites. Archaeological Computing Newsletter 47: 5-10.

ESTRADA BELLI, F., 1997b.Between the Volcanoes and the Ocean: Surveying and Excavating on the Pacific Coast of Guatemala. Context. 1996: 22-25.

ESTRADA BELLI, F., 1999a. The Archaeology of Complex Societies in Pacific Coastal Southeastern Guatemala: A Regional GIS Approach. British Archaeological Reports International Series 890. Oxford: BAR Publishing.

ESTRADA BELLI, F.,1999b. La Milpa: A Large Maya city, 400 B.C.- A.D. 830. URL http://www.bu.edu/lamilpa

ESTRADA BELLI, F., 2001 Maya Kingship at Holmul, Guatemala. Antiquity 75(2001): 685-6.

F. ESTRADA BELLI AND L.J. KOSAKOWSKY, 1996 El Proyecto Santa Rosa: Resultados de la Primera Temporada. In IX Simposio de Investigaciones Arqueológicas de Guatemala, J.P. Laporte and H.L. Hescobedo (eds.) pp. 565-582, Museo Nacional de Arqueología y Etnología, Guatemala.

F. ESTRADA BELLI AND L.J. KOSAKOWSKY, 1997. Survey in Jutiapa, Southeastern Pacific Guatemala, 1997. Mexicon 20(3): 55-59 ().

ESTRADA BELLI, F., L.J. KOSAKOWSKY, AND M. WOLF, 1998. El Lugar de Santa Rosa en el Mapa Arqueológico de Guatemala: Desarrollo de Sociedades Complejas en la Costa Sureste de Guatemala. In XI Simposio de Investigaciones Arqueológicas de Guatemala, J.P. Laporte and H.L. Hescobedo (eds.), Museo Nacional de Arqueología y Etnologia, Guatemala. pp. 319-338 (

ESTRADA BELLI, F., L. J. KOSAKOWSKY, M. WOLF AND D. BLANK, 1996. Patrones de asentamiento y uso de la tierra desde el Preclásico al Postclásico en la costa del Pacifico de Guatemala: La arqueología de Santa Rosa, 1995. Mexicon 18(6): 110-115 ,

ESTRADA BELLI, F., L. J. KOSAKOWSKY, B. THOMAS, A. E. LEWIS, J. SCHULTZ, M. WOLF AND K. BERRY, 1998. Preclassic to Postclassic Patterns of Settlement, Land Use and Trade in Pacific Coastal Guatemala: The Archaeology of Santa Rosa

1995-96. XIII UISPP Congress 1996 - Proceedings - Sections 13/18, eds. G. Bermond Montanari, R. Francovich, F. Mori, P. Pensabene, S. Salvatori, M. Tosi, C. Peretto, pp. 479-485, A.B.A.CO. Edizioni, 1998, Forlì, Italy.

ESTRADA BELLI, F. L. J. KOSAKOWSKY, B. THOMAS, A. E. LEWIS, J. SCHULTZ, M. WOLF AND K. BERRY, 1997. La Arqueología de Santa Rosa 1996. In X Simposio de Investigaciones Arqueológicas de Guatemala, J.P. Laporte and H.L. Hescobedo (eds.), pp. 193-210. Museo Nacional de Arqueología y Etnologia, Guatemala (

ESTRADA BELLI, F. AND G. TOURTELLOT III, 2002. (in press) A Terminal Classic Ritualized Landscape At La Milpa, Belize. In The 2000 Chacmool Conference Proceedings, The University of Calgary, Calgary, Canada

FIALKO, V., 1996 Yaxha y Nakum: Jerarquías y patrones de asentamiento en sus espacios intersitios.Mayab 26(11):15-24

FORD, A. AND S. FEDICK, 1992. Prehistoric Maya settlement patterns in the upper Belize River area: initial results of the Belize River Archaeological Settlement Survey JFA (19)1: 35-49.

FORD, A AND J.A. MONTES, 1999. Medio ambiente, uso de la tierra y desarrollo sostenible: la Reserva Arqueológica de El Pilar para la Flora y Fauna Mayas de Belice y Guatemala. Mesoamérica (20)37: 31-50.

KOSAKOWSKY, L.J., AND F. ESTRADA BElli, 1997. La Cerámica de Santa Rosa Una Vista desde la Costa Sur. In X Simposio de Investigaciones Arqueológicas de Guatemala, J.P. Laporte and H.L. Hescobedo (eds.), pp. 651-660. Museo Nacional de Arqueología y Etnologia, Guatemala.

KOSAKOWSKY, L. J., F. ESTRADA BELLI, AND H. NEFF, 1999. Late Preclassic Ceramic Interaction Spheres: The Pacific Coast as Core, Not Periphery. Journal of Field Archaeology. 26(4): 377-390.

KOSAKOWSKY, L.J., F. ESTRADA BELLI, P. PETTET, 2000. Preclassic through Postclassic: the Chronological History of Southeastern Pacific Guatemala. Ancient Mesoamerica II (2000) 199-215

KVAMME, K. L., 1992. Geographic Information Systems and Archaeology. In Computer Applications and Quantitative Methods in Archaeology. G. Lock and J. Moffat (eds.), pp. 77-84. B.A.R. International Series 577, Oxford.

LAPORTE J.P, AND H. E. MEJIA, 2001. Los sitios Arqueológicos de la Cuenca del Río Salsipuedes en el Sureste de Petén, Guatemala. Mexicon 23(3):65-72

MADRY, S. L. H. AND L. RAKOS, 1996. Line-of-Sight and Cost-Surface Techniques for Regional Research in the Arroux River Valley. In New Methods, Old Problems. H. D. G. Maschner (ed.). pp. 104-125. Center for Archaeological Investigations. Occasional Paper No. 23. Southern Illinois University, Carbondale.

MARCUS, J., 1993. Ancient Maya political organization. In Lowland Maya Civilization in the Eighth Century A.D. edited by J. A. Sabloff and J. S. Henderson pp. 111-183. Dumbarton Oaks, Washington, D.C.

MARTIN, S., AND N. GRUBE, 1995. Maya Superstates: How a few powerful kingdoms Vied for Control of the Maya LowlandsDuring the Classic Period. (A.D. 300-900). Archaeology 48(6):41-46

NEFF, H., J.W. COGSWELL, L.J. KOSAKOWSKY, F. ESTRADA BELLI, AND F. J. BOVE, 1999. A New Perspective on the Relationships among Cream Paste Ceramic Traditions of Southeastern Mesoamerica. Latin American Antiquity. 10(3): 281-299

STEPHENS, J. L. , 1843. Incidents of Travel in Yucatan. 2 vols. London: John Murray.

TOURTELLOT III, G. , F. ESTRADA-BELLI, J. ROSE AND N. HAMMOND, 2002. (in press) Late Classic Maya Heterarchy, Hierarchy, and Landscape at La Milpa, Belize In Heterarchy, Political Economy, and the Ancient Maya: The Three Rivers Region of the East-Central Yucatan Peninsula. V. L. Scarborough, F. Valdez, Jr., and N. Dunning (eds.).Tucson: Arizona University Press.

TOURTELLOT III, G., M. WOLF, F. ESTRADA BELLI & NORMAN HAMMOND, 2000 Discovery of two predicted Ancient Maya sites in Belize. Antiquity 74 (2000):481-2.

WOLF, M. K. GARDELLA AND F. ESTRADA-BELLI, 2002. Holmul's minor centers: Preclassic and Classic territorial nodes. Paper presented at the SAA 67 annual meetings, March 2002, Denver Colorado

From Artifact to Landscape: A Theoretical Approach to a Simulated Reconstruction of Historical Processes in Ancient Ethiopia

Rodolfo Fattovich

Istituto Universitario Orientale, Naples (Italy)
P.zza S. Domenico Maggiore, 12 - 80134 Napoli
fattovich@spinweb.it

Abstract. Modelling historical processes on the material evidence of past human activities must be a primary task of archaeology. Historical processes consist of the sequence of changes at different temporal and spatial scales in the complex system of interactions: a) between human population and the physical environment; b) between human population and the biotic environment (animal and plants); c) among different populations; d) among the members of a single population. The reconstruction and analysis of these processes requires a) to generate ecological, socio-economic, cultural, and cognitive models for the specific areas and periods; b) to correlate the models in a diachronic perspective; and c) to infer the factors that caused the changes through time. A computer simulation of the process should be the last step in this analysis. The Horn of Africa is a good laboratory to generate and test hypotheses about the dynamics of historical processes at a regional and global scale, because of the great environmental and ethnic diversity of the region. In this paper, a theoretical approach to a simulated reconstruction of the development of Aksum, the capital city of the ancient kingdom of Aksum in northern Ethiopia, will be suggested using the evidence the Istituto Universitario Orientale of Naples (Italy) and Boston University (USA) are collecting at Bieta Giyorgis, to the northwest of Aksum, as part of the joint IUO-BU Aksum Archaeological Project in progress since 1993.

Computer simulation is not just a new method to add to the social researcher's armoury, but a new way of thinking about society, and especially social processes

Nigel Gilbert[1]

1. Introduction

The aim of this paper is to present some personal reflections about a possible procedure to generate a simulated reconstruction of past historical processes on the basis of the archaeological evidence.[2]

The background to the paper is an ongoing research program, which involves varied lines of investigation: a) fieldwork and laboratory data processing; b) designing hypotheses about historical dynamics to be tested in the field; c) theorizing the role of archaeology in reconstructing past human ecosystems and their change through time.

In this paper the virtual reconstruction of ancient landscape is regarded as a stage in the full spectrum of analysis and interpretation of the archaeological record, from a single artifact/ecofact to landscape.

The basic assumptions are:[3]

a) Modeling real historical processes is a primary task of archaeologists, historians and anthropologists.

b) Archaeology can provide a solid evidence for reconstructing, modeling and explaining long-term social processes, and can generate interpretations of these processes that can be tested on the ground.[4]

Simulation of real historical processes can provide a better insight into the dynamics of economic, social, and cultural transformation, and thus a deeper understanding of present historical dynamics.

Ancient Ethiopia, and in particular the evidence from Bieta Giyorgis near Aksum (northern Ethiopia), have been chosen as a study case for the following reasons:

a) Ethiopia is a wonderful laboratory to test and generate hypotheses about the dynamics of historical processes at a regional and global scale. The region is characterized by an intense internal dynamism due to a great environmental and ethnic diversity, and her location at the interface of three major cultural regions: the Mediterranean world; the Indian Ocean world; and the African world.

b) Boston University (BU), Boston (USA), and the Istituto Universitario Orientale (IUO), Naples (Italy), are conducting since 1993 joint archaeological investigations at Bieta Giyorgis (Aksum) under the direction of K.A.

[1] Gilbert 1995.

[2] I want to thank Dr Andrea Manzo, IUO, Naples, for his precious comments and suggestions during the preparation of this paper. In any case, the responsibility of all statements is only mine.

[3] See Fattovich 1997.

[4] See Renfrew & Bahn 1991.

Bard and myself.[5] These investigations are providing fresh data to test different hypotheses and theoretical approaches on a direct evidence.

2. The archaeology of historical processes

2.1 Historical Process: a definition

Historical processes can be defined as the sequence of changes at a different spatial scale in *human ecosystems*, i.e. the system of interactions: a) between human populations and the physical environment; b) between human population and the biotic environment (animal and plants); c) among different populations; d) among the members of a single population.[6] These processes characterize the history of single populations through time and affect their biology and demography. They involve changes in the cognitive system, which can modify the collective behavioral patterns up to generate a new cultural identity of the population.

Historical processes can be described as a sequence of particular and unrepeatable events, as well as a progressive re-adjustment of single social, economic and cognitive systems to internal (e.g., cultural innovations; social strain) and/or external pressures (e.g., environmental fluctuations; economic/political changes in other regions; cultural influences from other regions). They depend on: 1) cultural processes, which reflect structural changes in the social, economic and cognitive system at a collective level,[7] 2) decision making processes, which reflect the perception the single members of a human group have of themselves, their society, and external world (i.e., natural environment and other peoples), and generate their reaction to stimuli in specific circumstances.[8] The investigation of a historical process thus requires both an ethic and an emic approach in order to correctly understand their dynamics.[9]

Historical processes can be examined at different spatial scales: 1) a global scale, for understanding the rise and decline of global systems of social-economic interactions through the time;[10] 2) a regional scale, for reconstructing the development of one people or the dynamics of peopling in a specific region through the time;[11] and 3) at a local scale, for outlining the changes in just one community.[12]

Finally, historical processes are characterized by 1) a long-term stability in some cultural patterns (*long durée*), and 2) punctuated changes, which can quickly modify the social, economic and cognitive system of the population.[13] Therefore, the reconstruction of historical processes requires a long-term perspective, over centuries or millennia, in order to identify the factors, which caused punctuated changes within long-term trends.[14]

2.2 The archaeological reconstruction of an historical process

Both archaeological and textual evidence can be used to reconstruct historical processes, as they provide two different perceptions of the processes.[15]

Archaeological evidence consists of the material remains of past cultures (artifacts; ecofacts; human remains) in their paleoenvironmental context. This evidence, though fragmentary,[16] provides a reconstruction of the structure of the ancient human ecosystems and their changes through time in an ethic perspective[17]

Textual evidence consists of literary texts, chronicles, administrative records, and ceremonial or votive inscriptions. These sources provide a narrative account of past events, and relevant information about social organization, economy and beliefs of ancient populations and their changes through time in an emic perspective[18] Written sources are often ideologically biased. They reflect the way the external world and events were perceived and interpreted, and information was selected and transmitted by contemporary people in order to legitimate social order and create a collective memory for consolidating cultural and political identity[19]

Archaeology in particular can contribute to a reconstruction and explanation of past historical processes, as this discipline focuses on the origins, and social, economic and cognitive development of the single populations, and their interaction within larger political and economic systems.[20]

Archaeology, as an anthropological discipline, is aimed at:

a) outlining the cultural evolution of mankind from the simplest forms of culture of earliest hominids to the

[5] See Bard *et al.* 2001.

[6] Fattovich 1990; see also Crumley 1994; Winterhalder 1994. These processes basically correspond to the social evolutionary trajectories suggested by Yoffee as a foundation to a "new social evolutionary theory" (Yoffee 1993).

[7] See e.g., Gelner 1988.

[8] See e.g., Goldschmidt 1990.

[9] Harris 1968; see also Crumley 1994

[10] See e.g., McNeill 1976; Abu-Lughod 1989; Frank & Gills 1996.

[11] See e.g., Trigger, Kemp, O'Connor, Lloyd 1983; Kemp 1989; Bard 1997.

[12] See e.g., Fattovich 1997.

[13] See e.g., Braudel 1958; van der Leew & McGlade 1997; Kristiansen & Rowland 1998.

[14] See Renfrew & Cooke 1979; van der Leew & McGlade 1997.

[15] See also Lightfoot 1995.

[16] See Schiffer 1987.

[17] See e.g., Renfrew & Bahn 1991

[18] See e.g., Topolski 1975

[19] Assmann 1992

[20] See Knapp 1987;; Sherratt 1993; Hayden 1993.

modern world, making evident the major transformations that characterized this evolution.

b) explaining how cultural diversity among different populations emerged.

Archaeology, as an historical discipline, is aimed at reconstructing:

a) the cultural history of the populations that occupied specific regions though the time.

b) the transformations that occurred in the culture of the single populations in the single regions because of internal causes and external (environmental; cultural) pressures.

c) the pattern of interaction among the populations that occupied a region along the time, and between them and the peoples living in other regions.

The reliability of archaeology to reconstruct past historical processes was tested on the evidence from Bieta Giyorgis, near Aksum, by directly correlating archaeological and textual evidence in order to asses how much they complement each other. The test was conducted 1) treating dated information from written sources as archaeological evidence; 2) interpolating the textual evidence in a well established archaeological sequence; and 3) checking whether they correspond to each other or not.[21]

The results of this exercise (see Fig..1) suggested :

a) In the Ethiopian case, archaeological and textual evidence are consistent.
This means archaeology can provide reliable "historical" information about the past, and thus a careful reconstruction of "prehistoric" social, economic and cultural transformations could be accepted as basically valid.

b) The reliability of archaeological information largely depends on the accuracy of excavation procedures and recording.

Only a rigorous control of the stratigraphy and chronology prevents the collected evidence from being open to subjective interpretations. Excavation procedures must be explicit in field reports, as they are the only means we have to properly evaluate the validity of a research and the interpretation of the results.

c) Analysis and synthesis must be constantly combined at any stage in the research - from fieldwork to laboratory study - in order to provide a firm reconstruction of past processes.

The analysis of diachronic (temporal) and synchronic (spatial) changes in the patterns of artifact assemblage (from stratigraphic units to sites) - in terms of site formation processes, typological sequences of artifacts, cultural and environmental contexts, settlement patterns, etc. - is crucial to generate a solid framework for the reconstruction of the process under investigation. A constant synthesis between the results of the analysis and other kinds of evidence (historical, linguistic, ethnographical, environmental, etc.) at a regional and interregional scale is necessary to outline more carefully the process. The reconstruction of the process can be then analyzed and interpreted in order to identify the particular factors, which affected its dynamics through time, and to suggest an explanation of these dynamics.

d) Research must be conducted with a multidisciplinary approach, ranging from geology and geo-archaeology to archaeology, bio-archaeology, ethnoarchaeology and ethnohistory. Investigations must be conducted in close collaboration with philologists and linguists, historians, social anthropologists and possibly geographers with a competence about the region under investigations.

In this perspective the different theoretical approaches processual and post-processual archaeology have emphasized can be integrated into a coherent and comprehensive system. These approaches provide us with analytical and interpretative tools to organize and "read" the data at different scales of complexity: a) Individual. b) Gender. c) Family. d) Society (from not hierarchical societies to hierarchical societies and states). e) People (from local community to ethnic nationality). f) World system.

Two approaches are particularly relevant to the archaeological reconstruction of historical processes: environmental archaeology and cognitive archaeology.

Environmental archaeology[22] deals with changes in Man-Environment interaction through time, and investigates:

a) the changes in the natural environment and their consequences, including draughts and famines, and catastrophic events, such as volcanic eruptions and earthquakes.[23]

b) the progressive exploitation and manipulation of the environment by the single populations to sustain more and more complex social and economic organizations, resulting in a man-made (cultural) environment.

c) The progressive adaptation of the single populations to the natural and cultural environment in order to guarantee their social and biological survival.

[21] Fattovich 1997.

[22] See Butzer 1982, 1990; Jochim 1990; Hastorf 1990; McIntosh, Tainter & Keech McIntosh 2000.

[23] See e.g., Dramis & Fattovich 1994.

ABSOLUTE DATING	CULTURAL PHASES	MONUMENTS	HISTORICAL SOURCES	CROSSES on COINS	CROSSES on POTTERY
850					*
825		OAZ V, ON XIII, TE I:			*
800	Post-	Compounds			*
775	Aksumite	BGS: Church, Cemetery			*
750		ON IX: Fire Place			*
725					*
700				*	*
675		ON XIII, IV-VI:		*	*
650*		Elite Residence		*	*
625**	Late	ON V: Domestic Area		*	*
600*	Aksumite	BGI: Church		*	*
575		TE I: Compound		*	*
550***			c.AD520-Byzantine ambassadors,	*	*
525***		BGS: Church	Cosmas, Procopius: war in S.	*	*
500*			Arabia	*	*
475****	Middle	ON VI, VII, VIII, X: Palace	c. AD 500 - Nine Saints	*	*
450****	Aksumite	ON V: Domestic Area		*	
425		TE I, GA I-II: Compounds	AD 368-431 - Palladius:	*	
400****			kinglet in Aksum	*	
375****			AD 356 - Athanasius: possibly	*	
350*			Christian kings	*	
325***			AD 330 - Frumentius	*	
300***		ON V, VI, VII, VIII: Palace	introduces Christianity		
275**	Classic	ON II: Lithic Workshop	c. AD 276 - Mani: Aksum,		
250***	Aksumite	ON XVI: Domestic Area	"world" power		
225*		OAZ II: Double Stele	c. AD 200-230 - Aksumite		
200*		TE: Cemetery	Intervention in S. Arabia		
175					
150**			c. AD 150 - Ptolemy: Aksum,		
125******			the King's Palace		
100**					
75****					
50		ON IV, V, VI, VII, VIII:	c. AD 40-70 - PME: metropolis		
AD 25		Palace	called the Aksumite		
0	Early	ON IX: Palace			
25 BC	Aksumite	ON XVI: Domestic Area			
50		ON II: Lithic Workshop			
75*		OAZ: Tomb 1, Tomb 2			
100*		OAZ IV: Monolith			
125					
150					
175*					
200***		OAZ I/VI, OAZ VII-X: Elite			
225		Cemetery			
250		ON IX, GN: Offering			
275	Proto-	Deposits			
300	Aksumite	ON II: Lithic Workshop			
325*		ON III, ON V: Evidence of			
350*		Walls			
375		ON IX: Palace or Temple			
400*					
425					
450	Pre-	OAZ I: Evidence of Walls			
475	Aksumite	OAZ III: Monolith			
500					

Fig. 1 Aksum cultural phases, monuments, historical sources, sequence of coins, and pottery with crosses.

Cognitive archaeology deals with the reconstruction of the mental templates and processes, which form the cognitive system of the ancient populations.[24] The cognitive system is the normative system, which underlies the behavior of the individuals in the daily life, and largely affects the decision-making processes. This system includes:

a) Science, i.e. the description and explanation of the world based on the direct observation of natural phenomena.

b) Religion, i.e. the belief in deities transcending and ruling the material world.

c) Magic, i.e. the symbolic manipulation of supernatural and superhuman forces to accomplish practical tasks.

[24] See e.g,, Gardin & Peebles 1992; Renfrew & Zubrow 1994.

d) Ideology, i.e. the interpretative framework sustaining the political relationships within the single communities or among different communities.

3. Simulation of historical processes

3.1 Predictive simulation of social and economic processes vs. simulated reconstruction of historical processes

Simulation is frequently used in social science to generate predictive models of social and economic processes, and to investigate the behavior of individuals as social actors. Different approaches have been followed to accomplish specific tasks, from system dynamics to artificial societies.[25]

The simulated reconstruction of historical processes differs from the predictive simulation of social and economic processes because we usually have some information about the early and final stages of the process, and a basic knowledge of the changes, which characterized the process through the time. Simulation can be therefore used as a method to test different hypotheses about the factors which affected the dynamics of the process.[26] Moreover, the results can be checked in the field.

In my opinion, a simulated reconstruction of a historical process requires:

1) To identify the general system of interactions between environmental and cultural factors, which characterize the historical dynamics and regulate the trajectories of the human ecosystems at a regional scale.

2) To outline the structure of the ancient human ecosystems and their changes through time.

3) To infer from (1) general rules and from (2) specific rules which may affect the choices of the social actors at the different hierarchical levels.

4) To simulate the decision making processes of the social actors using a) the rules which have been inferred in (3) and b) possible culturally induced forms of behavior of the members of the single human groups.[27]

5) To test in the field the process emerging from simulation.

[25] For an overview of the state of art in social simulation, see e.g., Gilbert & Troitzsch 1999.
[26] See Gilbert & Troitzsch 1999: 21-24.
[27] See e.g., Benedict 1934.

3.2 An archaeological approach to the simulated reconstruction of historical processes

Simulation is not a new method of investigation in archaeology, the first efforts dating to the 1970s, but with unsatisfactory results.[28] In the 1990s simulation became again a fashionable method in archaeological research with the application of agent-based programs such as Sugarscape, Swarm, etc., and more empirically grounded simulations.[29] At present, however, no simulation of a historical process, connecting in a simulation program the changes in the spatial organization of the evidence (from assemblage to region) with the changes in the cultural system (i.e., social, economic and cognitive system) has been attempted, as far as I know.[30]

In my opinion, a simulated reconstruction of a historical process *might* be generated through a chain of operations, from data collecting to agent-based modeling, in order to formalize and correlate the data, to infer from them rules, and to apply these rules in agent-based models for simulating decision making processes.

The following, tentative procedure is suggested (Fig. 2):[31]

1. Data collection (survey and excavation).

This is the basic stage of the procedure, and requires a rigorous control of the stratigraphic sequence. At present, the Stratigraphic Unit method is most suitable to investigate the vertical and horizontal relationships among features and soil strata in the archaeological deposit.[32]

2. Data Processing[33]

This stage consists of data description and recording, dating, and analysis.

a. Description and Recording
The first operation is the formalization of the collected data in a data base. Data Base programs like Access, DB III, DB IV, Excel can be used for data storage and management. Each entry is described on the basis of a specific attribute list, and distinct Data Bases must be created for the single entries. Data base recording different classes of materials sharing some attributes can be correlated.

b. Dating
Programs such as Calib and Oxcal can be used to calibrate radiocarbon dates in order to fix the absolute

[28] See e.g., Thomas 1972; Zubrow 1973; Hodder 1978.
[29] See e.g., Doran 1996; Kohler & Gumerman 2000.
[30] For a preliminary attempt to simulate the behaviour of hunters-gatheres linking GIS and agent-based models see Lake 2000.
[31] Dr. Andrea Manzo contributed to the description of the single stages of the procedure.
[32] See Harris 1989.
[33] Clark, 1968; Renfrew & Bahn 1991.

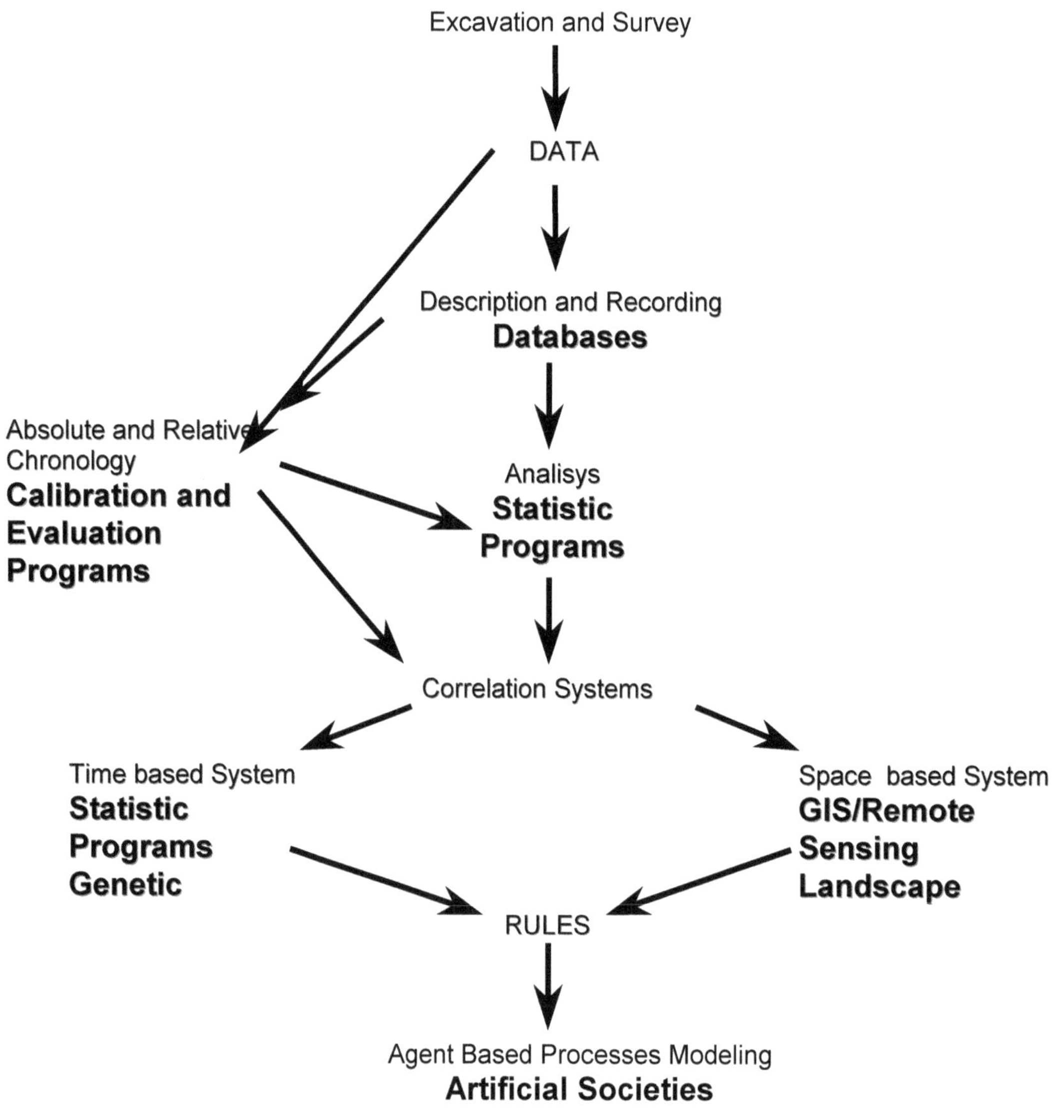

Fig. 2. Chain of operations from data collection to agent based modeling

chronology of evidence. Calibrated dates can be checked on the stratigraphic sequence, cultural phasing, and other possible dating evidence (e.g. coins, imported materials) as well as other dating procedures (e.g., dendrochronology, thermoluminescence, etc.). This control can be performed with the most recent versions of some calibration programs.

c. Data Analysis

Statistic programs such as SPSS or Systat can be used to process the Data Bases,[34] in order to obtain: a) frequencies; b) assessments of distribution (also performed by the Data Base programs); c) test of reliability of the data set (e.g. chi-square tests), which can show the occurrence of relations among variables (e.g. Pearson chi-square), or detect the occurrence of clusters in the data (e.g. Hierarchical Cluster Analysis, K-Means Cluster Analysis). Statistic programs can analyze changes of variables or sets of variables along the time scale established with programs of calibration and date

[34] See e.g., Shennan 1988; Doran 1988; Drennan 1996.

evaluation, or in the space, analyzing data from contexts at different spatial scale (from intra-site to world system). Remote sensing provides a base to generate thematic maps for territorial analysis.

3. Data correlation
Programs, such as GIS, Genetic Programming and applications of SPSS (e.g., PCA), can establish correlations among the different categories of data at a different scale in a synchronic and diachronic perspective. These programs can make evident: a) spatial relations among variables (e.g., settlement types and environmental context, settlements and traditional road network); b) temporal changes in these relations (e.g. correlation between the diffusion of a ceramic type and an environmental situation; conditions favoring the success of certain types of artifacts).

4. Modeling

The results of data processing with correlation systems can generate spatial and temporal models. GIS interfaced with remote sensing can provide territorial models for different phases and periods.[35] Genetic Programming and statistic programs can generate models of relations among factors which caused environmental and cultural changes.[36] In particular, virtual reality and digital reconstruction of ancient landscape can provide a more articulated view of the general organization of the territory in the past. Simulation programs can be already applied to these models to explore possible responses of the population to environmental and cultural stimuli.[37] The interpretation of these models can suggest predictive rules the behavior of social actors (e.g., human responses to environmental and cultural change; economic exploitation of the territory; production, use, and circulation of goods, energy and information).

5. Agent-Based Simulation

Artificial Societies programs, such as e.g., Swarm, Sugarscape, and Ascape,[38] can be used to simulate human social phenomena, including the interaction with the environment. The rules obtained from the interpretation of the models in (4) could be used as rules of behavior for agents in an environment which has been modeled on the real (or reconstructed) one. The initial conditions of agents and scenario in the Artificial Societies could be modeled on a known social and environmental situation. The possibly changing conditions of the environment and human behavior could generate a better understanding of the evolution to new conditions. Finally, the interpretation of this process could suggest how human ecosystems change through time and generate new hypotheses to be tested in the field.

4. Implementation

The implementation of the suggested procedure is in progress in the Laboratory for Archaeology of the Department of African and Arabian Studies, IUO, Naples, as part of the IUO/BU Archaeological Project at Bieta Giyorgis, Aksum.

At present, the investigation is conducted at two levels:

a) Analysis of the traditional Ethio-Semitic state and society in order to infer general rules for a simulated reconstruction of the development of the kingdom of Aksum (northern Ethiopia and Eritrea).

Processing and modeling the data from Bieta Giyorgis in order to infer specific rules, which may have affected the dynamics of development of Aksum as a capital-city.

4.1 Modeling traditional Ethio-Semitic state and society

Ethiopia is characterized by a strong contrast between temperate highlands up to 3000-4000 m in elevation and arid or semiarid lowlands. The environment exhibits a great variety of climatic and ecological conditions, that may sharply change in contiguous regions.[39] The country is also exposed to environmental hazard, such as climatic fluctuations, causing droughts and famines, geo-dynamic activity, causing frequent volcanic eruptions and earthquakes, invasions of desert locust swarms, and epidemics.[40]

People include Semitic, Cushitic, Omotic, and Nilotic-speaking groups, each one divided into many linguistic sub-groups. These populations developed different adaptive strategies, ranging from nomadic or semi-nomadic pastoralism in the lowlands to sedentary vegeculture, hoe farming, and ox-plow agriculture on the highlands, as well as different social and economic organizations, from not hierarchical societies in the western lowlands to highly centralized states on the highlands.[41] Moreover, populations with different adaptive strategies and social and economic organizations are often very close to each other, with a complex system of interaction at a micro- and meso-regional scale.

Regional historical dynamics were characterized by:

[35] Aldenderfer & Maschner 1996.
[36] See Manzo & Perlingieri 2001.
[37] See e.g., Van West & Kohler 1998; Behrens 1996; Dean *et al.* 2000.
[38] See Gilbert & Troitzsch 1999; Epstein and Axtel 1996; Parker 2001.

[39] See e.g., Dainelli 1935.
[40] Ethiopian Mapping Agency 1988: pl. 19-21; Gouin 1979; Zein & Kloos 1988.
[41] Almagià 1935; Conti Rossini 1937; Zanetti 1996.

a) A progressive interaction between the single populations at a different regional scale. Warfare was often a relevant component of this interaction.

b) Their inclusion into larger networks of trade and cultural contacts with the Mediterranean regions, Middle East, and African hinterland.

c) The constant re-adjustment of the single social and economic organizations to environmental hazard and political or economic pressure from the outside, in order to maintain their basic cultural identity.

A crucial factor in these dynamics was the rise and expansion of the Ethio-Semitic Christian State from the northern highlands to the south since the 1st millennium AD.[42] The Ethio-Semitic Christian State originated from the "Kingdom of Aksum", which consolidated on the northern highlands in the early 1st millennium AD.[43]

The state was organized as a "feudal" polity including small-scale polities with chiefs under a paramount king. After the introduction of Christianity in the 4th century AD, monasteries and churches became important religious and political centers. The expansion of the state was mainly due to the need to get control over resources and trade routes. New territories were incorporated into the state and dominated as part of an empire. [44]

The traditional state was characterized by a social hierarchy with the king, chiefs, lower rank nobles, and priests at the top, and farmers at the base. Artisans always formed a marginal (and spited) component of the society. Donations of land were used by the king to award nobles and churches, and to get political support.[45]

The king and the elite were directly involved with contacts with the external world, and controlled long distance trade. They were exposed to external cultural stimuli, and time by time adopted foreign symbols of power and ideology, as well as technological innovations to exhibit their status. Farmers were more conservative, and maintained their basic traditions. They were more exposed to environmental hazard and time by time adopted different buffer mechanisms to cope with this hazard.[46] These two components of the traditional society also operated in an autonomous way at the decision making level. Political choices depended on the elite, while the farmers were completely free in organizing subsistence activities and finding solutions to cope with the environmental hazard.[47] Artisans had a relevant role in implementing and spreading technical and stylistic innovations. Cultural external influences were always adapted to the local patterns and transformed into "Ethiopian" cultural elements.

[42] See e.g., Marcus 1994.
[43] See e.g. , Munro-Hay 1991; Phillipson 1998.
[44] Tamrat 1972; Munro-Hay 1991; Crummey 1999.
[45] See Crummey 1999.
[46] See also Halstead & O'Shea 1989.
[47] See Crummey 1999.

Economy, law, and religion linked the different components of the society in one system. Economy was based on the household and its productive resources. Products were consumed inside the household, exchanged in periodical markets, and given as a tribute to the elite. In turn, the productivity and sustainability of the household depended on the land-tenure system.[48] Law consisted of two distinct, interacting systems: a) royal law, providing the formal framework to the organization of the society, and b) customary right, providing the people with specific rules of social behavior in their daily life.[49] Religion (Christianity) provided the ideology linking elite and farmers in one cognitive system, and was the background to the traditional law.[50]

The preliminary analysis of the traditional Ethio-Semitic Christian state and society suggests the following remarks, which could be used as rules for a simulation:

a) A structural instability of the state due to political competition inside the elite, resulting in a tendency to fragmentation when central power was weak, or commercial, political and territorial expansion when central power was strong.
b) Warfare was a relevant component of the social dynamics of the state, as a result of internal competition and external expansion, and characterized the ideology and self-perception of the elite.[51]
c) External contacts and responses to changes in the external geo-political system were under the control of the elite, and innovations spread from the top to the base of the society.
d) Subsistence economy was conservative, and responses to environmental pressures were under the direct control of the farmers.
e) Market, land-tenure, and tribute were crucial institutions in the political and economic organization of the state.
f) Decision-making processes occurred at two distinct levels: elite and farmers. At the elite level, decision-making processes were affected by internal competition and external contacts. At the farmer level, decision-making processes were mainly affected by the perception of the environment and economic considerations.[52]
g) Religion and traditional law were the base of the cognitive system.

Finally, two patterns of personality, characterizing the elite, are recurrent in Ethiopian history. The first one, typical of the chiefs, is well outlined in a source of the mid-1st century AD describing an early Aksumite king as "a stickler about his possessions and always holding out for getting more, but in other respects a fine person and well versed in reading and writing Greek."[53] The second

[48] See Simoons 1961; Bauer 1985; Diriba 1995; Crummey 1999.
[49] See Tzadua & Strauss 1968; Conti Rossini 1916.
[50] See Tamrat 1972.
[51] See also Conti Rossini 1935.
[52] See Wolde-Mariam 1991; McCann 1995.
[53] Casson 1989: 52-53.

one is represented by the holy man, modeled on the early Christian monks.[54]

4.2 A study case: Bieta Giyorgis

The application of the procedure I have suggested in section 3.2, from collecting and recording artifacts/ecofacts to landscape modeling, is described in detail in another paper delivered at this workshop.[55]
The preliminary interpretation of the processed evidence points to the following possible rules for a simulation:[56]

a) The settlement pattern seems to be constant from the 7th century BC to the 13th century AD, and is directly linked to the geomorphology of the area. In some phases the sites are concentrated in some sectors of the hill. In Post-Aksumite times, particularly, the sites are located mainly in the SW and SE sectors.

b) The rural landscape was enough stable, although the extension of cultivated areas may have changed with fluctuations in the demographic density.

c) Elite residential palaces were a dominant component of the landscape along the whole Aksumite history.

d) Major transformations in the landscape were due to ideological changes. Beginning in Proto-Aksumite times open air cult places were probably substituted with cult buildings, and in Christian times churches were erected on earlier temples and shrines. The continuity in the location of the cult building points to a basic stability in the symbolic landscape of the region.

e) Changes in subsistence economy are not directly related to other cultural changes, save for a correlation among the increasing production of teff, increasing number of cattle and diffusion of gray ware.

f) Environmental variables are almost unrelated to economic and cultural changes

g) External stimuli were the main factor of change in Aksumite culture.

The analysis of the artifacts, moreover, indicate that:

a) Warfare was a relevant ideological value of the elite, as military symbols occur in elite tombs.

b) Christianity was introduced initially at the court and slowly spread to rural base in about 150 years.

c) External elements were quickly transformed into distinctive local elements.

These remarks suggest that the model of the traditional state and society, we are inferring from the analysis of later textual evidence, can also fit the Aksumite state, and might be used for a simulated reconstruction of the development of the ancient kingdom. In particular, they point to:

a) Changes involved mainly the elite who was more exposed to contacts with the external world, and moved from the top to the base of the social hierarchy.

b) The rural base of the society was stable along the whole history of ancient Aksum, the only significant change being an increasing in the cultivation of teff in the 7th century AD.

c) Changes in religion and related ideology were the major factors of transformation in the Aksumite society.

Conclusions

The preliminary analyses we are conducting on the evidence from Bieta Giyorgis seem to support the initial assumption that archaeology can provide reliable data to reconstruct a historical process and to generate rules for a possible simulation of the process. In any case, the conclusion of this paper is necessarily open, as the final test (i.e., simulation) has not yet been performed. We are planning to do it as soon as possible, when more rules will be defined on the evidence from Aksum.

References

ABU-LUGHOD J. L., 1989. *Before European Hegemony. The World system AD 1250-1350*. New York. Oxford University Press.

ALDENDERFER M. and H. D. G. MASCHNER, eds. , 1996. *Anthropology, Space, and Geographic Information Systems*. New York: Oxford University Press.

ALMAGIÀ R. , 1935. "Geografia antropica ed economica." *L'Africa Orientale*: 195-283. Roma: Reale Società Geografica Italiana.

ASSMANN J., 1992. *Die kulturelle Gedachtnis. Schrift, Erinnerung und politische Identitat in fruhen Hochkulturen*. Munchen: C. H. Beck'sche Verlagbuchhandlung.

BARD K. A., ed., 1997. *The Environmental History and Human Ecology of Northern Ethiopia in the Late Holocene*. Napoli: Istituto Universitario Orientale.

BARD K. A., CRESCENZI L., DIBLASI M. C., D'ANDREA A. C., FATTOVICH R., FORTE M., HARRIS M. S., JOHNSON G. H., KOCH M., MANZO A., PERLINGIERI C., SCHMID T., TILIA S., 2001. "The Joint Archaeological Project at Bieta Giyorgis (Aksum, Ethiopia) of the Istituto Universitario Orientale, Naples (Italy), and Boston University, Boston (USA): Results, Research Procedures and Preliminary Computer Applications," paper delivered at the Italy-United States Workshop *"The Reconstruction of Archaeological landscape*

[54] See Kaplan 1984.
[55] See Bard et al. 2001.
[56] See also Manzo & Perlingieri 2001.

through Digital Technologies," Boston, MA, November, 1-3, 2001.

BAUER D. F., 1985. *Household and society in Ethiopia.* East Lansing, MI: Michigan State University Press.

BEHRENS C. A., 1996. "A Formal Justification for the Application of GIS to the Cultural Ecological Analysis of Land-Use Intensification and Deforestation in the Amazon." Aldenderfer M. and H. D. G. Maschner, eds., *Anthropology, Space, and Geographic Information Systems*: 55-77. New York: Oxford University Press.

BENEDICT R., 1934. *Patterns of Culture.* Boston: Houghton Mifflin Company.

BRAUDEL F., 1958. "Histoire et sciences sociales. La long durée," *Annales E.S.C.*, 13 (4): 725-753.

BUTZER K. W., 1982. *Archaeology as Human Ecology.* Chicago: Chicago University Press.

BUTZER K. W., 1990. "A Human Ecosystem Framework for Archaeology." E. F. Moran, ed., *The Ecosystem Approach in Anthropology*: 91-130. Ann Arbor: The University of Michigan Press.

CASSON L., 1989. *The Periplus Maris Erythraei.* Princeton: Princeton University Press.

CLARK D. L, 1968. *Analytical Archaeology.* London: Methuen & Co.

CONTI ROSSINI C., 1916. *Principi di diritto consuetudinario dell'Eritrea.* Roma: Tipografia dell'Unione Editrice.

CONTI ROSSINI C., 1935. "L'Etiopia è incapace di progresso civile." *Nuova Antologia*, 17/9/1935: 171-177.

CONTI ROSSINI C., 1937. *Etiopia e genti d'Etiopia.* Firenze: Bemporad.

CRUMLEY C. L., 1994. "Historical Ecology: A Multidimensional Ecological orientation." C. L. Crumley, ed, *Historical Ecology*: 1-16. Santa Fe: School of American Research press.

CRUMMEY D., 1999. *Land and Society in the Christian Kingdom of Ethiopia from the Thirteenth to the Twentieth Century.* Urbana: University of Illinois Press.

DAINELLI G., 1935. "Le condizioni fisiche dell'Africa Orientale." *L'Africa Orientale*: 69-191. Roma: Reale Società Geografica Italiana.

DEAN J. S., G. J. GUMERMAN, J. M. EPSTEIN, R. L. AXTEL, A. C. SWEDLUND, M. T. PARKER & S. MCCAROLL, 2000. "Understanding Anasazi culture change through agent-based modeling." Kohler T. A. and G. J. Gumerman, eds., *Dynamics in Human and primate Socieities: Agent-Based Modeling of Social and Spatial processes*: 179-205. New York: Oxford University Press.

DIRIBA G., 1995. *Economy at the crossroads. Famine and food security in rural Ethiopia.* Addis Ababa: CARE International.

DORAN J.,1988. "Expert systems and archaeology: what lies ahead?" L. N. Ruggles and S. P. Q. Rahtz, eds., *Computer and Quantitative Methods in Archaeology 1987*: 237-241. *Oxford: B.A.R. International Series.*

DORAN J., 1996. "Artificial Societies and Cognitive Archaeology." *Archeologia e Calcolatori*, 7: 1231-1245.

DRAMIS, F, and R. FATTOVICH, 1994. "From Past to Present: Research Perspectives in Environmental Archaeology on the Ethiopian Plateau." B. Zewde, R. Pankhurst and T. Beyene, eds., *Proceedings of the Eleventh International Conference of Ethiopian Studies*, I: 9-14. Addis Ababa: Institute of Ethiopian Studies.

DRENNAN R. D., 1996. *Statistics for Archaeologist.* New York: Plenum.

EPSTEIN J. M. and R. AXTEL, 1996. *Growing Artificial Societies.* Cambridge, MA: The MIT Press.

ETHIOPIAN MAPPING AGENCY, 1988. *National Atlas of Ethiopia.* Addis Ababa: P.D.R.E.

FATTOVICH R. , 1990. "Processi storici e microevoluzione umana: riflessioni sulla possibile integrazione tra discipline storiche ed antropologiche," *Rivista di Antropologia*, 68: 5-35.

FATTOVICH R., 1997. "Archaeology and historical dynanics: The case of Bieta Giyorgis (Aksum), Ethiopia". *Annali Istituto Universitario Orientale di Napoli*, 57: 48-79.

FRANK A. G. and B. K. GILLS, eds., 1996. *The World System.* London: Routledge.

GARDIN J.-C. and C. S. PEEBLES, eds., 1992. *Representations in Archaeology.* Bloomington: Indiana University Press.

GILBERT N., 1995. *Simulation: an emergent perspective.* CRESS, Department of Sociology, University of Surrey (UK), *online publication.*

GILBERT N. and K. G. TROITZSCH, 1999. *Simulation for the Social Scientist.* Buckingam: Oxford University Press.

GELLNER E., 1988. *Plough, Sword and Book. The Structure of Human History*, Chicago: The University of Chicago Press.

GOLDSCHMIDT W., 1990. *The Human Career. The Self in the Symbolic World.* Cambridge, MA: Blackwell.

GOUIN P., 1979. *Earthquake History of Ethiopia and the Horn of Africa.* Ottawa: International Development Research Centre.

HARRIS M., 1968. *The Rise of Anthropological Theory. A History of Theories of Cultures.* New York: Thomas Y. Crowell Company.

HARRIS E., 1989. *Principles of archaeological stratigraphy.* London: Academic Press.

HASTORF, C. A., 1990. "The Ecosystem Model and Long-Term Prehistoric Change: An Example from the Andes." E. F. Moran, ed., *The Ecosystem Approach in Anthropology*: 131-158. Ann Arbor: The University of Michigan Press.

HAYDEN B., 1993. *Archaeology. The Science of Once and Future Things.* New York: W. H. Freeman and Company.

HALSTEAD P. and J. O'SHEA, 1989. "Introduction: cultural responses to risk and uncertainty." P. Halstead and J. O'Shea, eds., *Bad year Economics: Cultural Responses to risk and uncertainty*: 1-8. Cambridge: Cambridge University Press.

HODDER I., ed., 1978. *Simulation Studies in Archaeology*. Cambridge: Cambridge University Press.

JOCHIM, M., 1990. "The Ecosystem Concept in Archaeology." E. F. Moran, ed., *The Ecosystem Approach in Anthropology*: 75-90. Ann Arbor: The University of Michigan Press.

KAPLAN, S., 1984. *The Monastic Holy Man and the Christianization of Early Solomonic Ethiopia*. Wiesbaden: Franz Steiner Verlag.

KEMP B., 1989. *Ancient Egypt. Anatomy of a Civilisation*. London: Routledge.

KNAPP A. B., ed., 1987. *Archaeology, Annales and ethnohistory*. Cambridge: Cambridge University Press.

KOHLER T. A. and G. J. GUMERMAN, eds., *Dynamics in Human and primate Socieities: Agent-Based Modeling of Social and Spatial processes*. New York: Oxford University Press.

KRISTIANSEN K. and M. ROWLAND, eds., 1998. *Social transformations in archaeology. Global and local perspectives*. London: Routledge.

LAKE M. W., 2000. "MAGICAL Computer simulation of mesolithic Foraging:" Kohler T. A. and G. J. Gumerman, eds., *Dynamics in Human and primate Socieities: Agent-Based Modeling of Social and Spatial processes*: 107-143. New York: Oxford University Press.

LIGHTFOOT K. G., 1995. "Culture contact Studies: Redefining the Relationship between Prehistoric and Historical Archaeology," *American Antiquity*, 60 (2): 199-217.

MANZO A. and C. PERLINGIERI, 2001. "Genetic Programming and Archaeology," paper delivered at the Italy-United States Workshop *"The Reconstruction of Archaeological landscape through Digital Technologies,"* Boston, MA, November, 1-3, 2001.

MARCUS H. G., 1994. *A History of Ethiopia*. Berkeley: University of California Press.

MCCANN J., 1995. *People of the Plow. An Agricultural History of Ethiopia, 1800-1990*. Madison: The University of Wisconsin Press.

MCINTOSH R. J., J. A. TAINTER and S. KEECH MCINTOSH, 2000. *The Way The Wind Blows. Climate, History amd Human Action*. New York: Columbia University Press.

MCNEILL W. H., 1976. *Plagues and Peoples*. Doubladay: Anchor Press.

MUNRO-HAY S., 1991. *Aksum. An African Civilisation of Late Antiquity*. Edinburgh: Edinburgh University Press.

PARKER M. T., 2001. "What is Ascape and Why Should You Care?" *Journal of Artificial Societies and Social Simulation*, 4 (1), online publication <http://www.soc.surrey.ac.uk/JASSS/4/1/5.html>

PHILLIPSON D.W., 1998. *Ancient Ethiopia. Aksum: Its Antecedents and Successors*. London: British Museum Press.

RENFREW C. and P. BAHN, 1991. *Archaeology: theory and methods*. London: Thames & Hudson.

RENFREW C. and K. L. COOKE, 1979. *Transformations. Mathematical Approaches to Culture Change*. New York: Academic Press.

RENFREW C. and E. B. W. ZUBROW, eds., 1994. *The ancient mind. Elements of cognitive archaeology*. Cambridge: Cambridge University Press.

SCHIFFER M., 1987. *Formation Process of the Archaeological Record*. Albuquerque: University of New Mexico Press.

SHENNAN S., 1988. *Quantifying Archaeology*. Edinburgh: Edinburgh University Press.

SHERRATT A., 1993. "The relativity theory" in N. YOFFE, A. SHERRATT eds, *Arhaeological Theory: Who sets the agenda?* Cambridge: Cambridge University Press, pp. 119-130.

SHERRAT A. 1994, *Core, periphery and margin: perspectives on the Bronze Age*, in MATHERS, STODDART 1994, pp. 335-345.

SIMOONS F. J., 1961. *Northwest Ethiopia. Peoples and Economy*. Madison: The University of Wisconsin Press.

TAMRAT T., 1972. *Church and State in Ethiopia 1270-1527*. Oxford: Clarendon Press.

THOMAS D. H., 1972. "A computer simulation model of Great Basin Shoshonean subsistence and settlement patterns." D. L. Clarke, ed., *Models in Archaeology*. London: Methuen & Co. LTD.

TOPOLSKI J., 1975. *Metodologia della ricerca storica*. Bologna: Il Mulino.

TRIGGER B. G., B. KEMP, D. O'CONNOR and A. LLOYD, 1983. *Ancient Egypt: A social history*. Cambridge: Cambridge University Press.

TZADUA P. and P. L. STRAUSS, 1968. *The Fetha Nagast. The Law of the Kings*: Addis Ababa: Haile Selassie I University.

VAN DER LEEW S. E. and J. MCGLADE, 1997. *Time, Process and Structured Transformation in Archaeology*. London: Routledge.

VAN WEST C. and T. A. KOHLER, 1998. "A Time to Rend, A Time to Sew: New Perspectives on Northern Anasazi Sociopolitical Development in Later Prehistory." Aldenderfer M. and H. D. G. Maschner, eds., *Anthropology, Space, and Geographic Information Systems*: 106-131. New York: Oxford University Press.

WINTERHALDER B. P., 1994. "Concepts in Historical Ecology: The View from Evolutionary Theory." C. L. Crumley, ed., *Historical Ecology*: 17-42. Santa Fe: School of American Research Press.

WOLDE-MARIAM M., 1991. *Suffering under God's Environment. A Vertical Study of the Predicament of*

Peasants in North-Central Ethiopia. Berne: Univedrsity of Berne.

YOFFEE N., 1993. "Too many chiefs? (or, Safe texts for the '90s)." N. Yoffee and A. Sherrat, eds., *Archaeological Theory: Who sets the Agenda?*: 60-78. Cambridge: Cambridge University Press.

ZANETTI U., "Les Langues de l' Ethiopie et de la Corne de l'Afrique." X. van der Stappen, ed., *Aethiopia. Pays, Histoire, Populations, Croyances, Art and Artisanat* : 214-219. Luxembourg: Gordon and Beach Arts International.

ZEIN A. Z. and H. KLOOS, eds., 1988. *The Ecology of Health and Disease in Ethiopia.* Addis Ababa: Ministry of Health.

ZUBROW E., 1973. "Adequacy Criteria and Prediction in Archaeological Models." Ch. L. Redman, ed., *Research and Theory in Current Archeology*: 239-255. New York: John Willey & Sons.

Real Space Beyond Solid Models: Spatial Metadata in Ethnoarchaeology

Monica Foccillo[2], Andrea Manzo[1], Cinzia Perlingieri[1], Rosario Perlingieri[2],

[1]Istituto Universitario Orientale,
P.zza S. Domenico Maggiore, 12 - 80134 Naples, Italy
ybeyene@iuo.it
[2]novaricerche srl

Abstract. The solid modeling is used in this work as only a starting point to highlight the information potential of computer-generated environments in archaeology. The main purpose is to develop a research plan for a critical information use to really help archaeological interpretation. A virtual-modeling application, strictly associated with a GIS-based approach, was used with data from an ethnographic survey of ceramic manufacture and settlement habits, together with archaeological data from the surroundings of Aksum (Ethiopia). This process, based on digitized data and related multidisciplinary datasets and textual description, generates a provisional multi-level model of settlement structures in their original environment.

1 Introduction

The model showed in this pages represents a section of an Aksumite palace dated back to the first centuries A.D., and it is the major impact result of our research project. But our final goal is to go beyond the model and make of three dimensional space reconstructions real means of analysis and to demonstrate, through our direct experience, how solid models can be tools of actual impact in terms of information acquisition as well as of communication. Is there any real space beyond the reconstruction of this solid model? That is the provocative question that we asked ourselves.

The model is the result of a very long process of analysis and study on the archaeological data from the excavation at Bieta Giyorgis, Aksum (Tigray, Ethiopia) (Bard et al., 1997; Fattovich et al., 2000; Bard et al. this volume). It represents a hypothesis of the construction techniques, internal coatings and furnishings, as well as of the function of the rooms at its interior and the household goods and equipments.

It is many years that archaeologists have discovered computer graphic databases, applied to the spatial analysis, not only as drawing aid, but mostly because the construction of vector and raster three dimensional computer models of ancient sites and monuments offers us a powerful self-guided view of our data through time and space (Chirica and Cojocaru, 2000; Csáki and Redö, 1996; D'Andrea and Niccolucci, 2000; Forte and Beltrami, 2000; Forte and Guidazzoli, 1996; Gillings, 1999; Reilly, 1992; Reilly & Rahtz, 1992). Computer models allow us to accurately measure and analyse ancient monuments, for instance, to determine ancient appearances of deteriorated or multi-phase monuments, at a further stage, to re-create the processes of ruined structures and the organization of ancient settlements under particular environmental conditions (Earl, 1999; Eisler et al., 1988; Reilly and Richards, 1988; Wood and Chapman, 1992).

Data visualization and solid modelling, among archaeologists, have been for long only synonymous with advanced state of the computer graphics art. Rarely archaeologists, even in the recent past, have had direct experience of these advanced systems and haven't fully appreciated the uses to which they can be put. For many scholars, their only contact with this technology is through the generated pictures, and computer graphics has been regarded exclusively as a sophisticated presentation tool, completely separated from the actual use of information.

More recently, indeed, a new point of view is developing around this topic and the facilities of picture generation and visualization proper of these systems are starting to be considered as key properties in many fields of archaeological studies (Clubb, 1999). That is why the considerable curiosity and interest that mostly the visualization potential of three dimensional data has gained, is replacing by other important attributes lately recognized as very useful to archaeological research.

2 The case of Bieta Giyorgis

At the very beginning of this project, we asked ourselves, if it was really worthy and useful reviewing our archaeological and ethnographic data about ancient Aksumite buildings under the new light of modelling techniques. We were stressed on the thought of a flirtation with this attractive technology and on the risk of a big waste of time simply to add new appealing application to traditional output of our results. But working on it, we must say, we have been requested with an unexpected twofold effort, of *communication* with non-archaeologists on one hand, and of getting a *global vision* of our own work on the other.

We had to translate our knowledge of the fascinating ancient Ethiopian culture to work out a reconstruction clear and accessible to many and this has to do with how much we are interested in an actual impact of our work, if outside of circles of scholars. There is sometimes an intolerable gap between our scientific experience and what of this experience reaches the outside of scientific workshops and conferences.

On the other hand we realized how much of our work we leave someway behind lacking a full global vision of the diverse information generated by every single disciplines contribution that have always been involved in the Aksum project. An attempt of combine different contributions at the Aksum multi-disciplinary project has been made by the authors also through the application of genetic algorithm and other statistic approaches (see Manzo and Perlingieri in this volume, De Falco et al. nd.a)

databases, bear much easier the evidences of the facts beyond numbers. Virtual models produce an extra-information about how space was used, if not perceived, and then constitute a valid base to link different sets of data. They allow us in place back artifacts in what is the supposed original context. Actually through this system, that is conceived as a flexible archival, it is possible to establish links with many different groups and categories of data. It incorporates solid modelling within an integrated multi-media system dedicated to the management of the archaeological information. We can synthesize that this method reflects the same principles of the intra-site GIS based approach (Agresti et al., 1996; Baena et al., 1996; Huggett, 2000; Lamprell et al., 1996), but in the more powerful means of the three-dimensional representation (see also Harris and Lock, 2000).

Different datasets linked to the 3D model:

- Ceramic Analysis
- Construction Techniques
- Archaeozoology
- Palaeoethnobotany
- Ethnography
- Imported Materials
- Palinology

Fig. 1. Virtual model of a Classic Aksumite store room and related sets of archaeological data

What we learned of this experience, by contrast, is that applications of solid modelling in archaeology are beginning to show real benefits for archaeology in the definition and analysis of ancient three-dimensional space.

We refer mainly to the possibility of building up complex archives and structures managing enormous volumes of data and information. The impressive potential of modelling technology is at least as important and interesting to archaeologists as is the possibility of recording and interpreting primary archaeological information.

The goal of this project is to build a comprehensive database of artifacts, features and attributes, based on the spatial interrelationships among three dimensional objects. This is based on the simple assumption that visualization in general terms actually helps the comprehension of significances and relations of lost spaces and contexts. That is what usually happens dealing with long and complex sets of data: graphs visualizing

We are in absolute need of integrate the different sets of data and archives of data in a single system that allows us to make fast and simultaneous consultations, to identify similarities, connections, interactions that help us in reaching ever more complex stages of inference and interpretation of the archaeological evidence. The three-dimensional model and the related data sets give the possibility of pushing the re-play button, of going very easily from the interpretation (i.e. the model) to the rough data and *viceversa*, with an evident improvement of the archaeological method from the kuhnian point of view (Kuhn, 1962).

Our virtual-modelling applications, in association with a GIS-based approach, are focussing on the definitions of integrated methodologies able to give answers at all the aspects at every stage of the investigation. The theoretic-methodological structure of this project is aimed to the realization of a complex system in which the spatial information is associated with all the categories of datasets coming from the disciplines involved in a multidisciplinary project like the Aksum Archaeological

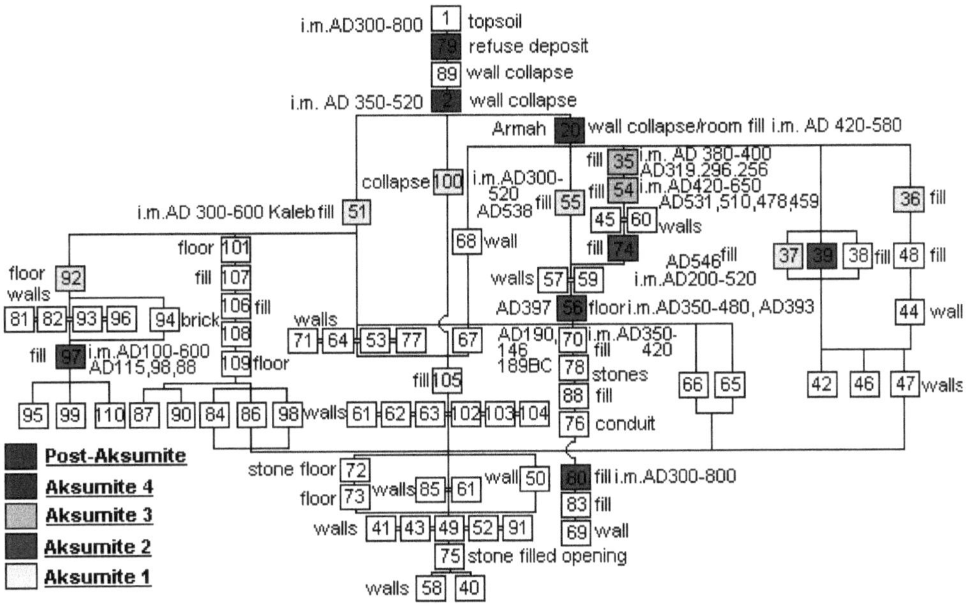

Fig. 2. ON VII-VIII/X, Stratigraphic Matrix

Project (Bard et al. this volume). The investigating process starts with the systematic collection and classification of all the archaeological items and it passes through further types of analysis offered by different disciplinary spheres (Fig. 1).

A comparative study, in both archaeology and ethnography, of excavated ancient structures, modern inhabited houses and recently abandoned compounds enlighted the strong liasons between ancient and modern construction techniques and made possible an attempt to reconstruct the elements, in terms of time, space and behaviour, that determined or affected the organization, use and abandonment of the ancient structures excavated at Ona Nagast. To get this point we tried this integration of traditional methodologies, statistics and virtual reality. The sites at Bieta Giyorgis, excavated to a high standard, provide us a well documented complex of information about the archaeology of the Aksum area. Utilizing this material in the pursuit of research goals has been one of our major challenge and necessitated since the beginning the adoption of new tools and methodologies for the data to be integrated, processed and evaluated at the best. In this work it will be presented a synthesis of the principal sets of data that have been used to reconstruct a sector of an ancient domestic context at Ona Nagast:

a. spatial analysis of the buildings found at the settlements excavated at Bieta Giyorgis,
b. systematic study of depositional processes and site formation processes,
c. systematic analysis of artefacts, mainly ceramics, and ecofacts,
d. ethnographic investigation of domestic life and material and social organization among rural communities in the surroundings of Aksum and Bieta Giyorgis.

3 From archaeological features to living spaces

The excavation at the site of Ona Nagast put in light very well preserved palatine structures, and allowed us to identify several rooms of a peripheral area of the main architectural structure excavated. The great part of the structure is left unexcavated under a cultivated terrace.
The segments in this area are part of a wide residential complex encompassing many rooms, courtyards, corridors and still visible architectural features like stone thresholds and steps. The phases of construction and modification of these structures, as well as the phases of deposition of archaeological material, cover a long period since phase Aksumite I to phase Aksumite III since the 1st century B.C. up to the mid 6th century A.D.

In archaeological terms the complex is constituted by two main typologies of archaeological features: structural standing elements, i.e. walls and all the features that separate and define rooms and residential areas; and stratigraphic deposits at the outside and/or at the inside of the structural elements, as a result of both natural and cultural processes of site formation (Fig. 2).

Once removed a strongly disturbed superficial layer in the trenches VII-VIII-X, several strata of structure collapse came in light, with construction materials mixed with sandy soil, alternated with layers of collapse with dense accumulation of rock fragments and artifacts in a matrix of sandy soil. The stratigraphy shows clearly the collapse of a roof and a second floor, maybe more than one, during the last phases of deterioration of the structure after its definitive abandonment. The structure at ON is attributed to the phases Aksumite II and III dates also confirmed by the association of the imported Mediterranean pottery found in the same archaeological contexts.

As seen in modern ruins of recently abandoned houses, the time of collapse and deterioration after the abandonment is very short, a few years, this explain why, even in case of temporary reuse of the structure as a deposit, or animal enclosure, the ceramic materials and other features are found fairly homogeneous in the lower layers of the deposit.

After the abandonment of the residence, the deterioration of the wooden supports and the weight of the materials of construction causes the collapse of the upper floors. This kind of collapse includes construction materials, and very likely furniture, fragments of pottery, other goods and objects; this layer, in an archaeological deposit, is clearly distinguishable from the lower ones. After this phase of collapse, the original walls still standing, will be furthermore weakened and, exposed to the action of atmospheric agents, also will collapse adding more layers of deposit even clearly distinguishable from the previous ones, including mostly stones with much less of other objects.

A brief description of the archaeological contexts and the deposition processes at the inside of the structures at ON, will give us the basic elements for observing and considering the presence and the frequency of the objects found in association with the deposits, and the functional meaning of the rooms.

The sequence at Ona Nagast, trenches VII-VIII-X

Walls excavated at ON VII-VIII-X, define three main rooms:

Room 1

The first phase of construction inside Room 1 is represented by the walls SU 58 and 40; subsequently the openings in the wall SU 58 have been filled up, SU 74 has been deposited, and the platforms SU 45 and 60 have been built, strengthening in such a way the base for the building of a new floor; at the interior of this room have been subsequently deposited SU 35 and SU 54.

SU 35 is a deposit of rocky fragments by structure collapse, in a matrix of sandy clayey soil. This unit represents a late phase of filling formed by the collapse of stones from construction materials and, most likely, by rests of an upper floor. Artefacts in this layer include ceramics, fragments of stone bowls, animal bones, glass, metal fragments, grinding stones, blocks of dressed stone and fragments of fired clay bricks. The most part of the stone elements has been found in a vertical position, fluctuating in the matrix of the deposit, witnessing a clear collapse of walls, roof, and upper floors. The pottery associated with SU 35 has been dated back to the phase Aksumite III, and a radiocarbon date from SU 35 has been calibrated at 35/335 A.D.

SU 54, lying under SU 35, is a deposit of filling between walls SU 40, 49/58, and SU 45/43. This layer consisted of stone rubbish mixed with a considerable amounts of ceramic fragments, bones, grinding stones, glasses and personal ornaments in a matrix of sandy clayey soil mixed with gravel. This layer is most likely associated to the collapse of the roof covering and of the paving of the upper floors, or as well, can be the result of a gradual accumulation of domestic discards over a long period of time. For this SU we got a radiocarbon date calibrated to the 190-550 A.D.

Stratigraphic Unit 74, a very thick deposit of stone rubble with moderate concentration of archaeological materials (ceramics, animal bones and glasses), in a matrix of sandy clayey soil mixed with gravel. This layer could result by a structure collapse of an older building or could have been intentionally re-filled during a later phase of construction. The homogeneity and the low concentration of materials included in it, seems to confirm the second hypothesis.

Hypothesis of later phases of structural restorations and modifications, could be inferred by the presence of some small rooms found in the immediate vicinities of the principal rooms, These are SU 37, 38, and 39.

SU 37 is a very thick deposit found at the inside of a long and narrow room between walls SU 46, 43, 42 and 41. It consists mainly of stone rubble and great concentrations of big fragments of pottery, very well preserved, many of which were restored in many almost complete bowls. These fragments were included in a slime-sandy clayey soil. Together with the ceramics were recovered numerous fragments of glass, grinding stones and pestels. This layer could have been intentionally deposited during a period of reconstruction as a deposit levelling the space between two old walls.

SU 38 is a thick deposit of slime-sandy clayey soil, found in the long narrow space between the walls SU 40, 43 and 41. It is mainly characterized by stone rubble and a considerable amount of artefacts, and similarly to SU 37, it could have been deposited during a period of reconstruction to fill up a useless space between two walls in disuse.

Stratigraphic Unit 39, a very clayey soil, is most likely a deposit levelling the space between two old walls, like the SUs described above. In it were found quantities of grinding stones and pottery sherds, deposit levelling the space between two old walls

Room 2

The most ancient phase of construction in this room, is represented by the wall 69. The wall SU 50 abuts SU 49, a more ancient wall, and the wall SU 53 abuts SU 50. Apparently the contiguous rooms called Room 2 and Room 3, were originally an open courtyard, delimited at the East and West sides by walls SU 50 and 61. The later construction of wall SU 53 has divided the original courtyard in two different rooms connected by a door with a threshold made up of big flattened stones. This

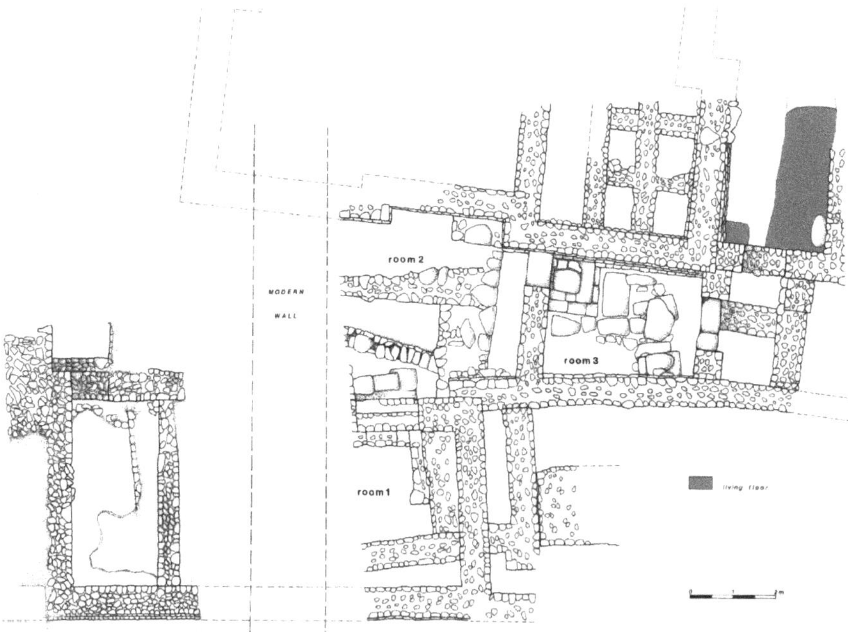

Fig. 3. ON VII-VIII/X, Excavation map

"restoration" of the environments seems to points out to a change in the organization of the house space and in the type of utilization of this space through time.

The stratigraphic sequence at the inside of Room 2 lies under a thick deposit consisting of three upper layers, SUs 1, 2 and 20. SU 55, that lies under SU 20, is a deposit of mainly stone rubble and great concentrations of pottery sherds, animal bones, glasses, fragments of bronze and iron, grinding stones and blocks of smoothed and shaped stones in a clayey soil mixed with gravel. This layer could be the result of an initial process of slow deterioration and collapse of the original walls or could have been deposited during a long process of occupation during which structure residuals and remains have been accumulated. Anyway this layer is most likely the result of a formation process happened after the first phase of construction/occupation of this building.

SU 56, lying under SU 55 constitutes with no doubt an earlier occupation phase. This deposit is a compacted clayey sandy soil mixed with gravel and fine inclusions of deteriorated rock, in which several bricks, made of fired clay were found. Associated with this layer was found a considerable quantity of pottery, animal bones, glasses, bronze and lithic objects. Under SU 56 was found a very thick deposit, SU 70, a compact sandy soil with inclusions of deteriorated rock including a surprising concentration of remains: pottery, animal bones, fragments of glass and stones. This layer was most likely an intentional deposit to level walls in disuse (SU 69) and of a drain pipe (SU 76).

Room 3

This room is defined by walls SU 61, SU 52/50, and 53, and represents the most ancient phase of construction of the whole area. The careful observation of the construction techniques might suggest that wall SU 50 was built later at the top of SU 52. Later was built wall SU 53 on the top of SU 52/50. A new threshold and a series of steps (SU 77) was later built joining the wall SU 52. This series of steps exhibits much less accurate construction technique if compared to SU 71 and it seems of having been built during a later phase of reconstruction after wall SU 52.

SU 51 is a deposit resulted by the collapse at the inside of the room among walls 61, 52 and 53. Is characterized by stone rubble mixed with gravel including a good quantity of artifacts (ceramics, animal bones and fragments of glass) in a sandy clayey soil. This layer lies at the base of SU 2. A coin of the king Kaleb (500-550 A.D. ca) was found in association with this layer. SU 71 is a series of three steps at the base of the threshold SU 64, the stone slabs, roughly shaped, are tied up together with a clayey mortar mixed with gravel. Another series of steps, SU 77, built as well with roughly shaped stone slabs abuts wall SU 52 and it lays on the floor SU 72/73.

The architectural features found in all the rooms are homogeneously characterized by stone slabs of different shape and size, with differently rough smoothing. This points out on an apparent absence of strong level of standardization in the construction techniques.

The floor SU 72 is characterized by stone slabs of different size and shape it lays close to SU 73, a paving constituted by compact clayey sandy soil with inclusions of deteriorated rock.

At East of Room 3 and walls 52/50, was found SU 36. This deposit is mainly constituted by stone rubble from collapse mixed with artifacts (mostly ceramics and animal bones) in a matrix of sandy clayey soil. At the East of wall SU 52, there is the layer SU 48, it lays all along under SU 36 and consists mainly of stone rubble, gravel, deteriorated rocks and ceramics in a matrix of clayey soil. The frequency of the artifacts inside this layer is very low and likely this deposit represents the upper part of the foundation of wall SU 52.

Very interesting is SU 68, a series of stone steps that most likely are part of a stairs going to an upper floor. It is very difficult to say if the stairs was placed at the interior or at the exterior of the house. The slabs used for the steps are roughly shaped and held together with a clayey mortar. The steps joins wall SU 67 to the N side and wall SU 63 to the S. Both abut wall SU 61. This stairs connected the rooms of the lower floor with those of the upper internal floor of the central residential building of the architectural complex at Ona Nagast, that is supposed to be at the side of walls SU 61 and SU 62.

The sequence at Ona Nagast, trenches VII-VIII, have put in light structures with walls defining three main rooms. The structure of the rooms show different phases of building and re-building, paving, and intentional filling. The collapse of the upper floor sealed the base of the walls and living surfaces under thick rubble strata. The architectural features found in all the rooms are homogeneously characterized by stone slabs of different shape and size, with differently rough smoothing. This points out on an apparent absence of strong level of standardization in the construction materials.

Most features found in the excavation have their parallel in the ethnographic evidence in terms of both materials employed and construction techniques. The modern traditional residences in the Aksum region are of two basic types: tukul, huts with circular plan, and hidmo, houses with square plan. Both are built with the same basic technique: the walls are made up of stones, simply compacted with soil, the roof, made of several overlapping layers of stones (not always), mud, ashes and straw (that has a waterproof effect), is sustained by a framework of wood poles and tables. Single households compounds show different patterns of structures at their inside: some have both tukuls and hidmos, some have only tukuls, some have only one hidmo. The tukul is a single room, therefore different tukuls represent different areas and have different functionality; the hidmos contain several rooms on different floors (basically two).

4 From artifacts to household goods

The origin of the archaeological strata and the mixing due to collapse phases do require a big caution during the interpretation. But together with the artifacts included in the archaeological deposits, the strata (i.e. the context) constitute powerful means for the formulation of hypothesis concerning the nature and function of the rooms found at the lowest level of the building and, where possible, they help in distinguishing layers of collapse from upper floors.

It is worthy here to consider an argument only apparently of secondary importance, that is actually an intriguing item in the ceramic studies, and one of big utility in terms of comprehension of site formation processes, and for the interpretation of the spatial organization of settlements: why is the pottery found in fragments? Through which dynamics it is spread inside a deposit? What is the destination of old vessels once broken? It is a rare coincidence find pieces belonging to the same vessel and reach in fit them in the complete shape, the pottery from pits for discards and from different depositional layers in a settlement, are usually fragments (Arnold, 1988; Shott, 1996).

In evaluating the assemblages it should be considered that pottery used in the household, once broken, is subjected to several kinds of processes, as ethnographic research at Bieta Giyorgis has demonstrated. The pottery, once broken, is subjected to several kinds of processes if it is re-used and if it is discarded and abandoned as well:

1. if it is re-used
- restored, fitting it with rope or vegetable fibres through small holes made ad hoc along the fractures;
- (if big enough) reused as dish or wide open bowl, once smoothed the edge of the fracture;

2. If discarded and/or abandoned, it can be:
- mixed with soil and thrown in deposits filling interstices between walls at the interior or at the exterior of the houses;
- thrown to the outside of the house thus subjected to a casual and unintentional process of trampling that shatters and scatters them furthermore; (even the floors in the rooms and courtyard are periodically swept to remove garbage and discards).

A container of pottery or terracotta when subjected to a mechanical shock very rarely is crashed in small fragments, it breaks into a few big pieces, but anyway it is irremediably of no use, especially if is a vessel for the conservation of liquids or is a cooking pot. Some fragments can be fitted for other uses, but most of them are abandoned and exposed to further stresses.

Moreover, in evaluating the meaning of the presence of single ceramic shape is an assemblage it is also important to consider that the frequency of a certain ceramic type represents a mathematic function of his frequency proportional to other ceramic types in a specific interval of time, but is also a mathematic function of his duration of use. The technical characteristics, i.e. resistance, temperature of firing, and finishing of the surfaces, do affect the life and the duration of the vessel, and more, they are affected by the domestic context. For instance, jars and amphorae for the conservation of liquids, are in a context, and have a use, that we can say define passive, they are in a stable position, on the ground or on low supports, and rarely are moved; on the contrary small jars for transportation of water or cooking pots have an "active" role in the domestic organization, then suffer of frequent displacing, mechanical stresses and thermal shocks, they are therefore much more exposed to risks of fractures.

After evaluating the depositional formation process of the context and after establishing that it does not reflect

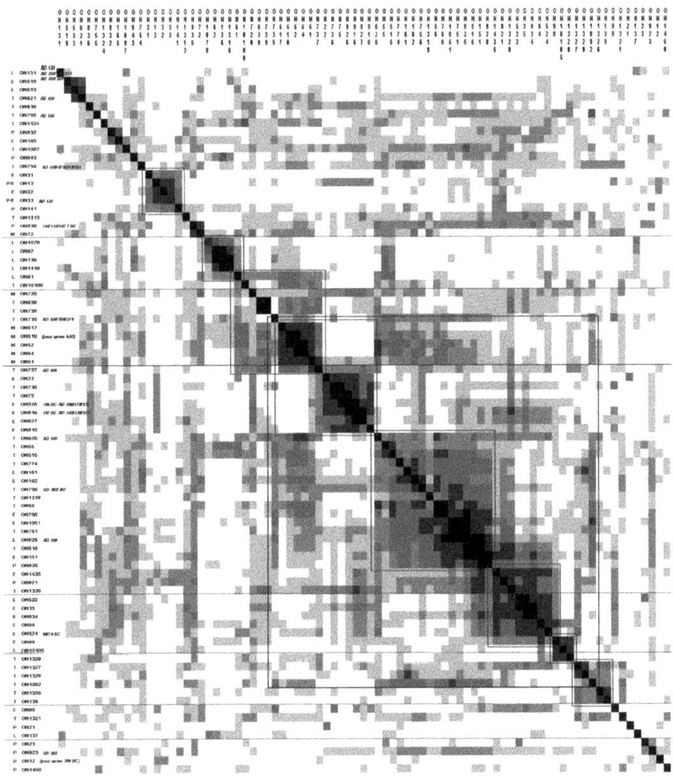

Fig. 4. matrixes relative to the distribution of types in the sequence

disturbances and mixings but an actual ancient assemblage, it is interesting to compare diverse types of pottery vessels that are used in a traditional household and those that we find in the archaeological deposits.

All these elements, if considered in the archaeological study of different phases and comparable settlement contexts, help in building a more and more rich and precise outline of the original goods and strategies of domestic space organization.

Unfortunately, very rarely archaeological deposits are clear, more frequently, the interpretation is complicated by the alternation of different phases of use, re-use, abandonment and collapse. Nevertheless, in the case of ON VII-VIII, some inferences on the distribution of artifacts are possible.

On the base of the matrix relative to the distribution of the vessels in the rooms at ON VII-VIII, we notice the higher percent of brown ware jars in SU 35, SU 54 and SU 55 in contrast with the other shapes, this suggest a considerable presence of storing vessels rather than cooking pots in these stratigraphic units. On the contrary, the low frequency of brown ware jars in stratigraphic units 56, 70, 80, 51, 92 and 97, and the massive presence of orange ware small bowls seem to witness a different use of the rooms associated with these layers.

The different nature and the variability of the archaeological datasets from the excavation at BG, imposes, for his complexity, the application of mathematical methods able to manage an enormous quantity of data.

Once included our data in a database software, interrogations allow us in quantify the presence, the frequency and the combinations of the morphological attributes. The combinations have been considered and further verified with a statistic process from which it has been generated "similarity matrixes" relative to the frequency of the different attributes of each class of pottery and, later, to the frequency of the typological classes themselves in all the stratigraphic units of the sequence. In this way, the sequence resulted from the excavation appears marked by peculiar technological characteristics or by peculiar stylistic trends, etc. This "picture" generated by the mathematical elaboration is of great utility not only to distinguish different chronological-cultural phases inside every sequence, but also to compare sequences from different areas and to fill the gaps of a given sequence with the data coming from another one.

An archaeological deposit, in fact, is often not clear and/or badly preserved, moreover, it can result from a discontinuous occupation; therefore it is a precious help having the possibility of work on and compare more than one stratigraphic sequence to comprehend the development of the system of production and use of the pottery in its context of origin. Thanks to this methodology most of the lacking information of the incomplete units at Bieta Giyorgis have been integrated.

The theoretic principle that lays beyond the interpretation of the statistic data is that: stratigraphic units showing a statistic configuration similar to each other have to be considered as belonging to the same cultural phase, vice versa, statistic configurations that show strong differences are the clear index of a chronological or functional differentiation, then they have to be distinguished.

The matrixes of data, relative to the sequences of attributes for each class of pottery, has been subjected to cluster analysis (grouping). This process is based on the "Pearson's method" of the dendrogram matrix and on the "complete linkage" as a system of internal comparison. These procedures have been arranged using the software *Systat*. The matrix of percent values, obtained after the first elaboration, has been secondarily transformed in dendrogram (it is the most easier representation of the analysis of cluster to read and interpret) using the coefficient of Pearson for the distance and the complete linkage. The procedure is synthesised as follows: 1- matrixes based on the Pearson' method, 2- dendrogram of the clusters according to the coefficient of Pearson. We have preferred to display here a modified version of the original matrix, this to make clearer and more easily appreciable the values (Fig. 4).

The analyses of cluster have been elaborated in two different steps, on the base of two separate sets of data. The first group of matrixes relative to each class of forms, is based on the distribution of the single morphological attributes *per* form, their associations represent the "types". In second instance the data relative to the resulted "types", *i.e.* the aggregations of attributes seen before, have been subjected to a second analysis of cluster generating the general matrix relative to the distribution of the types in the whole archaeological sequence.

It is worth to note that the cluster analysis doesn't produce groupings of absolute reliability: the data resulting have to be interpreted in their probable errors and gaps and have to be compared with their original archaeological contexts and the other evidences coming from other sets of data.

As an example, using a traditional method of study the preponderant percent of several features, at the inside of a given archaeological unit, suggests the attribution of the unit itself to a certain cultural phase; in general terms we are able to discard those elements that we consider "intrusive" interpreting them with our direct experience. In a mathematical process, like the cluster analysis, these elements are counted together with the others and risk to throw the statistic elaboration balance off; usually they simply don't aggregate with the other elements (whose stronger association is evident in the cluster) and are "dispersed" in the matrix. Nevertheless they can someway alter the statistic picture and can lower the coefficient of aggregation.

This is the main motive why it is important, before a statistic analysis aimed at reconstructing living spaces operate a selection of the data excluding those coming from the archaeological units that are incomplete, not clear, numerically insufficient. In the case of Bieta Giyorgis have been excluded all the upper units in many sites (the surface collections always give second deposition mixed material with no stratigraphic and numerical significance) and every unit resulting from a not reliable deposit (it is the case of the site ON IV).

5 The model

After the collection, elaboration and interpretation of the archaeological data in the light of ethnoarchaeological evidence, we have all the rough material we need to built up a model.

More powerful computers and increasing sophistication in software enable great realism to be introduced into the computed views of the model.

The realization of this kind of model is articulated in two different phases: modelling of geometry (slide con and "characterization" of the model through the application of raster type textures. To model the geometry, in this case has been used a bi-dimensional CAD relief imported in the solid modelling program. Through the same process is possible to model different kind of objects, like vessels, rotating their section around their axis (Fig. 5). The phase of texturing the model is particularly delicate, actually on it depends the realistic effect of the final scene. This procedure allow us to simulate virtually all the materials of the reality starting from JPG images of the materials themselves. Realism is achieved even by including some elements of the social context and function of the objects in the visualization and thereby bring the visualization to life (for instance, figures in ancient clothes, busy in their activities).

Fig. 5: right, picture of some Proto-Aksumite bowls; up, 3D model of a Classic Aksumite open bowl

And at the end natural lighting effects greatly contributes to reach the desired realistic effect. The rendered views of the model contain accurately modelled perspective, and the ray tracing techniques were employed to compute shadows. Ray tracing is one of the most popular techniques, it facilitates shading due to direct and indirect illumination and both reflection and refraction of light. Rays are cast from an eye through the pixels on the screen into the three dimensional space containing the model, when a ray intersects with an object calculations are performed to determine colour and all the other surface properties. The cumulative colour of these rays is added to the pixel value to define the final effect.

Final Remarks

As we have seen, the computer graphic systems at our disposal offer us many different tools to display always better our data (Fig. 6). What we like to enlighten in this occasion is that in all the process we made we searched for the best ways and strategies to put in light all the available information starting, as a basic primary step, with the archaeological evidence. What we think as archaeologists is that as we improve in computational processes we have to make progress in data acquisition techniques, in data analysis, and on this base use computer graphic systems to visualize and present hypotheses in the best light possible. What we would consider an actual success of our project it would be to conclude that archaeology takes a real advantage of the technological abilities of the virtual reality to create an experience, near to reality, of ancient environmental and social contexts.

Defining movements through artificial landscapes and living spaces is seen as powerful means of re-interpreting and demonstrating social behaviour and cultural meaning. We think that virtual reality can be considered and used much more than as sets of nice pictures; models and single objects can be linked to text, image, sound, and databases permitting authomatic research or educational virtual tours of ancient sites in which users can learn about history, construction details or daily life with a click of the mouse. Worlds programmed in VRML can be sent over the Internet or run on CDs providing an interactive and exciting research experience, it is a valid alternative to traditional paper-based publications: a 3D computer model can be used as well as up-to-date visual index to an excavation report becoming the report by itself (see also Eisler et al., 1988; Heyworth et al., 1996; van Leusen et al., 1996; Wansleeben and van den Dries, 2000).

The same models can also be used as instruction materials for public schools or museums directly engaging students or the general public in a participatory learning experience utilizing not only a fascinating technology but also the latest archaeological evidence (see also Bossons and Ord, 2000; Ruggles, 1988; Cornforth et al., 1992). And this comes to the twofold aim of this project that is to utilize virtual reality as an "exchange box" to which all data types are linked and create a globally integrated and interactive system of virtual worlds for research, teaching, fieldwork and museum exhibitions to offer new and wide access to information and cultural knowledge; or in other words, in which scientific data by a few are entered to be uttered as easy learning means to many.

References

AGRESTI E., A. MAGGIOLO-SCHETTINI, R. SACCOCCIO, M. PIEROBON and R. PIEROBON-BENOIT, 1996, "Handling Excavation Maps in SYSAND", *Analecta Praehistorica Leidensia*, 28, 31-36.

BAENA F.J., F. QUESADA and M.C. BLASCO, 1996, "An application of GIS intra-site analysis to Museum Display", *Analecta Praehistorica Leidensia*, 28, 465-471.

BARD K.A., R. FATTOVICH, A. MANZO, and C. PERLINGIERI, 1997. "Archaeological Investigations at Bieta Giyorgis (Aksum), Ethiopia: 1993-1995 Field Seasons", *Journal of Field Archaeology*, 24, 387-403.

BOSSONS P.D. and D.E. ORD, 2000. "Computer Networks in Higher Education. A Case Study: Staffordshire University", Lockyear K., T.J.T. Sly and V. Mihailescu-Bîrliba, *Computer Applications and Quantitative Methods in Archaeology 1996, BAR International Series*, 845, Oxford, 149-155.

CHIRICA V. and A. COJOCARU, 2000. "Multidimensional analysis of the archaeological discoveries from the multiphase Palaeolithic site at Mitoc-Malu Galben", Lockyear K., T.J.T. Sly and V. Mihailescu-Bîrliba, *Computer Applications and Quantitative Methods in Archaeology 1996, BAR International Series*, 845, Oxford, 49-53.

CLUBB N.D., 1999. "Have we Failed to Provide a Strategic Vision for Information Systems in Archaeology?", Dingwall L. *et al.*, *Computer Applications and Quantitative Methods in Archaeology 1999, BAR International Series*, 750, Oxford, 219-226.

CORNFORTH J., C. DAVIDSON, C.J. DALLAS and G.R. LOCK, 1992. "Visualising ancient Greece computer graphics in the Sacred Way Project", Lock G. and J. Moffett, *Computer Applications and Quantitative Methods in Archaeology 1991, BAR International Series*, S577, Oxford, 219-225.

CSÁKI G. and F. REDŐ, 1996. "Documentation and modelling of a Roman imperial villa in Central Italy", *Analecta Praehistorica Leidensia*, 28, 433-438.

D'ANDREA A. and E. NICCOLUCCI, 2000. "L'archeologia computazionale in Italia: orientamenti, metodi e prospettive", A. D'Andrea e F. Niccolucci, *Atti del I Workshop Nazionale di Archeologia Computazionale, Archeologia e Calcolatori*, 11, 13-31.

DE FALCO I., A. IAZZETTA, A. MANZO, C. PERLINGIERI, and E. TARANTINO nd.a. "Genetic programming in archaeology. Extracting rules from excavation data: the case of Bieta Giyorgis", *Archeologia & Calcolatori*, in press.

EARL G.P., 1999. "Visualising Danebury: Modelled Approaches to Spatial Theory", Dingwall L. *et al.*, *Computer Applications and Quantitative Methods in Archaeology 1999, BAR International Series*, 750, Oxford, 259-263.

EISLER J., J. PEIŠA and K. PREUSS, 1988. "A digital model of archaeological excavations as the starting point of a database of primary information in Egyptology: method-procedure-experience", Rahtz S.P.Q., *Computer and Quantitative Methods in Archaeology 1988*, *BAR International Series*, 446, Oxford, 109-132.

FATTOVICH R., K.A. BARD, L. PETRASSI, and V. PISANO, 2000. *The Aksum Archaeological Area: a Preliminary Assessment*, Naples.

FORTE M. and R. BELTRAMI, 2000. "A proposito di Virtual Archaeology: disordini, interazioni cognitive e virtualità", A. D'Andrea e F. Niccolucci, *Atti del I Workshop Nazionale di Archeologia Computazionale*, *Archeologia e Calcolatori*, 11, 273-300.

FORTE M. and A. GUIDAZZOLI, 1996. "Archaeology, GIS and desktop virtual reality: the ARCTOS project", *Analecta Praehistorica Leidensia*, 28, 439-451.

GILLINGS M., 1999. "Engaging Place: a Framework for the Integration and Realisation of Virtual Reality Approaches in Archaeology", Dingwall L. *et al.*, *Computer Applications and Quantitative Methods in Archaeology 1999*, *BAR International Series*, 750, Oxford, 247-254.

HARRIS T.M. and G.R. LOCK, 1996. "Multi-dimensional GIS:exploratory approaches to spatial and temporal relationships within archaeological stratigraphy", *Analecta Praehistorica Leidensia*, 28, 307-316.

HEYWORTH M., S. ROSS and J. RICHARDS, 1996. "Internet archaeology: an international electronic journal for archaeology", *Analecta Praehistorica Leidensia*, 28, 517-523.

HUGGETT J., 2000. "Looking at intra-site GIS", Lockyear K., T.J.T. Sly and V. Mihailescu-Bîrliba, *Computer Applications and Quantitative Methods in Archaeology 1996*, *BAR International Series*, 845, Oxford, 117-122.

KUHN T. S., 1962. *The Structure of Scientific Revolutions*, Chicago.

LAMPRELL A., A. SALISBURY, A. CHALMERS and S. STODDART, 1996. "An Integrated Infromation System for Archaeological Evidence", *Analecta Praehistorica Leidensia*, 28, 37-41.

REILLY P., 1992. "Three-dimensional modelling and primary archaeological data", Reilly P. and S.P.Q. Rahtz, *Archaeology and the Information Age A global perspective.* London, 147-171.

REILLY P. & S. RAHTZ, "Introduction: archaeology and the information age", Reilly P. and S.P.Q. Rahtz, *Archaeology and the Information Age A global perspective.* London, 1-21.

REILLY P. and J. RICHARDS, 1988. "New perspectives on Sutton Hoo: the potential of 3-D graphics", Ruggles C.L.N. and S.P.Q. Rahtz, *Computer and Quantitative Methods in Archaeology 1987. BAR International Series*, 393, Oxford, 173-185.

RUGGLES C., 1988. "Software for the Leicester Interactive Videodisc project", Rahtz S.P.Q., *Computer and Quantitative Methods in Archaeology 1988. BAR International Series*, 446, Oxford, 523-529.

VAN LEUSEN M., S. CHAMPION, J. LIZEE, and T. PLUNKETT, 1996. "Toward a European Archaeological Heritage Web", *Analecta Praehistorica Leidensia*, 28, 507-515.

WANSLEEBEN M. and M.H. VAN DEN DRIES, 2000. "ArchWEB: a web site for Dutch arcaheologists", Lockyear K., T.J.T. Sly and V. Mihailescu-Bîrliba, *Computer Applications and Quantitative Methods in Archaeology 1996*, *BAR International Series*, 845, Oxford, 71-79.

WOOD J. & G. CHAPMAN, 1992. "Three-dimensional computer visualization of historic buildings – with particular reference to reconstruction modelling", Reilly P. and S.P.Q. Rahtz, *Archaeology and the Information Age A global perspective.* London, 123-145.

The Remote Sensing project for the archaeological Landscape of Aksum (Ethiopia)

Maurizio Forte

CNR-ITABC, Istituto per le Tecnologie Applicate ai Beni Culturali, Roma
maurizio.forte@mlib.cnr.it

Abstract. Pioneer relevance of this international archaeological project is to identify a new methodological approach specifically addressed towards a VR spatial interaction starting from spatial data-entry of GIS, fieldwork, and remote sensed data. The project involves the application of advanced digital technologies for a detailed reconstruction of the archaeological landscape: analysis and classification by remote sensing and GIS, interpretation and presentation of the results through virtual reality and visual information systems. A relevant aspect is constituted by a multidisciplinary approach, starting with the acquisition of the data on the ground and, then, creating predictive classified maps of the geomorphological and anthropic landscape, including all the paleoenvironmental factors. Remote sensing applications have embraced: supervised and unsupervised classifications, data fusion, DGPS survey and micro-DEM's creation. Most part of the remote sensing classifications have been confirmed by the comparison with anthropological and geoarchaeological interpretations. In the project aerial photos, Landsat TM, SPOT XS satellite imagery have been used; in the near future an Ikonos multispectral image will be processed in collaboration with NASA.

1 Introduction

Final and fundamental goal of the Askum project is the reconstruction of the archaeological landscape on the basis of ethno-anthropological, paleoenvironmental, GIS, remote sensing and spatial data (Bard K.A., R. Fattovich, A. Manzo & C. Perlingieri 1997; Bard. K.A., M. Coltorti, M. DiBlasi, F. Dramis & R. Fattovich 2000). So the main key words could be: spatial, virtual, remote sensed and ecological approach (see my paper on the mindscape in this volume). In one word I could call this work a cybercultural landscape's reconstruction. The amount of archaeological data regarding Aksum is relevant, because of the several investigations undertaken in the past and in the last years (Littmann E., D. Krencker & T. von Lüpke 1913; Fattovich R. & K.A. Bard 1993; Phillipson D.W. 1998; Forte M., K.A. Bard, R. Fattovich, M. Foccillo, A. Manzo & C. Perlingieri 2001).

Figure 1. General map of Aksum

According to this methodological approach we distinguish principally: a) an observed landscape; 2) a classified landscape. The observed landscape corresponds to the diachronic visualisation of all the raster sources of information (aerial photos, satellite images, landscape's pic tures), the classified landscape is the interpreted landscape after digital and spectral processing of data.

Starting from 2000, activity of the project has been strongly increased by the collaboration of CNR-ITABC of Rome; in particular new directions of research were addressed towards GIS and remote sensing applications, digital topographic survey and DGPS applications.

Actually the project is supported by Agenzia 2000 CNR throughout a cooperative project between CINECA (Supercomputing Centre, Bologna), IUO (Istituto Universitario Orientale, Naples), University of Padua (Department of Archaeology) and CNR-ITABC (Rome). A short overview of the geomorphological features of Aksum comprehends: 1) Beta Giyorgis and Mai Qoho hills and the plain of Abba Liqanos. 2) a peripheral zone, related to the Northern and Southern plain of Aksum. 3) A boundary zone, comprehending the hills surrounding the plain of Aksum (figs.1-2).

Figure 2. View of the landscape in the North of Bieta Giyorgis

The whole project includes systematic researches and excavations in multidisciplinary fields such as paleo botany, archaeozoology, etnoarchaeology, ethnography, anthropology, geology, geomorphology, palinology, finally GIS, remote sensing and virtual reality (see the other papers in this volume and Forte M., K.A. Bard, R. Fattovich, M. Foccillo, A. Manzo & C. Perlingieri, Trabassi, 2001).

In this paper we will describe remote sensing applications constituting the fundamental base of data, whether for the investigation and predictive monitoring of the geo-archaeological landscape, or for the virtual reconstruction of the archaeological landscape. The pioneer relevance of this international archaeological project is to identify in a new methodological approach specifically addressed towards a VR spatial interaction from GIS data-entry, fieldwork and remote sensed data (Forte M., K.A. Bard, R. Fattovich, M. Foccillo, A. Manzo & C. Perlingieri, Trabassi 2001). For this aim specific software and interface written in C++ are in progress using Terravista and Vtree (thanks to a collaboration with Terrex Inc. and Tesscom). Further developments of the projects are planned with the acquisition of a multispectral Ikonos image in collaboration with NASA.

2 Remote sensing and digital reconstruction

In the case of Aksum the problem of the cartographical base was really heavy, because of the lack of data fit to the advanced goals of the project. In fact, at the beginning of the project, only 1:25.000 and 1:100.000 cartography and some aerial photos were available. Therefore the contribute of remote sensing was addressed either to integrate the previous data either to constitute a new georeferenced base for a digital spatial processing.

Figure 3. 3D Texture mapping of XPOT XS on cartographic DEM (1:25.000).

In this situation it is evident that the contribution of satellite remote sensed images has constituted a fundamental start up for having an acceptable and updated cartographic-territorial support.

In short, remote sensing applications have been finalised to the following aims:

- Georeference and rectification of '60 aerial photos (fig.5).
- Stretching and contrast enhancements of the panchromatic data.
- Digital filtering for the areas of interest.
- Creation of photomosaics including aerial photos and satellite imagery (fig.23).
- Multilayered overlay of the raster data in order to compare colours, components and geoarchaeological features (figs.12, 14, 20, 23).
- Multispectral supervised and unsupervised classifications for the interpretation of soils, vegetation and association of sites and terrain (figs.14, 21).
- Data fusion between panchromatic images (aerial photos) and multispectral satellite images (fig.18).
- DGPS survey (differential GPS) in the SO of Beta Gyorgis Hill for the creation of a centimetric DEM (figs.6-9).
- Texture mapping of all the satellite raster data (before and after multispectral classification) and aerial photos over the DEMs (like fig.3).
- Integration of all the spatial data in a GIS (Arcview 3.2, figs.4, 12, 20).
- Virtual reconstruction of the archaeological landscape with OpenGl graphic libraries (see Forte, Kay in this volume).

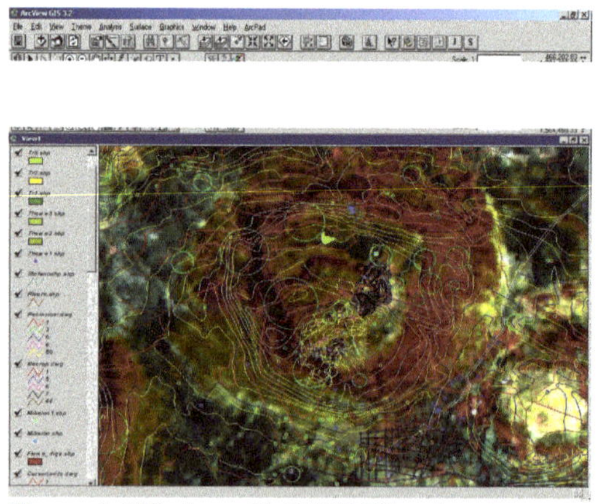

Figure 4. GIS view with SPOT XS and vector cartography in overlay

The methodology used in 2000-2002 has involved the creation of remote sensed preliminary models of the areas of interest so as to analyse the territory before the planning of archaeological fieldwork (May-June 2001).
This phase of processing of the "observed landscape" before the fieldwork has been very useful for preparing the 2001 expedition and for planning the survey with DGPS and total laser stations.

According to these premises a methodological sequence of the Aksum project was started with the following approaches:

a) Analogic (autoptic analysis and identification of archaeological features on aerial photos and cartography).
b) Digital-analogic (on rectified and gereferenced aerial

photos).

c) Digital-spectral (image processing with remote sensing software, unsupervised and supervised multispectral classifications).
d) Digital-virtual of the observed landscape (three dimensional reconstruction of the territory with DTMs created by cartographic contour lines).
e) Analogic (survey, fieldwork, archaeological excavations).
f) Digital-DGPS (acquisition of topographic data in real time with DGPS; exploration and topographic survey with geo-archaeological sampling of soils and anthropological classification of the territory).
g) Virtual (after the fieldwork and data acquisition on the basis of remote sensing and cultural interpretations, one starts the virtual reconstruction of the archaeological landscape interacting in real time with all the spatial data and 3D digital libraries).

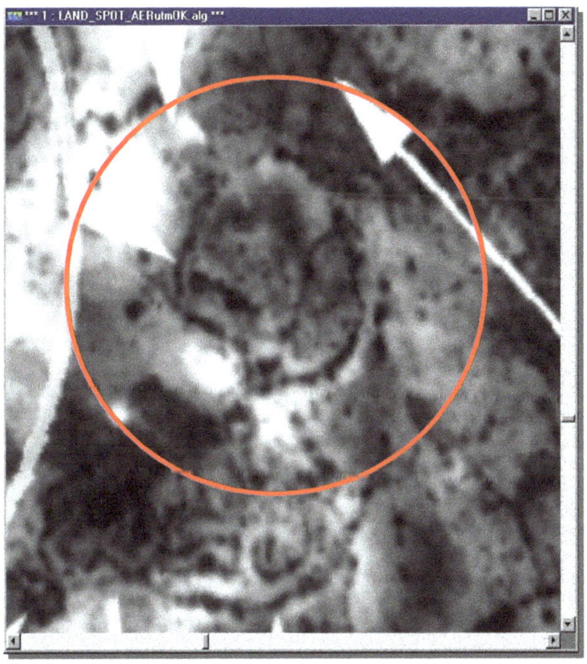

Figure 5. Aerial photointerpretation of possible archaeological features.

This description of a methodological protocol is a preliminary attempt for "virtualizing" all the phases of work from the ground to the lab and vice versa.

Moreover, the application of a methodological protocol in the field of the remote sensing applications in archaeology is the best approach to the multidisciplinary investigation of the territory.

3. DGPS survey

The Global Positioning System (GPS) is a worldwide radio-navigation system formed from a constellation of 24 satellites and their ground stations. Differential GPS involves the cooperation of two receivers, one that's stationary and another that's roving around making position measurements (figs.7, 9). There are several measuring techniques that can be used by most GPS Survey Receivers. We can say that DGPS in the last years has changed the approach of the archaeological survey in terms of time, precision and spatial data acquisition. A good campaign of DGPS during the archaeological fieldwork is able to construct a very detailed mapping (2D and 3D) of points, sites, territories, structures, integrating ancient preexistences with the modern landscape; that is in a very short period it is possible to produce a wide GIS base (DEMs included, like fig.22). Then, in kinematic way, hundred thousands points can be recorded in a few days or weeks, multiplying the potentialities of investigation and of digital processing.

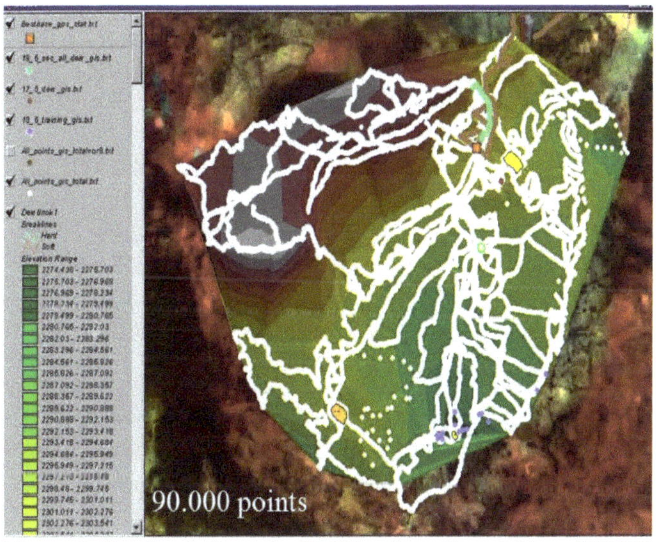

Figure 6. The dots indicate the 90.000 points of acquisition of DGPS in kinematic way.

Figure 7. DGPS in kinematic way (rover).

Use of GPS in archaeology is focussed mainly in two phases: applications of metric precision (typically in a range of 30-100 mt) or sub-meter and centimetric precision (DGPS single or double frequency).

Three main fields of applications can be identified in:
- cartography/survey of topographic points;
- remote sensing;
- DEM (digital elevation model).

In cartography and in topographic survey, the use of GPS is finalised to the acquisition of geographic coordinates of points (sites, settlements, etc.). Acquisition of georeferenced points permits to input the data directly in a GIS, after the conversion of coordinates in a local datum.

In the multispectral remote sensing the DGPS applications can be addressed towards identification and classification of training regions, namely the areas that must be sampled on the ground for classifying in a satellite image all the regions with the same spectral characteristics. High precision of this method allows a very detailed identification of the training region and, therefore, a better spectral (supervised) classification.

This centimetric precision of sampling corrects a very common error of the spectral classifications, that is the wrong boundaries of the training regions, or the choice of ambiguous pixels (for instance belonging to more soils and more than one investigated class). This error, typically made by interactive identifications of the training regions using the colours of the pixels on satellite images, produces a redundancy effect, namely too many classes of interest are identified.

Moreover, in the field of the creation of DEM, the use of differential GPS in cinematic method permits to acquire on the ground thousands or hundred thousands of elevation points in a very short time. Main DGPS techniques used are the following:

Figure 8. Agricultural terraces visualised by DEM created with DGPS.

Static - Used for long lines, geodetic networks, tectonic plate studies etc. it gives high accuracy over long distances but is comparatively slow.

Rapid Static – It offers high accuracy on baselines up to about 20km but it is much faster than the Static technique. Typical use is for creating local control networks, network densification etc.

Kinematic - Used for detailed surveys measuring many points in quick succession. With this method it is possible to record spatial coordinates and elevation data in real time during the fieldwork. In terms of methodological approach it has changed the praxis of the archaeological topography. In the cinematic DGPS we have a space of work of circa 10 km. This type of survey is particularly useful for GIS applications, because definition and drawing of polygons, points and lines can be made continuously.

Figure 9. DGPS in static way (reference).

RTK - Real Time Kinematic uses a radio data link to transmit satellite data from the Reference to the Rover. This enables coordinates to be calculated and displayed in real time, as the survey is being carried out.

4. Survey at Bieta Gyorgis

One of the key problem of our research was the lack of an adequate cartographic base (with the exception of 1:25.000 map) and of the total absence of cartographic control points in the main areas of investigation; in fact Bieta Gyorgis has only artificial terraces and small rural villages not mapped in cartography. This problem has involved a GPS systematic survey for positioning all the archaeological sites excavated and surveyed. GIS applications then, have integrated GPS data with raster and vector layers.

On the top of the Bieta Gyorgis hill the survey has involved these DGPS methods (model Leica 510, single frequency):

- Static;
- Kinematic-rapid (sampling of 2 and 15 seconds);
- stop & go.

90.000 points have been acquired (figs.6-10).

The 2001 campaign of DGPS planned the following objectives:
- General survey for correcting previous non differential GPS acquisitions and for creating a new and affordable cartographic basis (static method).
- Creation of a micro-topographic DEM of part of the top of the Bieta Gyorgis hill (kinematic method, for a

planimetric area of 107 hectares, fig.6).

- Creation of a micro-DEM of the archaeological site of Tukul Emeni, excavated from the American team of the Boston University. This site was located on a small hill and, in 3 hours of work, has been surveyed and reconstructed in 3D interpolating the elevation points (fig.22).
- Definition of the training regions for multispectral classification of the SPOT XS and Landsat TM satellite images (stop & go method, fig.17).

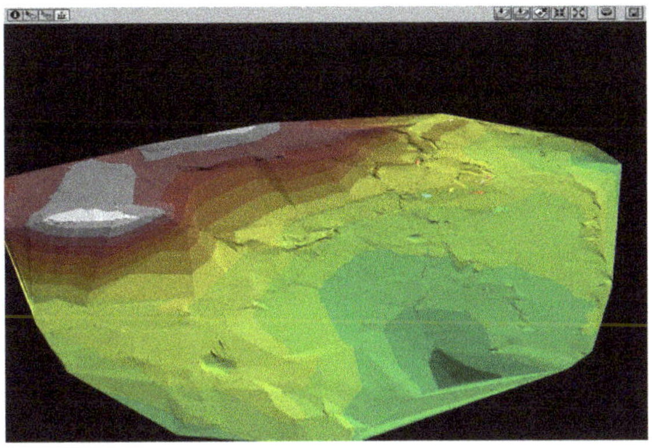

Figure 10. 3D DEM visualization of DGPS points interpolated by TIN method.

In the construction of a DEM by DGPS data, it is amazing the faculty to acquire a big amount of topographic data in a short period (about 107 hectares surveyed in 9-10 days of work with an elevation range of 2266-2345 over the sea level). Then, the interpolation of the data by grid and by TIN has given important details concerning agricultural boundaries and artificial terraces (figs.8, 10); in many cases these structures insist on ancient pre-existences). In this way it is possible to compare the DEM created by cartographic 1:25.000 contour lines with the DEM created by DGPS (fig.20): in the first case the surface of the hill seems flat, while the second detailed DEM demonstrates that the hill is very sloped and possible relations between sites and landscape morphology can be investigated (figs.8, 10).

Use of DGPS in this context has substantially changed the perspectives of archaeological survey *in situ*. In fact the repositioning of all the sites and areas of interest has allowed the correction of the previous surveys detecting a mean error of localisation between 100 and 200 m (towards South). Then, the integration of the centimetric DGPS survey with the total laser station, has completely recreated the archaeological cartography.

Finally, detailed identification of the training regions, because of the centimetric precision of the DGPS positioning, has produced a very interesting supervised spectral classification (fig.21). For example, for clay and calcareous soils, the classification is very detailed: in fact the pixels with the same spectral features are isolated and circumscribed in very small areas (1 pixel = 20 m) without spectral redundancy (fig.17). Using higher resolution images (like Ikonos), the result would be even more detailed. In the above reported table the American team of geologists has suggested a preliminary identification of the soils of Bieta Giyorgis on the basis of the FAO classification (tab.1, fig.21).

Characterisitics	Soil types			
	Bakahel	Macaeo	Ogunma	Walka
(FAO classification)	Leptosol	Cambisol (ferralic)	Cambisol	Vertisol
Colour	Light brown	Red	Brown	Black
Depth	Very shallow	Shallow	Shallow	Variable
Organic matter content	Very low	Low	Low	High
Texture	Loam	Sandy clay loam	Clay loam	Clay
Cracking	Slight	Slight	Slight	Severe
Drainage	Good	Good	Moderate	Poor
Moisture retention	Low	Low	Moderate	High

Tab.1

5. Multispectral classifications

Spectral components of the satellite images allow a detailed classification of the landscape according to the spatial resolution and the digital features of the data (fig.23). Classification is a statistical process which embraces homogeneous pixels into areas of interest based upon a concept referred to a spectral pattern recognition. Several classification algorithms or classifiers are known in image processing, such as: parallelepiped, minimum distance to means, and maximum likelihood. This last classifier is the most affordable and used and it has given good results also in the case of Aksum.

The two principal types of classification are supervised (human assisted), or unsupervised (clustering) techniques. In general, a typical sequence of classification involves:

- image display,
- image interpretation,
- ground truth,
- training regions,
- unsupervised classification,
- supervised classification,
- accuracy assessment.

The unsupervised classification does not require training sites and it identifies the multispectral data into a number of classes based on the same intrinsic similarity within each class (fig.14). So, values within a given cover type should be close together in spectral space, as opposed to data in different classes being comparatively well separated.

On the other hand, supervised classification uses existing region or class statistics to label each cell in an image as belonging to a particular class. Statistics of the training regions can be drawn by hand, converted from a vector image, or by GPS.

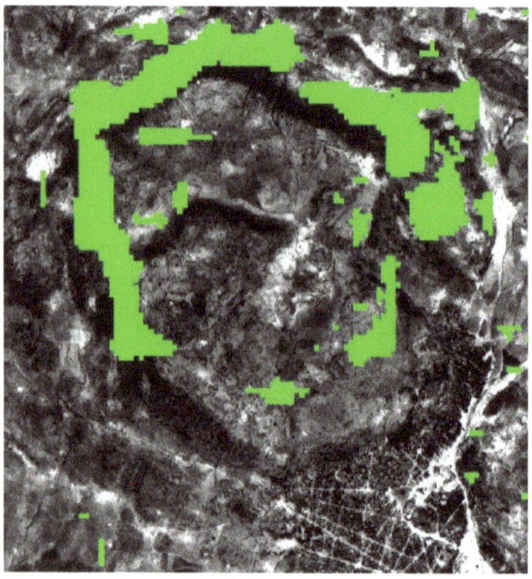

Figure 11. Vegetation index calculated by SPOT XS and superimposed on aerial photo (1964).

Figure 12. Map of the archaeological sites surveyed in the project from 1993 (on SPOT XS).

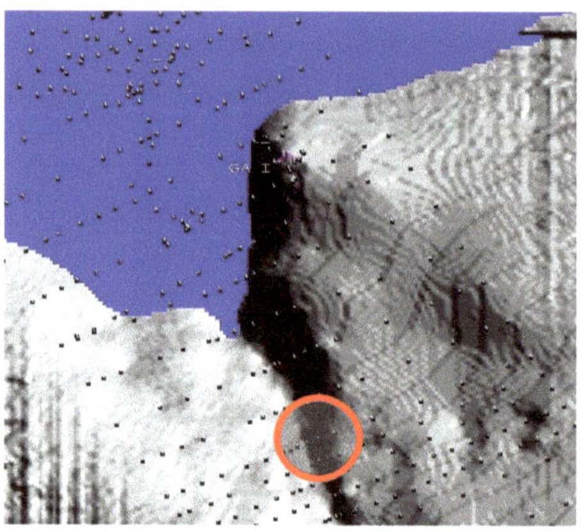

Figure 13. Simulation of a flood in the area of the site of Tukul Emeni shown by the arrow.

6. Remote sensing

Because of the lack of detailed cartography and for integrating panchromatic data, the need of satellite remote sensed data was fundamental, in particular in order to classify the landscape by spectral signatures and color components. For Aksum we have used two satellite images: a Landsat Thematic Mapper TM image of 11/22/1984 and a SPOT XS of 3/23/1993 (fig.23); in the near future an Ikonos image will be used thanks to the collaboration with NASA. The first step of the processing was the creation of a multispectral and multilayered georeferenced photomosaic including Landsat TM, SPOT XS and the aerial panchromatic photos (fig.23).

A preliminary satellite processing was made by Magaly Koch and Thomas Schmid by selecting different landcover types (soil, rock and vegetation) in the region of Bieta Giyorgis (Fattovich R., K.A. Bard, L. Petrassi & V. Pisano 2000). 15 land-cover types have been classified, and four different classifiers were tested, *Mahalanobis Distance, Minimum Distance, Maximum Likelihood, and Paralellepiped* classifier. In particular, *colluvium* and red soil classes were mainly detected by Mahalanobis and Minimum Distance classifiers. On the contrary, in the central plane of Bieta Giyorgis were identified light soils (high reflectance values) with spectral signature close to pyroclastic material (trachyte; light blue class) found in neighbouring hills (e.g., Mai Qoho, figs.12, 14, 23). Even if these interpretations are very preliminary, in the TM image it is possible to see three types of plutonic domes:

1) not breached domes where erosion has not yet uncovered the igneous rocks, but the shape of the dome is clearly visible;
2) breached domes (Mai Qoho) where erosion has started to disclose pyroclastic rocks;
3) denuded domes (Bieta Giyorgis) where the core of the dome is being exposed by erosion (see Bard et alii in this volume).

In short, remote sensed raster bases of Aksum are constituted of:

- Landsat TM 7 bands (30 mt of resolution);
- SPOT XS, 3 bands (20 mt of resolution);
- Aerial photos (1951, 1964, 3 mt of resolution);
- DEMs (by DGPS and by contour lines).
- All the remote sensing classifications have been involved after the aerial georeferencing and photo-interpretation in order to map the main features of the archaeological landscape.

The processing of the SPOT XS (20 m of resolution) has involved the combination of the three RGB bands in the sequence 1, 2, 3 (figs.3, 4, 12); in this combination red pixels show very well the extension of the eucalyptus forest. The unsupervised classification (20 classes, fig.14) in pseudo colours was very useful for identifying preliminarily the main classes of soils, rocks and vegetation on the top of Bieta Giyorgis. In fact it is possible to see in dark blue the shadow determined by the sunlight (therefore corresponding to not classifiable regions), mainly in the western part of the hill, then, in light green the eucalypts, in orange, light blue, fuchsia, and other blends, further classes of soils, rocks and vegetation (fig.14). For classifying in detail the classes of vegetation, a vegetation index was calculated: fig.11 shows the superimposition of the vegetation index over an aerial photo of 1964. It is evident that a wide extension of the eucalypts in the last ten years hide part of the archaeological landscape visible, otherwise, in the sixties.

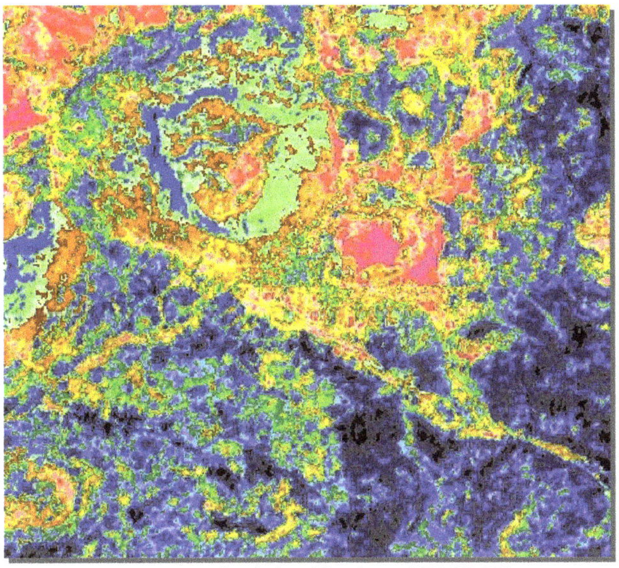

Figure 14. Unsupervised classification of SPOT XS.

For solving the above mentioned ambiguities of spectral classes, a supervised classification is needed, starting from known training regions. For this aim in the 2001's campaign, the most relevant training regions were surveyed on the ground using a differential GPS (DGPS, figs.17, 21); namely for each area of interest a perimeter has been drawn on terrain by DGPS, so that to obtain a vector polygon. For all the training regions a picture and a sample of terrain were taken, in order to compare the results of the supervised classifications with geo-archaeological analyses. On the basis of these training regions concerning the main soils and relating sites of Bieta G., the following classes were identified using Maximum Likelihood Enhanced and Maximum Likelihood Standard supervised classifications (for each colour there is a sequence of sites/structures, fig.21):

- **green**: NB_SB1-3, TR 1, TR 7, TR 4, BA, GA, OAZ3, TR 3, TR 6, G(uadguad) A(gazien)
- **yellow**: TR 3, GN, TR 1, TR 5, TR 4, TR 7, O(na) E(nda) A(boi) Z(ewge) X-XII.
- **blu**: OAZ9, OAZVT2, OAZ2
- **red**: OAZVT2, OAZ1, OAZ2
- **dark red**: OAZ7, TR 10, TR 7, TR 8, TR 9, ONA NAGAST AREA
- **light blu**: TR 3
- **orange**: TR 3, T(ukul) E(meni) (figs.14, 17, 20)

Supervised and unsupervised classifications have allowed to investigate relevant relations between soils and sites, distinguishing specifically 6 areas of soils and 5 significant clusters of archaeological sites. In figure 21 the soil's supervised classification by DGPS's training regions identify areas with green pixels specifically in the North-North-east part of the hill, with yellow pixels in the West-North-West, with red, dark red pixels in the middle and in the South, with light blue pixels in the middle towards North; finally an other class of orange pixels is statistically very small but scattered all over the hill. The correspondence of these classifications with preliminary geological interpretation of the soils (table 1) in walka, bacahel, macaeo, ogunma (in Ethiopian language) opens interesting perspectives of further investigations on the relations between soils, land use and presence of sites and settlements. In particular, on the basis of these classifications and remote sensed signatures, we have circumscribed settlement location in four main areas (in figure 21 defined by the circles): 3 in the North and 1 in the centre-South. This supervised classification demonstrates very well that different spectral signatures (measured on the ground by training regions) are corresponding to different gemorphological features and compositions of the soils (geological analyses of the terrain will give more detailed information about this problem); therefore, for each main class of soils we have a particular concentration of sites/structures/settlements (figs.20-21).

Finally, additional very important results were reached integrating remote sensing and archaeological fieldwork: an ancient road crossing the hill from the East is visible from the SPOT XS; this road passes very close to a monolith-stele located in a plain in the middle of the hill (fig.21). The archaeological interpretation of this anepigraph monolith-stele is relevant, because it is not linked with a tomb, as in all these Aksumite typologies, but it stands alone, with the very probable function of boundary (in fig.21 the arrow shows the stele and the dot line the road). It is interesting that more than one path crosses the area of Tukul Emeni, just where the monolith-boundary was erected in late Pre-Aksumite times. Moreover, rock-inscriptions, most likely used as property landmarks, were located at the base of the southern side of the hill.

According to my hypothesis (see fig.21), comparing these archaeological data with remote sensing applications, we could define two principal areas or districts of settlement in pre-Aksumite and Aksumite period, separated by a stele-boundary: area 1 in the North and area 2 in the South (these two areas are separated by the stele and the road). On the basis of old surveys and recent excavations we identify preliminarily a more destined "sacred-symbolic" landscape in the North (area 1) and a peopled/industrial landscape of settlements in the South (area 2). An open-air site, dating to late Pre-Aksumite or initial Proto-Aksumite times, has been discovered at Gunda Nebri, on the northeastern side of the hill. An other probable ceremonial site, Tukul Emeni, was excavated by Mike Di Blasi in 2001 and it is located in the North-East part of the hill, close to the stele's area (fig.22). Structures of a possible monumental temple, dating to Proto-Aksumite and Early Aksumite times, were otherwise found at ON in the southern part of the hill. More detailed results coming from the next investigations will can confirm these hypotheses.

7. Data fusion

Data fusion is the combination of multiple types of data into a single view. For example, sharpening a Landsat TM image with a SPOT Panchromatic image is a data fusion, or combining panchromatic bands of aerial photos with spectral components of satellite images such as the case of SPOT XS. The Brovey transform is a method to fuse different data together using one image; for example Landsat TM for spectral or colour information, and the other image for spatial or sharpness (in our case the SPOT XS).

In the case of Aksum we have used three RGB bands of the SPOT XS (B1, 0.545_um; B2, 0.645_um; B3, 0.84_um) and the panchromatic bands of the aerial photos of the sixties, so that to have a good resolution of the aerial images with the spectral colour component of the satellite imagery (SPOT and Landsat). Even if this fusion combines data belonging to different periods, the result is interesting: for example it is remarkable the difference of the vegetation cover (eucalypts) between the older aerial photos and (see also the vegetation index in fig.11) the newer satellite SPOT. In fact in the sixties no eucalyptus was present and the archaeological visibility on the terrain was excellent; on the contrary the satellite imagery in transparency shows very well that the red spectral component referring to the eucalypts covers a wide extension of the hill (we have calculated almost 50% of the hill). In fact this kind of tree is not endemic and was imported from the Europeans in the eighties for populating the area of Aksum. Unfortunately, the strong increment of the density of the eucalyptus forests over the hill has hided part of the ancient settlements (or possible archaeological sites) well visible in sixties, like the data fusion shows (figs.11, 18).

An other benefit of the data fusion is the property to zoom the satellite imagery in order to see details present in the aerial photos, comparing several layers of spatial information (even if these information concern different years and seasons).

Several hundred features have been identified in the aerial photos of the area, and two main categories of features were distinguished: a) linear features (including quadrangular or polygonal structures), and b) circular (fig.5).

The remains of two possible dams, associated with three walls, were found along the Guadguad Agazyen (fig.19) on the northern edge of the hill. An artificial terrace wall was also built along the north-eastern side of the basin. Also in this case we have reconstructed the micro-landscape with DGPS: the draft 3D visualizations show very well this particular water system, probably aimed for the land use management. These structures can be dated between the mid-1st millennium BC and mid-1st millennium AD.

8. Virtual and 3D visualization

As I have written in the paper dedicated to the mindscape in this volume, I follow an ecological or ecosystemic approach in the interpretation of the landscape (see Forte, "Mindscape: ecological thinking and virtual archaeological landscapes"). Because all the relations within the environment are autopoietic, the creation of an artificial world of simulation can constitute the basis for observing and investigating all these interrelations in a perceptive and interactive way. For this aim we are working for constructing a specific VR software for spatial georeferenced data and for the inclusive interaction in a virtual archaeological landscape. Even in the remote sensing, VR properties of visualization can increase quality and level of the interpretation's process thanks to the inclusive and stereoscopic way of interaction in real time. In this project we have tested two solutions: the Virtual Theatre at the CINECA (Supercomputing Centre, Bologna, fig.15) and the Visionstation (thanks to an agreement with the company Elumens, fig.16). The Virtual Theatre consists of a wide stereo hemispheric screen and a 3D surrounding sound: in this environment the experience of visualization is collective, even if there is only one inter-actor. The Visionstation is an hemispheric display system (fig.16) fit for a semi-immersive visualisation connecting the display with a PC or a workstation. The visualisation is made by combining software API and lens, creating spherical images and projecting pixels equally through the hemispherical screen surface.

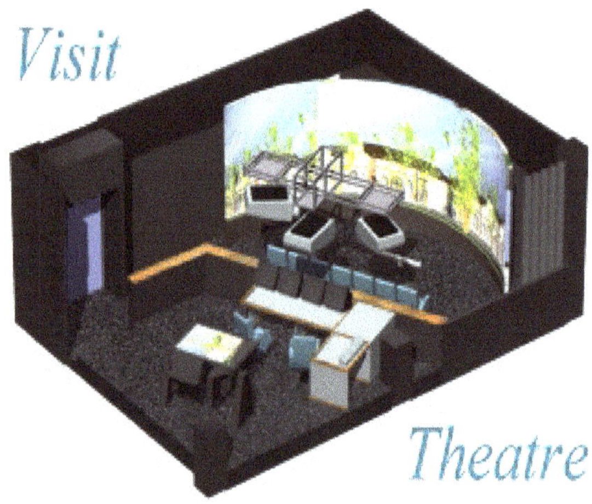

Figure 15. Virtual Theatre at CINECA (Supercomputing Centre, Bologna)

Figure 16. Visionstation by Elumens

Figure 17. Training region by DGPS: on the left the supervised classification of TR 5, on the right a picture of the sampled terrain

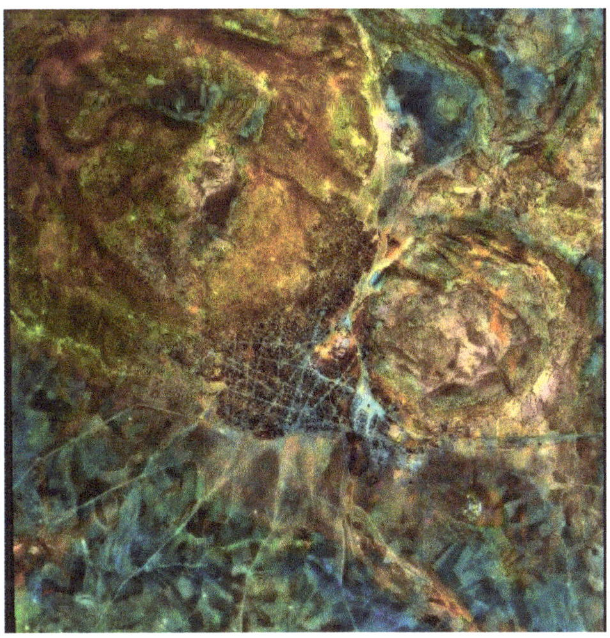

Figure 18. Data fusion between SPOT XS (bands 1, 2, 3, 20 m of resolution) and a panchromatic aerial photo (3 m of resolution)

In both the environments the focus of interaction is the capacity to perceive the virtual in an holistic and inclusive view (Forte 2000, 2002). All the details are visualized dynamically in 3D with a significant involvement of the users: in this way it possible to perceive and to acquire information and data otherwise not available. This approach is very useful in all the three phases of the research: observing the landscape, classifying the landscape, reconstructing the landscape according to the eco-interrelations. In particular, virtual reality systems implemented and tested before of the fieldwork have allowed to simulate a walkthrough in the landscape before of knowing the territory and with the possibility to plan the activity of sites' investigations. Then, in particular, the real time navigation through the 3D landscape mapped with aerial photos, has given important results in terms of verification of archaeological features.

Finally, the virtual landscape reconstructed can always be compared with all the observed landscapes of the different spatial sources of information: a demonstration of a cybercultural ecological reconstruction.

9. Conclusions

This kind of project is always in progress because it is an amazing case study for testing new and revolutionary methodological approaches. In particular, it is really difficult to understand deeply potentialities of VR systems in the spatial realm: we investigate universes of information and we try to classify it, but it is difficult to reach a satisfying interpretation, mainly in diachronic terms. The great challenge is to investigate the ancient landscape within a digital environment of interrelations and according an holistic approach (without context there is no communication, Bateson, 1979; Forte, 2002).

With spectral data fusion high resolution photos were integrated with multispectral component, so that a "virtual" multispectral images of 3 m have been obtained for the area of Bieta Giyorgis.

The vegetation index calculated by satellite imagery and superimposed to aerial photos has shown the extension of eucalyptus's forests in comparison with the diachronic visibility of the archaeological landscape from the sixties.

Remote sensing multispectral classifications have remarked the link between soils and archaeological sites, sampling the spectral signature of the main soils and land uses of the top of Bieta Gyorgis.

In this context the use of training regions by DGPS gives important results for a preliminary classification of soils, semi-buried structures and anthropogenic terrains. Possible connections between soils, terrains and sites can show or negate:

- relations between ancient and modern land uses;
- relations between anthropogenic/organic terrains with buried or semi-buried archaeological structures (organic components and artificial struc

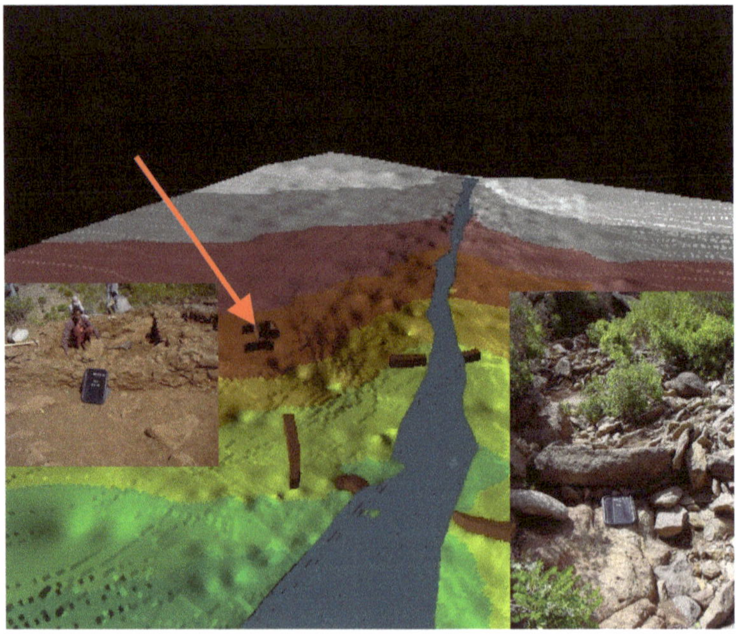

Figure 19. Reconstruction of an ancient dam using DGPS on an archaeological excavation (Guadguad Agazyen, in the area 1 of Bieta Giyorgis).

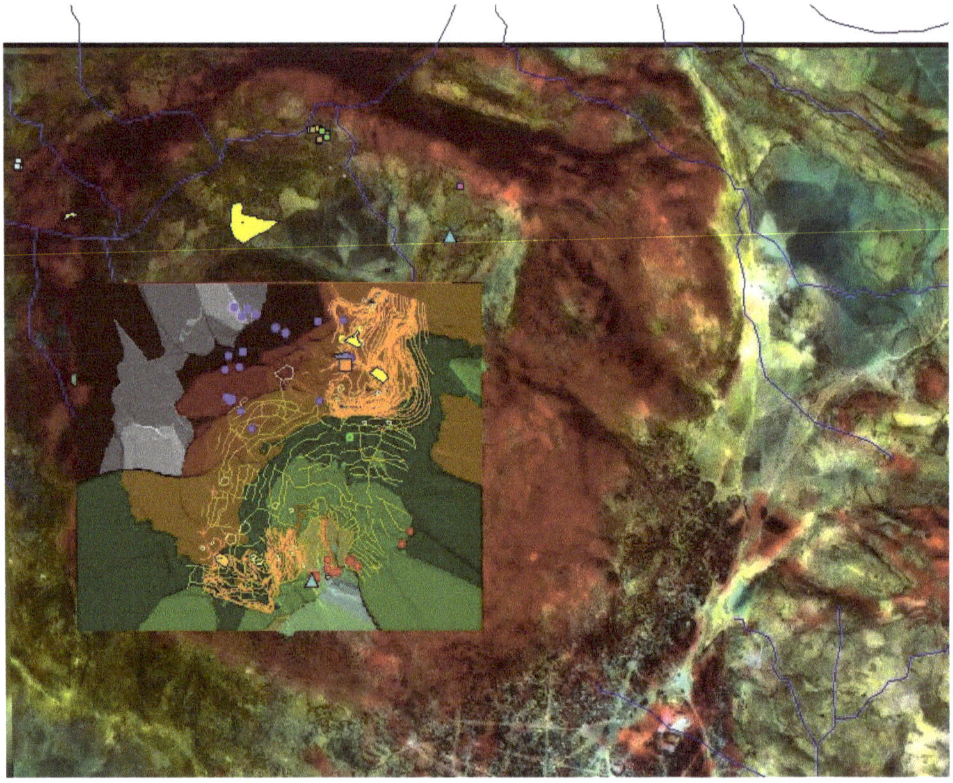

Figure 20. Data fusion of SPOT XS and aerial photomosaic; in the South part of Bieta Giyorgis overlay of the DEM (grid interpolation), of the archaeological sites and of the agricultural terraces.

tures can modify the soil reflectance);

- relations between settlement ancient topography, destination (i.e. religious, sacred, civil, according to the power, etc.) and soils;
- relations between archaeological sites and current vegetation;
- relations between sites and terrain elevation;
- relations between sites and slope, aspect, sunlight.

Starting from this methodological discussion, one of the most interesting challenges of the future is the possibility to investigate how the buried archaeological structures can change the reflectance of the soils and, in this way, can suggest predictive archaeological maps. For example the compared analyses of geophysical and satellite remote sensed data could give significant answers.

Moreover, the use of DGPS has changed the perspectives of research or in spectral analyses whether in the geoarchaeological mapping. Principal results can be summarized in:

- definition of the spectral training regions from the ground.
- Creation of a micro-DEM of the southern part of Bieta Giyorgis of centimetric precision.
- Creation of the micro-DEM of the site of Tukul Emeni during the archaeological excavations.
- Micro-topographical surveys integrating the work of the DGPS with the theodolite's positioning (correcting also all the previous surveys).

Having a so detailed DEM it has been possible to simulate floods using our models. As we can see in fig.13, concerning the archaeological ceremonial site of Tukul Emeni, a flood (in dark blue) can't touch the site (in the circle) because it was constructed in an elevated position. This is the most frequent situation identifiable over the hill of Bieta Giyorgis for this type of sites.

The quality of the DGPS's DEM is visible in figs.6, 8, 10: in fact, in the cartographic DEM, the hill seems quite flat, on the contrary of the reality.

DGPS topographic survey has covered an area of 107 hectares with 90.000 points of acquisition (x, y, z) in about 10 days of work. DGPS survey of training regions has identified 15 areas of archaeological and geological interest; final results will be published with pedological and geological analyses but these preliminary remote sensed classifications are very encouraging. In fact, possible relations between sites and other components of the landscape can be determined by: typologies and land uses, geological features, erosive phenomena, elevation, geomorphology, streams and water resources, site catchments, crop marks, sunlight. In the future of the project we will attempt to combine all these factors in order to identify special clusters of data.

In any case, it is important to underline that supervised and unsupervised classifications have shown significant relations between soils and sites, distinguishing specifically 6 areas of soils and 5 significant clusters of archaeological sites (figs.17, 21). The correspondence of

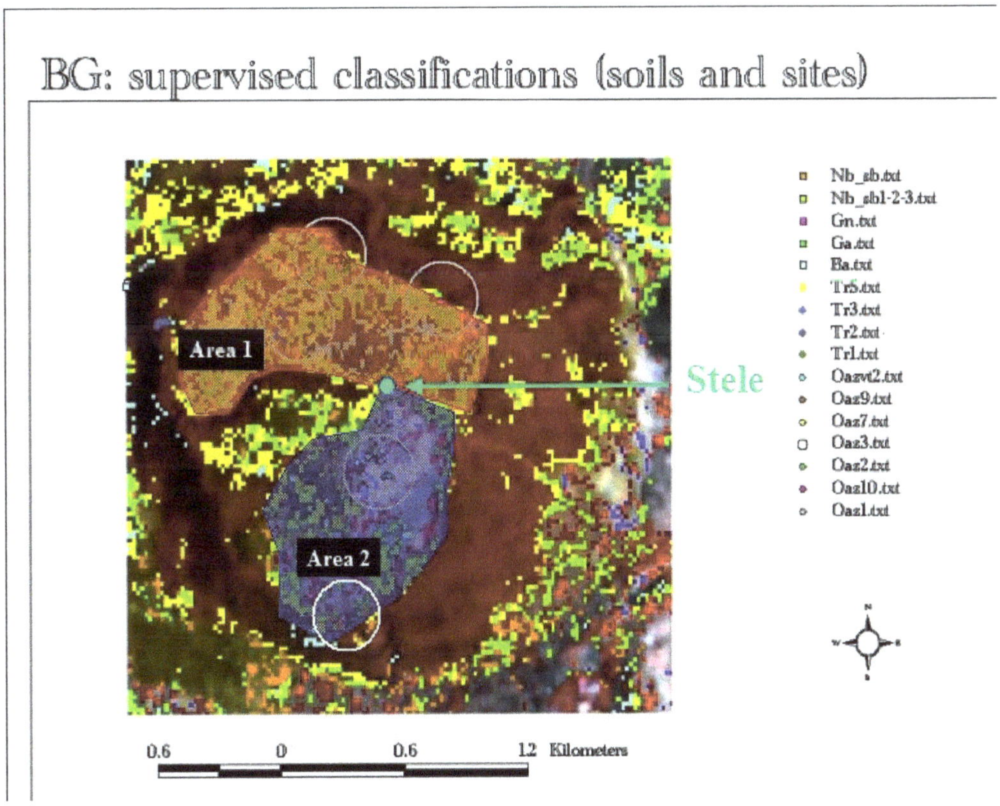

Figure 21. Supervised classification of SPOT XS with 6 sampled regions. The arrow shows the position of the stele-boundary and the dot line shows the path of the ancient road. Two main areas (1 in the North and 2 in the South) are corresponding to different land use and destination of the landscape.

Figure 22. Archaeological site of Tukul Emeni (shown by the arrow): micro-DEM created by DGPS in 3 hours of work. and in the small picture on the left a view of the site during the fieldwork.

these classifications with a preliminary geological interpretation of the soils is indicated in the table 1: walka, bacahel, macaeo, ogunma (in Ethiopian language). This classification circumscribes the archaeological sites in four main areas: 3 in the North and 1 in the centre-South.

Supervised classification demonstrates that different spectral signatures (measured on the ground by training regions) are corresponding to the above mentioned gemorphological features and composition of the soils: so for each class of soil there is a particular concentration of sites/structures/settlements (fig.21).

Finally, a remarkable archaeological interpretation derives from these remote sensing applications: an ancient path (dating from Pre-Aksumite period) crossing the hill from the East is visible from the SPOT XS; it passes very close to a stone monolith located just in the middle of the hill (fig.21). Our archaeological thesis is that this monolith-stele, because it is not linked with a tomb, is interpretable as boundary-landmark. Therefore in fig.21 two principal archaeological regions or districts are distinguished according to anthropological and remote sensing interpretation: area 1 in the North and area 2 in the South.

In conclusion, all the artificial landscape features at Bieta Giyorgis (agricultural terraces and fields, cisterns, dams, roads, paths, etc.), investigated in the fieldwork and by remote sensing, represent a remarkable social connective space, index of a relevant and organized power landscape.

Figure 23

Acknowledgments

In the Aksum project many teams and researchers have collaborated and are still collaborating. A special thank is due to: R. Fattovich, S. Kay, A. Manzo, C. Perlingieri, R. Perlingieri, B. Trabassi and, as Institutions, to CNR and CINECA. For the software and hardware support we thank the companies Terrex, Tesscom and Elumens.

References

BARD K.A., R. FATTOVICH, A. MANZO & C. PERLINGIERI 1997. *Archaeological Investigations at Bieta Giyorgis (Aksum), Ethiopia*: 1993, 1994 and 1995 Field Seasons, Journal of Field Archaeology, 24: 387-403.

BARD. K.A., M. COLTORTI, M. DIBLASI, F. DRAMIS & R. FATTOVICH 2000, *The Environmental History of Tigray (Northern Ethiopia) during the Holocene: A Preliminary Outline*, The African Archaeological Review, 17 (2): 65-86.

BATESON G., 1979, *Mind and Nature. A Necessary Unit*, Dutton, New York.

FATTOVICH R. 1997, *Archaeology and Historical Dynamics: the Case of Bieta Giyorgis (Aksum), Ethiopia.* Annali dell'Istituto Universitario Orientale di Napoli, 57 (1-2): 48-79.

FATTOVICH R. & K.A. BARD 1993, *Scavi archeologici nella zona di Aksum. C. Ona Enda Aboi Zaguè (Bieta Giyorgis),* Rassegna di Studi Etiopici, 35: 41-71.

FATTOVICH R., K.A. BARD, L. PETRASSI & V. PISANO 2000, *The Aksum Archaeological Area: a Preliminary Assessnent*, Napoli: Istituto Universitario Orientale.

FORTE M., 2000, *About Virtual Archaeology: Disorders, Cognitive Interactions and Virtuality*, J. Barceló, M. Forte, and D.H. Sanders, Virtual Reality in Archaeology, BAR International Series, 843, Oxford, 247-259.

FORTE M., 2002, *Archaeology and territorial "Musealisation": the virtual-cognitive reconstruction of archaeological landscapes*, in Virtual Archaeology, Proceedings of the VAST Euroconference, Arezzo, 24-25 November 2000 (ed.by F. Niccolucci), BAR International Series, Oxford, 2002, 29-39.

FORTE M., K.A. BARD, R. FATTOVICH, M. FOCCILLO, A. MANZO & C. PERLINGIERI 2001, *The Aksum Project (Ethiopia): GIS, Remote Sensing Applications and Virtual Reality*, in Z. Stancic e T. Veljanovski (eds.), Computing Archaeology for Understanding the Past - CAA 2000: 241-251, Oxford: British Archaeological Reports.

LITTMANN E., D. KRENCKER & T. VON LÜPKE 1913, *Deutsche-Aksum Expedition*, I-IV, Berlin: Druck und Verlag von Georg Reimer.

PHILLIPSON D.W. 1998, *Ancient Ethiopia. Aksum: Its Antecedents and Successors*, London: British Museum Press.

Mindscape: ecological thinking, cyber-anthropology and virtual archaeological landscapes

Maurizio Forte

CNR-ITABC, Istituto per le Tecnologie Applicate ai Beni Culturali,
Via Salaria, km.29,300, C.P.10, 00016, Monterotondo St., Roma, Italy,
e-mail: maurizio.forte@mlib.cnr.it

Living systems are units of interactions; they exist in an ambience[...]. A cognitive system is a system whose organization defines a domain of interactions in which it can act with relevance to the maintenance of itself, and the process of cognition is the actual (inductive) acting or behaving in this domain. Living systems are cognitive systems, and living as a process is a process of cognition. This statement is valid for all organisms, with and without a nervous system [...].
Humberto R. Maturana , Autopoiesis and Cognition, 1980

Information visualization focuses on information, which is often abstract. In many cases information is not automatically mapped to the physical world (e.g.geographical space). This fundamental difference means that many interesting classes of information have no natural and obvious physical representation. A key research problem is to discover new visual metaphors for representing information and to understand what analytical tasks they support
Gershon, N., Eick, S.G., and Card, S., 1998, Information Visualisation. "Interactions": March/April, 1998, 9-15.

Abstract
Knowledge and diachronic interpretation of an archaeological landscape depends on factors of perception, self-referring, interaction ("feedback") and cultural learning. The paper describes a new approach for the reconstruction of archaeological landscapes using virtual reality technologies within an eco-environment of simulation. The philosophical and epistemological approach derives from the ecological school of Bateson, Maturana, Varela and Ingold: the interpretation of the landscape depends on its relations, without context there is no communication. Therefore for interpreting the landscape we have to create the "map" in batesonian sense (*The map is not the territory*), the map is the virtuality, the "difference". the territory is the reality.
In theoretical sense we distinguish three levels of landscape to investigate and to interpret: mapscape (virtual landscape by GIS and spatial data), taskscape (activities and relations of the landscape), mindscape (perceived landscape, digital ecosystem).
In conclusion, we will face up methodological and technological problems of the virtual reconstruction using OpenGl software and libraries in order to project, to reconstruct and to navigate the ancient landscape in real time. The new rules of this digital ecosystem will be autopoietic, in the sense that they follow ecological and cyber-anthropological theories. Each interaction in real time within the virtual landscape will produce difference and, through the difference, new ways of learning.

1 Introduction

"Landscape ecology is a branch of modern ecology that deals with the relationships between man and his open and built-up landscapes" (Naveh, Lieberman, 1990). In the last 30-40 years "environment", "space", "place", "landscape" and "ancient landscape" were central in the international debate in epistemological, philosophical, psychological, technological and morphological terms (Gibson, 1950, 1979; Forman, Godron, 1986; Zubrow, 1994; Forte, 1995; Allen, Green, Zubrow, 1990; Basso, 1996; Gaffney, Ostir, Podobnikar, Stancic, 1996; Ashmore, Knapp, 1999; Forte, 2000; Gillings, Wheatley, 2000; Lock, 2000; Chapman, 2000; Dodge, Kitchin, 2001).

My cybernetic manifesto corresponds to the ecological school's thought of Gregory Bateson: we are our relations and there is no distinction between mind and body; this approach is very important in the study and in the digital reconstruction of the landscape. A recent book, very interesting, of Tim Ingold, "The perception of the Environment" (Essays in livelihood, dwelling and skill, 2000) represents a good base of discussion of the arguments treated in this paper. In fact our thesis is that the reconstruction of an archaeological landscape passes

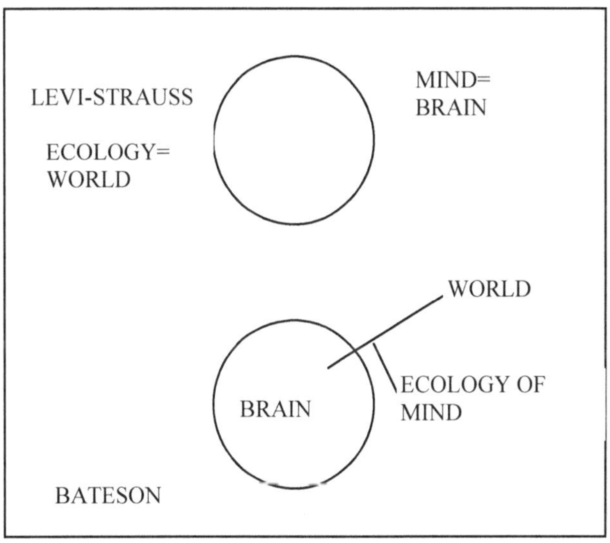

Comparison of Lèvi Strass's and Bateson's views on mind and ecology (Ingold, 2000, 18)

through a re-proposed perception of the ecosystem digitally reconstructed in a virtual reality's inter-related system. Without a new context, we can't have new differences, therefore no new information. Each cognitive process or act imply a learning through the difference (a preliminary ontological approach is discussed in Heiddeger, 1927) and, for increasing the difference, we have to multiply the behaviours in a "spatialised" digital ecosystem. In order to obtain this result a project of creation of a virtual system has to integrate ecological thought, spatial data and an adequate technological development. The challenge is to transform a virtual landscape in an open space of digital self-organised behaviours (active and passive) and implemented by the user/navigator. In this sense we can distinguish: a) the behaviour of the navigator in the virtual realm (interaction towards the environment); b) the behaviours between digital objects or elements of the system (interactions within the artificial life); c) the behaviour of the navigator with other agents (interaction towards the artificial life). Finally, in suborder, we could classify a visual action (fly, walk) and a "through-action", in the sense of inter-action with other virtual agents.

2 Cyber-anthropology and landscape

In the above mentioned context, cyber-anthropology embraces a cybernetic base applied to anthropological contexts according the ecological thought and the study of digital ecosystems. In fact the digital space of VR, in terms of inter-relations, connections, interactions, artificial life, artificial societies, is a cyber-ecosystem and the rules for interpreting must have a systemic-ecological base. Understanding these rules of learning means to interpret the environment: a cyber-anthropological dream in the realm of the neural network is to put an "intelligent" avatar in the virtual world and to ask: "can you describe what you are seeing ?". If the avatar could describe its environment or digital ecosystem, we would start to perceive this context from the inside, starting a new way of knowledge. In short, according on these premises we could enlarge this discussion in a wider field of anthropology of the cyberculture (Escobar, 2001, Abu-Lughon L., 1990; Lévy, 1997).

Cyberculture concerns specifically new technologies: artificial intelligence, multimedia, virtual reality and biotechnologies (Escobar 2001, 103); the relations between "bio" and digital ecosystems are relevant because they underline a different approach to the knowledge of new worlds and systems, part biological, part virtual.

Appadurai (1991) affirms the there are new differences produced by new technologies in the "global ethnoscapes". Escobar (2001, 107-108) writes that "bodies", "organisms" and "communities" have to be re-theorised because they are composed by elements born in three different domains: the organic, the technical and the textual (or cultural). Therefore cyberculture defines the ensemble of techniques, thoughts, attitudes, practices, developed with the increase of cyberspace (Lévy, 1997).
The construction of a digital landscape is a process of cyber-culture, cyber-anthropology, because it embraces organic, physical bases with technological issues. Cyber-anthropology studies new ways of learning and interactions in the cyberspace where we construct worlds, new experiences in the discover of the science of complexity. The science of complexity shows a pluralistic vision of the physical world: networks instead of structures, connections instead of distinct borders isolating immaculate systems, differences instead of homogeneity (Deleuze, 1967; Derrida, 1967 Deleuze, Guattari, 1987; Escobar, 2001.).

In the book of Ingold a key concept is the comparison between Lévi-Strauss's position and Bateson's position: "the mental world – the mind – the world of information processing – is not limited by the skin (Bateson, 1972, 429)". "Yet the ecosystem, taken in its totality, was nevertheless envisaged as two-faced. One face presents a field of matter and energy, the other presents a field of pattern and information; the first is all substance without form, the second is all form detached from substance [...]. Corresponding to this duality Bateson recognised two ecologies: an ecology of material and energy exchanges, and an ecology of ideas. And it was this second ecology that he christened the ecology of mind." (Ingold, 2000, 16).

3 Autopoiesis and systems

Features of an organism are always related to the links between the interaction of all the parts and from retroactions connecting the organism with the environment in which it lives. All the physical, chemical, geological factors are connected in a complex system. Therefore we have to think mind and nature as a unit (Bateson, 1972, 1979), there is no distinction between mind and body.

In this context the research of Maturana and Varela, an "autopoiesis" of living and cognitive systems (Maturana, Varela, 1980, 1987), emphasizes the fundamental equivalence between self-organised and cognitive capabilities, or for the organisms, whether for natural systems of superior order, or for the living organisms in the relationships with the environment as ecological systems. On this basis, according the ecological thinking of Bateson, Maturana, Varela and Ingold, human being, society and environment must be considered not like separated layers, but as a unique plot of relations. In "Autopoiesis and cognition" Maturana and Varela (1980) remark some of these issues, so in the following points we can summarize the key points useful even for the reconstruction of the landscape.

'How does the living organization give rise to cognition in general and to self-cognition in particular?'

- Organisms are adapted to their environments, and it has appeared adequate to say of them that their organization represents the 'environment' in which they live, and that through evolution they have accumulated information about it, coded in their nervous systems. Similarly it has been said that the sense organs gather information about the 'environment', and through learning this information is coded in the nervous system Living systems are units of interactions; they exist in an ambience.

- A cognitive system is a system whose organization defines a domain of interactions in which it can act with relevance to the maintenance of itself, and the process of cognition is the actual (inductive) acting or behaving in this domain. *Living systems are cognitive systems, and living as a process is a process of cognition.* This statement is valid for all organisms, with and without a nervous system (Maturana, Varela, 1980).

The approach of Maturana help us to comprehend that, considering the landscape a system, a living system, the reconstruction of a digital environment must keep, or try to keep, the same rules.

Theory of autopoiesis is fundamental for understanding feedback of behaviours and relations within an ecosystem: self-organised systems can develop different models starting from the same initial conditions. After some threshold these systems can have bifurcations towards multiple states or solutions; a given solution is random and it can't be planned. Whatever further evolution of the system will depend on the choice made in the bifurcation. These crossroads mark the way out towards the complexity: they represent innovation and diversification because they involve new solutions for the change. According Maturana and Varela self-organizing behaviours can compare some states (attractors), converting part of the energy of the system in a new ordered behaviour. So the structure is characterized by the collapse of the previous symmetry and by appearance of multiple choices. Finally, this "ontogeny" has created a difference within the system and it has produced new information and feedbacks. The same situation must be reproduced and analysed in a digital ecosystem.
In the last decade paper and publications concerning theoretical approaches about archaeological, political and social landscapes were published. Just for mentioning some key books and papers on these topics: "The ancient mind" (Renfrew, Zubrow, 1994); "Archaeologies of landscape" (Ashmore, Knapp, 1999), "A phenomenology of landscape" (Tilley, 1994); *Ethnologie et paysage*, in *Paysage au pluriel. Pour une approche ethnologique des paysa* (Lenclud, 1995) "Space and place. The perspective of experience" (Yi-Fu Tuan, 1977), "Semiotics of Landscape: Archaeology of Mind" (Nash, 1997), "Geofilosofia del paesaggio" (Bonesio, 1997), *GiS and Landscapes of Perception*, (Witcher, 1999) "Cartography and *Landscape Perception: A case study from central Italy* (Attema, in Barker, Mattingly, eds.,1999); "Social Mindscapes. An invitation to Cognitive Sociology" (Zerubavel, 1997), finally the Zubrow's paper in this volume, *The Archaeologist, the Neural Network, and the Random Pattern: Problems in Spatial and Cultural Cognition of Landscapes.*

In some of these contributions keywords such as space, place (see also Robinson, Zubrow, 1999), mind, are superimposed but the key words are perception and cognition: there are no spaces without places. Further relevant issues discussed in these publications are the frontier of GIS and spatial algorithms for the description of the archaeological landscape (Lock, 2000), in short: we need more for a holistic comprehension of the landscape.

In this paper we want to demonstrate that in order to approach the problem of the reconstruction and perception of the landscape we must start from the ecological thinking, or in epistemological terms, or in biological terms (Bateson, 1972, 1970; Maturana, Varela, 1980, 1987); then the archaeological-cultural landscape is a good metaphor for interpreting the past and the present. So the "digital phase" of virtual simulation must follow an epistemological creation of a landscape's mental model, a *mindscape.* What space does the archaeology occupy within the landscape ? Is the ancient landscape only a layer ?

Before of starting the landscape's reconstruction, it is very important to distinguish three principal fields of investigation:

- observed landscape,
- classified landscape,
- reconstructed landscape.

Observed landscape concerns the diachronic visibility of the modern landscape through remote sensing systems and analogic sources of information.

Classified landscape is corresponding to the visual data of the landscape after a digital classification (satellite, aerial, geophysical, spectral, etc.), namely a digital interpreted landscape.

Reconstructed landscape is the final processing related to the virtual reconstruction of the landscape through digital technologies: remote sensing, GIS, spatial processing and, lastly, virtual reality systems (Forte, 2000a, 2001).
Our theory is that a GIS platform is only the first step of the reconstruction but it is not enough because the interpretation of the landscape depends on the investigation of its relations, according to an ecological approach. The challenge is to apply these bases to a construction of a Virtual Reality system dedicated to the archaeological landscape.

A famous definition of landscape was written by Naveh and Lieberman (1990):

1. Ecotope (or site), smallest holistic land unit, characterized by homogeneity of the least one land attribute of the geoshpere-namely, atmosphere, vegetation, soil, rock, water, and so on, with non excessive variations in other attributes.
2. The land facet (or microchore), a combination of ecotopes, forming a pattern of spatial relationships and being strongly related to properties of at least one land attribute (mainly landform).
3. The land system (or mesochore) is a combination of

land facets that form one convenient mapping unit on reconnaissance scale.
4. The main landscape (or macrochore) is a combination of land systems in one geographical region.

"The landscape itself is a reticulate maze of criss-crossing lines of ancestral travel, with the most significant localities at its nodal points. Localities identified by particular landscape features – hills, rocks, gullies, waterholes, an so on- embody the ancestors' powers of creativity and movement in a congealed form." (Ingold, 2000, 141)

In the cyber-anthropological domain, and according to the above-mentioned premises, for the description of landscapes (and maybe in diachronic sense the landscape is always "old") I suggest to create three neologisms: mapscape, taskscape (term acquired from Ingold, 2000) and mindscape (fig.1).

4 Mapscape

Mapscape represents the act of studying cartographic relations and spatial overlay (analogic, raster and vector) in order to produce a theoretical model of the landscape starting from the contemporary viewpoint and in comparison with the ancient dwelling. Cartographic data and drawings of the modern landscape are mapscapes, a GIS overlay of raster and vector data is a mapscape; we could say that every metaphoric action for representing the "observed landscape" is a mapscape.

5 Taskscape

This neologism has been introduced by Tim Ingold (Ingold, 2000, 190-193) and remarks very well the bases of the ecological thought and, additionally, it constitutes a theoretical premise of our VR approach in terms of correspondence between tasks/activities and virtual behaviours/neural networks.

The definition of taskscape is interesting because Ingold doesn't intend the landscape in aesthetic sense (an object of contemplation), but as context of human and not human activities (taskscape). According to Ingold "the landscape as a whole must likewise be understood as the taskscape in its embodied form: a pattern of activities collapsed into an array of features". [...] "The activities that comprise the taskscape are unending, the landscape is never complete. Neither built nor unbuilt, it is perpetually under construction. [...] The landscape seems to be what we see around us, whereas the taskscape is what we hear. In other words, the taskscape exists not just as activity but as interactivity.

This thesis confirms that without an holistic approach embracing all the inter-relations of the landscape, it is impossible to reconstruct a virtual ecosystem.

6 Mindscape

The mindscape regards all the ecosystem's relations needed for constructing the "map" in ancient and modern minds; mindscape is the way with which it is possible to perceive the landscape (fig.1). We aware that only cultural and anthropological background can help us to imagine and to construct a mindscape in diachronic and territorial terms.

For example, according to Fred Myers (Myers, 1986, 49-50) the understanding of the landscape of the Pintupi of the Gibson Desert of Western Australia can be summarized in four precepts. "First, it is not a given substrate, awaiting the imprint of activities that may be conducted upon it, but is itself the congelation of past activity – on the phenomenal level, of human predecessors, but more fundamentally of ancestral beings. Secondly, it is not so much a continuous surface as a topologically ordered network of places, each marked by some physical feature, and the path connecting them. Thirdly, the landscape furnishes its human inhabitants with all the lineaments of personal and social identity, providing each with a specific point of origin and a specific destiny. And therefore, fourthly, the movement of social life is itself a movement in (not *on*) a landscape, and its fixed reference points are physically marked localities or "sites". In short the landscape is not an external background or platform of life, either as lived by the ancestors in the Dreaming or as relived by their ordinary human incarnations in the temporal domain. It is rather life's enduring monument." (Myers, 1986, 49).

Then, in the Richard Nelson's 1983 study of the Koyukon of Alaska, we read: "Reality is not the world as it is perceived directly by the senses; reality is the world as it is perceived by the mind through the medium of the senses. Thus reality in nature is not just what we see but what we have learned to see" (Nelson, 1983, 239).

According to Ingold (2000, 190-191), there is a strong distinction between land and landscape: "Land is not something you can see, any more than you can see the weight of physical objects. [...] Landscape is not land, it is not nature, and it is not space. Moreover, Gibson writes that "ambulatory vision takes place along what a "path of observation". A path is to be understood not as an infinite series of discrete points, occupied at successive instants, but as a continuous itinerary of movement. Thus the environment one sees is neither 'seen-at-this-moment' nor 'seen-from-this-point'. On the contrary, what one perceives is an environment that surrounds one, that is everywhere equally clear, that is in-the-round or solid, and that is all-of-a-piece" (1979, 195-97).

Starting from these premises, before of constructing a virtual landscape we have to create a mindscape of the past or several mindscapes of the past. Previously, this neologism has been used typically in sociological field for describing, generally, the mind's landscapes in terms of cognition/perception of the world. In our case we want to introduce this concept for showing the importance of the mental perception for describing and interpreting an archaeological landscape or, simply, a human landscape. So, the mindscape is a connective space of perceptual information, it is a self-organised world in which the human complex societies grow and develop. In order to interpret this environment an holistic and ecological approach is needed, according to the theories of Bateson (1973, 1979), Ingold (2000) Varela and Maturana (1980, 1987).

In sociology and in social anthropology, the concept of mindscape defines cognitive aspects of human learning and behaviour: "More than any of other sciences of the mind, cognitive sociology highlights the fact that what goes on inside our heads is also affected by our social environment. In probing the social dimension of our cognition, it reminds us that we think not only as individuals and as human beings, but also as members of particular communities with certain distinctive cognitive traditions that affect the way we process the world in our minds.... Focusing primarily on our cognitive commonality as human beings certainly helps cognitive scientists identify universal patterns in the way we process information, form concepts, or access our memory" (Zerubavel, 1999).

Mental and cognitive maps define an introspective space of the perceived landscape: not only a geographic-topologic horizon of the visible-visual, but horizon of significant, narrative, collective, memorised and anthropologised landscapes.

According to Ingold "Some things in the world are not represented in the mind, but some images in the mind have a counterpart in the real world. It is experience that mediates between the two worlds, providing both the raw material in the form of sensory data from which ideas are constructed and the opportunities to test them by empirical observation.

In a so complicated multilayered context of information a GIS creation is not enough; it represents the first important scientific step but we must apply more sophisticated approaches (see also Zubrow in this book, Forte, 2001; Latalova, Raczkowski, Tschan, 2000).

Our theory is that for reconstructing and trying to learn an archaeological landscape, it is fundamental to construct a virtual digital environment of simulation using neural networks and virtual reality systems. Neural networks are needed for constructing intelligent avatars and behaviours. Keywords and key concepts of this theory are concerning the following issues:

- Self-organization.
- Digital evolution of an ecosystem for simulating factors of interaction.
- Real time 3D navigation through the landscape.

For comprehending this cybernetic approach two definitions have to be suggested:

- The map is not the territory in the Reality.
- The map is the territory in the Virtuality (see the following paragraph and Forte, 2000, 2001).

Below some key concepts concerning the theory of self-organization and biological cognition of Maturana and Varela (Maturana, Varela, 1980).

Man knows and his capacity to know depends on his biological integrity; furthermore, he knows that he knows. As a basic psychological and, hence, biological function cognition guides his handling of the universe and knowledge gives certainty to his acts; objective knowledge seems possible and through objective knowledge the universe appears systematic and predictable. Yet knowledge as an experience is something personal and private that cannot be transferred, and that which one believes to be transferable, objective knowledge, must always be created by the listener: the listener understands, and objective knowledge appears transferred, only if he is prepared to understand. Thus cognition as a biological function is such that the answer to the question, 'What is cognition?' must arise from understanding knowledge and the knower through the latter's capacity to know (Maturana, Varela,1980, 5)

Man is a deterministic and relativistic self-referring autonomous system whose life acquires its peculiar dimension through self-consciousness; ethic and morality arise as commentaries that he makes on his behaviour through self-observation. He lives in a continuously changing domain of descriptions that he generates through recursive interactions within that domain, and which has no other constant element in its historical transformation than his maintained identity as an interacting system. That is, man changes and lives in a changing frame of reference in a world continuously created and transformed by him. (Maturana, Varela, 1980, 57)

In the realm of virtual reality and neural networks all the properties of interaction are self-learning, namely each action is aimed to understand the environment and to modify the behaviour on the basis of the surroundings.

A consistent difficulty of the interpretation of ancient landscapes is our contemporaneous superstructure, our mental map (Renfrew, 1982; Renfrew, 1984; Renfrew, Zubrow, 1994).

7 The map is not the territory

"The map is not the land" is a principle made famous by Alfred Korzybski (1941) according to which "...when there is thought or perception or communication of perception there is a transformation, a codification, between what is communicated, the Ding an sich, and its communication". According to Bateson the map is "a sort of effect which sums up the differences, which organises the information about the territory's differences". "The differentiation, for example, between map and territory, which the semanticists insist that scientists shall respect in their writings must, in cybernetics, be watched for in the very phenomena about which the scientist writes. Expectably, communicating organisms and badly programmed computers will mistake map for territory; and the language of the scientist must be able to cope with such anomalies.

According to Denis Wood "mapping and mapmaking "do not mean the same thing" (1992, 32)[...] Mapping is a capacity universal to humans, established along with other capacity of the human mind-brain through a process of evolution under natural selection. [...] Whereas mapping, like speaking, might be regarded as a "universal expression of individual existence, mapmaking, like writing, has to be seen as an unusual function of specifiable social circumstances arising only within certain social structures (Wood, 1992, 32; Wood, 1993; Ingold, 2000, 231).

According to Svetlana Alpers (Alpers, 1983) the "maps give us the measure of a place and the relationship between places, quantifiable data, while landscape pictures are evocative, and aim rather to give us some quality of a place or the viewer's sense of it. One is closer to science, the other is art. Anything on the map that evokes the experience of place or movement is dismissed by the scientific cartographer as mere decoration; anything in the picture that conveys factual information about spatial location is dismissed by the artists as mere topography." (Alpers, 1983, 124-26). For interpreting the ancient landscape we can't separate these two categories because we need information from the map, from the cartography and from the places; only the integration of spatial data with evocative information can give us a correct perspective of interpretation. According this viewpoint in the landscape's interpretation it is preferable to remark the difference between "map" (in the Bateson's sense) and territory, between virtual (map) and real (territory). The map is the alphabetised space, the "difference" between ontologies for interpreting information.

According Bateson "what is it in the territory that gets onto the map ? We know the territory does not get onto the map. That is the central point about which we here are all agreed. Now, if the territory were uniform, nothing would get onto the map except its boundaries, which are the points at which it ceases to be uniform against some larger matrix. What gets onto the map, in fact, is difference, be it a difference in altitude, a difference in vegetation, a difference in population structure, difference in surface, or whatever. Differences are the things that get onto a map". (Bateson, 1972, 457). [...] "What is the territory ? Operationally, somebody went out with a retina or a measuring stick and made representations which were then put upon paper. What is on the paper map is a representation of what was in the retinal representation of the man who made the map; and as you push the question back, what you find is an infinite regress, an infinite series of maps. The territory never gets in at all. The territory is *Ding an sich* and you can't do anything with it. Always the process of representation will filter it out so that the mental world is only maps of maps of maps, ad infinitum. All phenomena are literally appearances. Or we can follow the chain forward. I receive various sorts of mapping which I call data or information. Upon receipt of these I act. But my actions, my muscular contractions, are transforms of differences in the input material. And I receive again data which are transforms of my actions. We get thus a picture of the mental world which has some-how jumped loose from our conventional picture of the physical world" (Bateson, 1972, 460-61).

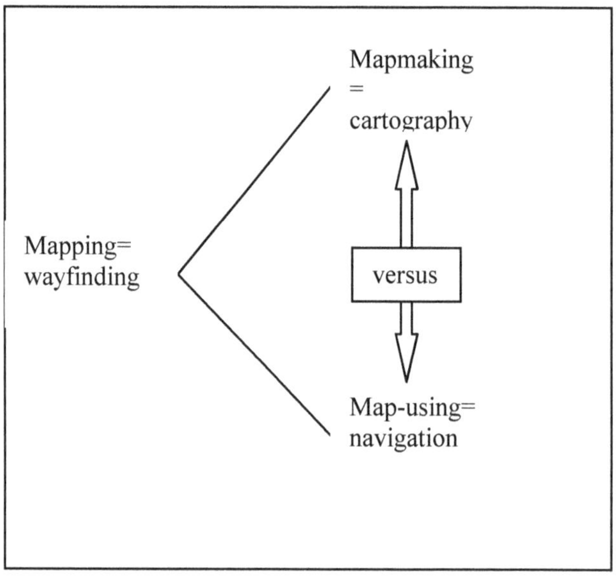

The relations between mapping, mapmaking and map-using (Ingold, 2000, 231)

In the realm of the virtuality the "map" is the creation/representation of information because it increases information through cybernetic means and codes. In the digital domain any process of knowledge passes through feedback, differences (in cybernetic sense) and interpretations of maps. A virtual reality system is a digital ecosystem, so the rules of the environment follow the "bio" relations between humans and digital environment, between actions and information. This ecological thinking represents the theoretical base of this project and of our approach, mainly according to the cybernetics of Gregory Bateson.

Perception of the environment and of the landscape involves all our senses: sight, scent, taste, hearing; the senses recreate contexts of learning (Latalova et alii, 2000; Barcelo, Sanders, Forte, 2000; Forte, 2000a, 2001). For this reason, evolution and implementation of virtual reality systems, immersive and semi-immersive, could constitute, in the near future, the best tool of advanced research for any perceptual multisensorial reconstruction. Innovative and inclusive ways of perception create advanced forms of virtual learning (Barcelo, Forte, Sanders, 2000): learning's acts through the "difference/feedback" we produce exploring a virtual environment (Bowen Loftin, 2002).

Main rules of virtual environment can be (Forte, 2001):

– Feedback or retroaction, the property to learn and to know the ecosystem/digital environment through actions, inter-actions and re-actions (answers). Perceptive phenomena provides an interaction level, or better an "exchange" or absence of behaviour between the actor and receiver; in this field we can identify multiple levels of interaction in real time.

- Difference. The learning is "through difference", the perception acts only on the difference. Receiving, grabbing information means always and necessarily to receive news of difference, and perception of the difference is always limited from a threshold. Differences too slight or presented too slowly are not perceptible: they don't feed the perception. The Bateson's theory of the knowledge is fit to explain the mechanisms of processing information: data are neutral objects, the knowledge of a spatial system is for interaction (difference) between the components or for the simulation of connected events. The more difference increases in the virtual interaction, the more learning increases.
- Reticular learning. This interaction and learning in VR is of reticular type (information spatially connected within a 3D cyber-net) because the use is immersed within reticules of information and visual data. Each interaction is carried out from its own system of virtual alphabetization; a new "alphabet" can suggest a different and fast perception of the visual and interactive space; in this way advanced directions of digital learning can be promoted so that the audience can receive in a very short time a relevant quantity of information. This mental re-composition is a new map.
- Connectivity. In VR all the information are interconnected in a 3D space; an ontology of connectivity entails mutual causality: actor and environment modify each other creating new information (see also Heidegger, 1927).

8 Space

The space represents the concrete symbol of the social system (Leroi-Gourhan, 1977, 374; Condominas, 1980); therefore the space is conceived not only as a physical place, as "container" in which a social system is structured, but also like a connective space (Levi Strauss 1966, 234). The space is conceived because we call it space.

Space is defined through places and facts, so the space would not exist without places: therefore a space is a context of natural and artificial places. The place is fundamental in the social collective and individual education because in the interaction between places and human activities: the landscape is narrative, imaginative and finally historical. Description and interpretation of the landscape concern the perception of space, but in a progressive sedimentation of information of different sources, but what the space is ?

An interesting example on the "power of the places" is without doubt the volume of Keith Basso *Wisdom Sits in Placet,* (Basso, 1996) an ethnographic study of the western Apache and of the rank of the place in the daily life. In this context the names of the places are not so fundamental rather than the stories to which are linked. In fact the stories have coded messages on the "correct" way of the behaviour. When one tales a stories, mental images are evocated linked with the places representing their local memory, or the inner landscape of the population living there.

A very interesting conference hold by Martin Heidegger in 1964 at Friburg faced the argument of body and (in) space (Heidegger, 1996); some of these thoughts are discussed below. Aristoteles defines the space with two different terms: *topos* and *chora*; *topos* is the space occupied by a body, is its place, *chora* indicates the space containing places. In Greek sense the space is considered starting from the body, as its place and container of places (Aristoteles' vision). Each body has its own place, and a landscape is constituted by bodies and forms.

According Condominas (1977, 1980, 1981) the space is determined from the ensemble of relation's systems established by the societies. In fact the societies manipulate the space or at technical level whether at symbolic level, namely the space becomes a product of a transformation process leaded on a given space. The elements structuring the social space are space and time, the relations with the environment, the goods trading, relations of kinship, relations of communication.

Each society is made of places and bodies, or of living, operating, interacting, bodies settling some places. Therefore we can't consider a society if it is not occupying a certain place and, more precisely, places of the space. The social space (occupied by the society) is not ever neutral and uniform: it is variegated, it is made of different places, electing some specific ambits of its territory to special destinations (Remotti, 1993, 31): for example ritual centres, landmarks, social attractors, territorial catalysers. This is the case of stele, artificial or natural monoliths (see the case of Aksum in Forte, in this volume), rank sites, time and space signs.

The human being is organizer of space that is extracted from the chaos for electing coded territories: we could define this process "territorialisation of the landscape". Territorialisation means construction of meanings, namely, the territory appropriates of the landscape and it identifies contexts: so the spaces become places.

On this subject is enlightening the definition of landscape of Lenclud: "the landscape is, first of all, ontologically, an expanse of space offering to the sight [...]. In any case it puts off to a support of perception [...]. There is no landscape without observer; it is necessary that a site can be seen because it can be defined a landscape. The landscape is a place, but a place isolated from the look; a site, but a contemplated site; a space, but a framed space; a datum, but a datum reconstructed by a visual analysis; a cutting of the world, but a cutting with a significance. [...] Therefore the landscape includes the existence of a space on which sight and conscience operate... (Lenclud, 1995, 4-5, 10-I).

For the contemporary anthropology the landscape is a cultural and social "process", modelled by a culture of a group (Hirsch 1995, 9). Culture is a transformation factor, the natural space is a mean, cultural landscape is a result. In the landscape anthropology all the economical and social systems tend to organize the exchange with the natural environment (in general the community rules the use of the resources).

In a holistic approach for the analysis and the representation of the archaeological landscape, the perception of the whole is interpretable through a mental map or, in our case, through a digital alphabetization of the content. Connective, natural, social, anthropic elements of the territory constitute the vision of the landscape and our cognitive horizon (Forte, 2000). In fact, according to the above mentioned theory of Alfred Korzybski, "the map is not the land" (Korzybski, 1941), because if the map was the same than the territory, no interpretation would be possible. The map in our case is represented from the virtual reality systems and from the 3D GIS applications; that is the map is not the territory in the reality but it is the territory in the virtuality.

The virtual reconstruction of the archaeological landscapes must be aimed to bring in an alphabetised cyberspace (the "map") what exists in an encoded space (the "territory"). How does the process of knowledge is in GIS for the landscape archaeology ? Which perceptive mechanisms could be within the landscape ? Korzybski's metaphor show in an extraordinary way the great difference between real space, territory, and descriptive space, or its representation, the map. Without a grammatical and metaphoric language of the map, no knowledge would be allowed in the "spatialized" and "territorialized" world.

The faculty to analyse in detail archaeological data, gives great potentialities of research, but it doesn't solve the crucial problem of interpreting the ancient space: the reconstruction of the landscape through mental maps (Renfrew, 1982, 1984, Renfrew, Zubrow, 1994). How was one living, was one seeing, was one imaging space, territory, landscape, in the ancient times? What elements were classifying it ? And in the same settlement system, were the inhabitants looking at the territory in a different way according to their position or hierarchy ? And the border's space what impact could create in ethnically different populations ?
All these questions and issues determines qualities of the ancient landscape, therefore in the following paragraph we will try to describe them very shortly.

9 Landscape's qualities

In the paper *Archaeology and territorial "Musealisation": the virtual-cognitive reconstruction of archaeological landscapes* (Forte, 2001) I have described some cognitive qualities of the archaeological landscape that can help us to use a correct anthropological methodology; I think it is useful add this discussion in this contribution.

Social-symbolic. The social-ritual knowledge and the awareness of the way things were perceived inside one's borders is fundamental in the formulation of the sense of belonging. The presence of social, anthropological or ritual indicator-descriptors (markers of the landscape), segments the territory in hierarchic terms (symbolic familiarisation). Memory and the sense for the border topologically delimit the sense of belonging.
Geometric-topologic. The landscape's geometry is the capability of describing surfaces and volumes (intervisibility, DEM, panoramas, horizons). These are the only descriptive factors that are taken into account in the majority of GIS spatial analyses.
Visual. Visuality is the property to see territories (one's own and of other's) without being conditioned by mental maps and hierarchic cognitive processes. It corresponds to the purely theoretical property of seeing without interpreting; for example, certain GIS algorithms are used to explore the field of vision on the basis of digital terrain models.
Mental. The perception of landscapes through mental maps constitutes the grammar of visual anthropology.
Expressive-artificial. "Landscape uses an artificial language, namely an object that is created by man and that incorporates the principles of organization and classification within human experience" (Nash, 1997).
Imaginary-narrative. The evocation of landscape as a narration, reproduces places of the mind, reruns those of memory and describes the condition of "power" also through mytho-history (landscape is what is narrated or could be narrated). The conservation of landscape's memory means the conservation of cultural identity (Ashmore, Knapp, 1999, 12-19). A narrative landscape describes the quality of a collective memory (for example, through legends, mythical places meaningful in the belonging to an elected space).
Semiotic. Every landscape is comprehensible thanks to diverse languages and grammars. The construction of an anthropic language contemplates the use of a common language to express the mind/mentality of the community. Semiotics builds the alphabet of the imaginary and of social and collective memory.
Hierarchic levels of power. The construction of anthropic landscapes follows rules of power, so that perception may occur through hierarchic passages (levels of importance) between settlement systems and between different sites (non symbolic political hierarchy).
Connected-interconnected. Landscape is a context in which the whole is greater than the sum of its parts. The connectivity of information represents the contexts and thus the communication of events.
Natural-not natural. The "natural" quality in the perception of a landscape is purely theoretical and provides for a neutral knowledge without an attribution of hierarchic values to information. According T. Ingold (Ingold, 2000, 191) "The landscape is not nature [...]. Suffice it to say that I have in mind the rather specific sense whose ontological foundation is an imagined separation between the human perceiver and the world, such that the perceiver has to reconstruct the world, in consciousness, prior to any meaningful engagement with it. The world of nature, it is often said, is what lies out there [...]. We live 'in here', in the intersubjective space

marked out by our mental representations."

Space-temporal. "The landscape represents a series of spaces that become places, thus establishing a territory.... In mentally perceiving the idea of place, our body becomes part of the landscape creating in this way a sense of space [...]" (Nash 1997: 1-2). The diachronic qualities of the landscape constitute by themselves another context, in other words, another interpretation resides in diachronic information.

Graphic-simulated. The archaeological landscape is reconstructed graphically or by computer without any dynamic interaction (graphic-visuality).

Virtually-simulated. Inside a virtual reality reconstruction you experience a passive interaction (the user navigates through and visualises a model) and also an active interaction (with behavioural exchanges between environment, user and events, Forte 2000). Both interactions reproduce contexts. In the evolution of a digital ecosystem, behaviours between actors (for example, natural events or anthropic actions) and receptors (the environment) have to be contemplated.

Virtual-Dynamic. The perception of a landscape can also take place through the dynamic perception/interpretation of virtual models (walking through and flying by). Moreover the conceptual model of landscapes comprises also actions and events.

Political-identifying. Borders of a territory establish a sense of belonging and identity to the landscape on the basis of relations and symbols of power.

10 Virtual reconstruction of the archaeological landscape: problems and components

The construction of the map in the sense of the semiotics of the landscape (cognitive landscape), is a process very complex that, starting from GIS and spatial data, should reach the cognitive-symbolic perception of the landscape, also through mental maps and multisensorial simulations. The "map" could represent the more advanced evolution of the GIS, a digital ecosystem able to reproduce perceptively the archaeological landscape through virtual reality applications. If the construction of the cognitive map is an acquired result in the conceptual model we have theorized, on the technological viewpoint we are still far from a satisfying realisation. On the contrary, on the basis of the digital processing one should ask: is this a cognitive map ? Is this a map ?

Considering the experience developed in GIS environments and in OpenGl three-dimensional applications (graphic libraries for virtual reality applications on PC and workstation), below some methodological and technological tools for the reconstruction of landscapes will be suggested (anyway all the activities are aimed to the construction of a Map"):

Paleoenvironmental analyses. Before of creating a digital spatial data-entry, it is necessary to start study and interpretation of all the paleo-environmental, pedogenetic, geomorphologic, anthropic, archaeological, climatic data. In this phase the "map" is constructed as a conceptual model with all the information available integrated in a digital system. It is fundamental that the data are spatial and georeferenced.

Construction of the actual physiography of the landscape (how it is).

Geometric-physiographic reconstruction is made with the DTM interpolation. Over the DTM aerial photos or satellite images will be georeferenced in order to contextualize modern and current landscape (for example, a short diachronic sequence can give a different visibility of traces or archaeological sites as we can survey today). The creation of the landscape physiography is important for contextualizing spatial data and for having a model or a base of reference for testing the simulations. The physiographic reading of the landscape in the sense of "how we can see it" (or "how we think to see it"), can reproduce our contemporaneous mental maps (if we know the territory), but it can't visualize the ancient landscape.

Construction of the ancient physiography (how it was).

Three-dimensional model of the ancient landscape can be obtained integrating actual data of DEMs with new micro-topographical acquisitions (with total laser station or DGPSs) and processing archaeological and geomorphological cartography (for example modifying the contour levels according to the landscape we want to construct). Overlay of the physiographic environments: the reconstruction of the archaeological landscape is integrated with the visualization of the current landscape through spatial data. In this case a re-reading of the different diachronic maps representing distinct spatial and perceptive layers (ancient and current) could be attempted.

Vectors and themes.

Vector data, associated with attributes (an attribute is corresponding to a graphic pattern like water, grass, soil, vegetation, etc.) become graphic objects in the OpenGL environment, with volumes, textures and thickness, for example the rivers with water, the roads with stones, linear structures like walls, fences, and so on (see Forte, Kay in this volume).

Vegetation.

Vegetation can be reproduced in two dimensions with textures mapped from graphic libraries or in three dimensions through quads or billboards or 3D models. Billboards and quads (single or cross shape) are 2D shapes representing trees, brushes, etc.. The creation of this kind of billboards can be obtained directly by photographs or by graphic rendering.

Textures

Textures can be used for creating a territorial model draped over the DTM, over the vegetation or linked and mapped on 3D structures (buildings, cities, monuments, etc.).

Modelling

Creation of three-dimensional models related to sites, terraces, buildings, urban and rural areas, finalised to the virtual peopling of the landscape. A 3D graphic library must be specifically created for each informative level of landscape to construct.

Graphic libraries

They concern the draping of the DEM, like vegetation, trees, forests, waters, soils, rocks and so on. Different texture resolution could show diverse qualities of the landscape (cultivations, soils, etc.). In this class of objects even three-dimensional or pseudo-trhee dimensional objects can be input, like billboards and quads.

Dynamic events

In the digital ecosystem the movement of objects and/or phenomena constitutes a base element of the simulation because it creates a relevant "difference" between interaction of the user and movement of objects; this "difference" catalyses the perceptive act. Some natural phenomena simulated by software are, for example, the meteorological conditions (rain, snow, wind, clouds) or actions as floods, landslides, movements of virtual objects. Project of dynamic events is complicated and must be planned so that to develop new perceptive contexts of collective interactions within the digital ecosystem.

Mental maps

In theoretical sense, mental maps should draw the perceptive-symbolic context, giving anthropological significances to landscape's geometries and landmarks. Some properties of the mental maps could be: the frontier, the border, the ritual-symbolic elements (natural or artificial), the sensorial horizons (sight, smell, hear), the geometric horizons (profile of mountains, rivers, valleys), imaginary or memory places, hierarchical settlement systems and all the artificial or natural landmarks of a territory. The mental maps are a symbolic and perceptive superstructures not easy to demonstrate, but they are the only mean for drawing a cognitive-computational cartography of the digital and virtual landscape (digital "territorialisation"). In short, this could be the grammar (the "map" in our sense) for decoding the landscape in anthropological and semiotic terms.

Virtual dynamic interaction

A key aspect of the simulations is the navigation in real time (with a frame rate from 15 up to 60 frames per second), namely a dynamic access to the 3D information. This type of action corresponds to the dynamic increment of the perception: throughout the movement, the model becomes more perceptible, because it puts into effect a process of difference between dynamic observatory and object/space simulated.

Spatial data we get from a GIS for the virtual reconstruction of the archaeological landscape can be: cartographies, vector and raster data, aerial photos, satellite images, Digital Elevation Models. Key element of the reconstruction is the topological model, namely every virtual reconstruction must be "spatialised", keeping the geographic information.

The issue is very complicated but a very interesting solution is determined by the OpenGL development, as we have described in an other paper of this volume (see Forte, Kay and Forte); so we are addressing our efforts in this direction, creating specific software in C++.

11 Projecting a VR GIS system for the virtual landscape

In technical sense, problems of the virtual reconstruction of landscapes will be discussed in an other paper in this volume (Forte, Kay). In this paragraph, even on the basis of the Aksum's case study, we will attempt to draw a general methodological template for projecting a VR system dedicated to the landscape.

For what is concerning the virtual reconstruction of the archaeological landscape, a key factor of the work is represented from the graphic interface and from the behaviours. The interface determines commands, menus, levels of interactivity, while the behaviours represent the actions planned in the 3D space and by the interactive choices of the user. Below we suggest a preliminary list of behaviours we would like to implement in a "standard" project of VR GIS.

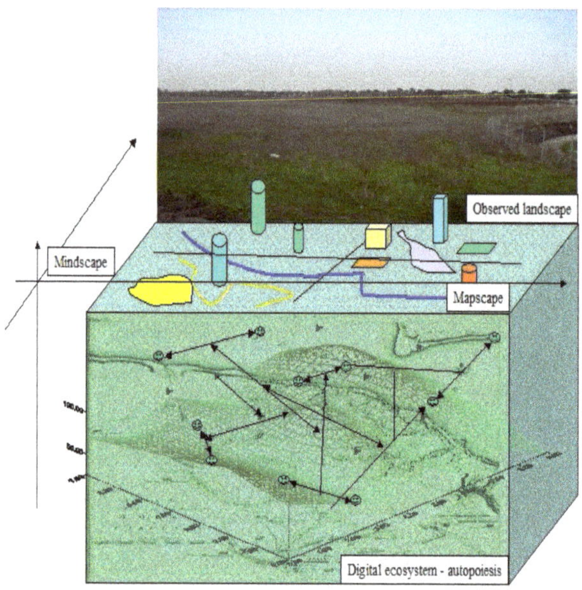

Figure 1. The landscape's reconstruction is the final result of integrated interpretations of taskscape, mapscape, mindscape and observed landscape. The basis of the knowledge process is the creation of a self-organised and virtual eco-system (autopoietic) in which, through inclusive interactions, we produce new ways of learning and of perception. This drawing attempts to describe the complexity of plots of mental maps needed for completing the vision/image/imagination of landscapes in our mind.

Behaviour 1: choice of the model
3D reconstruction can involve different landscape models: for example a DTM with only aerial photos and satellite imagery for showing the current terrain model, diachronic or morphological evolution of the landscape, ancient landscape and so on. At this stage of application the user can choose with which model wants to interact.

Behaviour 2: fly
Visualisation of the landscape from the top with the possibility to choose altitude and direction of the flight; this interaction attends to understand wholly the territory.

Behaviour 3: walk
Visualization of the landscape from the down, simulating to walk on the terrain.

Behaviour 4: choice of the 3D themes
3D themes populate the landscape in the same time or according to the choices of the user and the type of visualisation.

Behaviour 5: loading of the graphic libraries
This behaviour concerns the possibility to choose textures and graphic libraries according to typological bases. For instance, if we want to simulate agricultural cultivations with corns, we can choose between different types of maps and textures from the digital archives.

Behaviour 6: landscape anthropology
For increasing the cognitive value of the simulation we could input even behaviours of socio-economical and symbolic-cultural theme, created on the basis of the archaeological spatial analyses: settlement hierarchy, symbolic value of sites or territorial areas, roads and paths, control or border areas, areas associated with events, myths, legends. The visualization of these thematic maps will be through transparent coloured overlays (eventually linked with sound events) or markers superimposed on the landscape models. Sounds, music and effects could be linked with relevant areas in anthropological terms (i.e.ceremonial or funerary sites)

Behaviour 7: simulation of events
In the virtual landscape it is possible to plan pre-ordered events such as: floods, landslides, deforestation, forestation, artificial barriers, geodynamic or anthropic events, etc.. The visualization can involve pre-calculated movies of simulation, spatially georeferenced, or thematic maps.

Behaviour 8: husbandry, farming, herbage
The virtual landscape must be a living scene where the activities are evident. We can plan to input 3D characters or billboards.

Behaviour 9: peopling
The landscape is peopled of characters with specific behaviour chosen on the basis of their activities. Each character will be created in 3D so that the interaction in the 3D inclusive space can be at the best level.

Behaviour 10: agriculture
Ancient cultivations will be visualised in separated layers with the possibility to compare diachronic land uses.

Behaviour 11
The sound environment has to be developed in 3D (spatialised): music, sounds, voices, will be linked to territorial spaces or dynamic events. The spatialised sound helps very well to represent the idea of mindscape. Specific sounds could be linked with different colours.

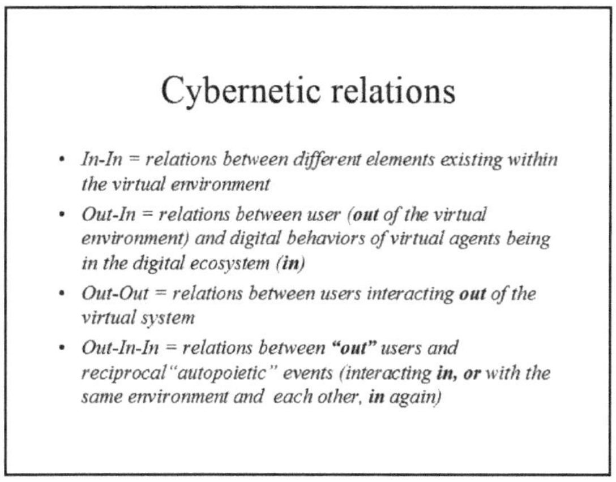

Behaviour 12: multimedia
In the 3D space multimedia windows will can be inserted in order to contextualize the digital environment with images, movies, animations, even for comparative analyses.

Behaviour 13: geomorphology and paleoenvironment
This action concerns the possibility to load maps and textures of soils, land uses and paleoenviromental data.

Behaviour 14: timeline
A virtual cursor representing a timeline will permit to move models through the time and visualizing different periods according to the availability of archaeological data.

Behaviour 15: narration
In the virtual landscape will be planned some special hyperlinks, that we define centre of narrative hearing. These points will be shown in the landscape with special symbols (and hot links) so that particular tales, narrations, comments on special places (in ecological, anthropological, territorial terms) can be heard. These attractive points will permit to see the landscape from special areas of interest so that to have the best visual perspective for interpreting and understanding the narration (ex. ceremonial sites, relevant or magic power of rocks, etc.). The creation of centre of narrative hearing will allow to read the ancient landscape even with the difficulties of interpretation given by the actual landscape.

Behaviour 16: query of the landscape
All the spatial data included in the virtual landscapes in terms of objects, areas, layers or themes will be processed as databases, so they will be queried during the navigation in real time and in 3D. We could start a query from the visual landscape or from specific tables.

Behaviour 17: ethnography/environment
In cases in which it is possible, spatial data about ethnography (ethnoscape), like movies, interviews, documents, will can be inserted in the virtual environment for allowing a comparison between different sources of information.

12 Conclusions

In this paper we have attempted to answer to some questions of this workshop: what does the landscape consist of (now)? How do we perceive the landscape and what do we perceive (now through our minds)? What did the ancient landscape consist of ? How did ancient people perceive their landscape (through the ancient minds)? How can we perceive the ancient landscape? How the Virtual and the sciences of complexity can help us to face up these issues ?

All these questions can't be solved only in terms of algorithms or theoretical geography (conceptual models, space analyses and so on), because they involve the fields of perception, of learning, of epistemology, of ecology of mind. And the ecology of mind, the cybernetic approach of Gregory Bateson and related ecological schools, constitute the theoretical and methodological basis of our work (see a good exegesis in Padovani, Manghi, 1998): we have to project the Virtual following ecological logics: all the rules of neural networks, virtual environments and artificial intelligence should be integrated in this plan. Integrating simulations and virtual environments it is possible to increase the cybernetic level of knowledge, because of the following properties: reciprocal behaviors/actions within the ecosystem (*in-in*), reciprocal behaviors between users and (*out-in*) ecosystem, reciprocal behaviors between different users (*out-out*), self-organized events of artificial life with growing of the ecosystem according to specific stimuli (*out-in-in*).

This epistemological approach starts from the ecological thinking and stresses the following things: the mindscape represent the difference between map and territory, the information of the cybernetic world. The more we increase the difference between map and territory, the more we learn, but it is not enough; the digital virtual ecosystem becomes an "autopoietic" system, namely it grows according inter-relations, interactions, that can't be preview. In this exchange of information we have a double level of learning: self-learning of the virtual agents in the digital ecosystem, and self-virtual-learning of the user exchanging behaviors with the digital environment. So, this digital mindscape exists in terms of relations between environments, agents and actions producing difference, therefore information.

In technological terms, and starting from the case study of Aksum, we are projecting a virtual ecosystem in OpenGL environment) integrating neural networks, a VR GIS, and a sequence of visual and inclusive digital behaviors. We intend visual behaviors the visual interaction of the user in real time (fly, walk) and inclusive behaviors (Bowen Loftin, 2002), the mutual interactions of 3D self-organized agents (user included) in the digital ecosystems. For example, each agent can react in different ways according to the events/behaviors of the surrounding, or the user can exchange information, actions and movements with the ecosystem (e.g. taking a stone, manipulating an object in 3D, hitting a target, using materials, talking with other avatars, querying the system, etc.).

The great challenge of the future is to develop digital ecosystems and to test their growth and behaviors in real time; on the basis of our recent projects and experiments it is possible to imagine a strong develop of the research in the following fields and applications:

- virtual collaborative and shared environments;
- projecting of avatars;
- integration of virtual reality systems with neural networks;
- creation of special display systems dedicated to the landscape;
- creation of artificial life in digital ecosystems;
- "autopoiesis" of digital agents;
- self-learning of virtual agents and final users;
- multisensoriality;
- inclusive spatialisation in immersive environments.

At the present time, we are constructing a virtual system of spatial data programming in C++ software and integrating actual and past information, remote sensing and GIS, 3d libraries and some virtual behaviours (see Forte, Kay, in this volume.

In conclusion, we are only at the beginning of a very long path for reconstructing and interpreting the ancient landscape but we think that the field of virtual reality, neural network and sciences of complexity represent the best level of research and application for the future, mainly addressing the progress towards a creation of integrated cybernetic systems. Maybe one dream of the cyber-anthropology could be realized: users, avatars and agents will interact in the same environment/cyberspace and with the same cybernetic language ("please tell me what you are seeing there...around you...in your digital surroundings"). Mental landscapes, mindscapes, mental maps, cybernetic maps are the border-line of this digital era, so the next question could be: Yes, this is the virtual reality, but where is the reality ?

Finally, let me to conclude with a cybernetic thought of G.Bateson: "The elementary cybernetic system with its messages in circuit is, in fact, the simplest unit of mind; and the transform of a difference travelling in a circuit is the elementary idea. More complicated systems are perhaps more worthy to be called mental systems, but essentially this is what we are talking about. The unit which shows the characteristic of trial and error will be legitimately called a mental system" (Bateson, 1972, 465).

References

ABU-LUGHON L., 1990, *The Romance of Resistance*, "Samerical Ethnologist", n.17.

ALLEN, K.M.S., GREEN, S.W., and ZUBROW, E.B.W. (Eds.) 1990, *Interpreting Space: GIS and Archaeology*, London, Taylor & Francis.

APPADURAI, A., 1990, *Disjuncture and difference in the global cultural economy*, "Public culture".

ALPERS, S., 1983, *The art of describing: Dutch art in the seventeenth century*, Penguin, London.

ASHMORE W., KNAPP A.B. (Eds.) 1999, *Archaeologies of*

landscape, Blackwell Publishers, Malden, 1999.

ATTEMA, P.A.J., 2000, *Cartography and Landscape Perception: A case study from central Italy*, in Gillings M., Mattingly D., Van Dalen J. (Eds.), 2000.

BARCELO J., FORTE M., SANDERS D., 2000 (Eds.), *Virtual reality in archaeology*, Oxford, BAR Publishing (BAR International Series S 843).

BASSO, K.H., K. 1996, *Wisdom Sits in Places: Landscape and Language among the Western Apache*, University of New Mexico Press, Albuquerque, 1996.

BATESON G., 1972, *Steps to an ecology of mind*, The University of Chicago Press, Chicago, 1972.

BATESON G., 1979, *Mind and Nature. A Necessary Unit*, Dutton, New York.

BONESIO, L., 1997, *Geofilosofia del paesaggio*, Mimesis, Milano.

BOWEN LOFTIN R., 2002, *Psychophysical Effects of Immersive Virtual Reality*, Proceedings of the IEEE Virtual Reality Conference 2002, 145-148.

CHAPMAN H. 2000, *Understanding wetland archaeological landscapes: GIS, environmental analysis and landscape reconstruction; pathways and narratives*, in G. Lock (ed.) 2000, 49-59.

CONDOMINAS, GEORGES, 1977, *Pour une définition anthropologique du concept d'espace social*, ASEMI, vo. VIII, 1977.

CONDOMINAS G., 1980, *L'espace social. A propos de l'Asie du sud-est*, Flammarion, Paris.

CONDOMINAS G., 1981, *Spazio sociale*, in Enciclopedia, XIII, Einaudi, Torino.

DELEUZE G., 1967, *Différence et répétition*, PUF, Paris.

DELEUZE G., GUATTARI, F., 1987, *a Thousand plateaus*, Minneapolis, Minnesota University Press.

DERRIDA J., 1967, *l'écriture et la différence*, Seuil, Paris, 1967.

DODGE M., KITCHIN R., 2001, *Mapping cyberspace*, Routledge, New York.

ESCOBAR A., 2001, *BENVENUTI CYBERIA. NOTE SULL'ANTROPOLOGIA DELLA CULTURA*, IN "AVATAR", NOVEMBRE 2001, MELTEMI, ROMA, 100-117.

FORTE, M. 1995, *Scientific visualization and archaeological landscape: the case study of a terramara, Italy*. In G. Lock and Z. Stancic. (eds) Archaeology and Geographical Information Systems, London: Taylor & Francis, pp. 231–238.

FORTE M. 2000, *Archaeology and virtual micro-topography: the creation of DEMs for reconstructing fossil landscapes by Remote Sensing and GIS applications*, in "Beyond the map. Archaeology and spatial technologies", (ed. Gary Lock), Oxford, 2000, 199-213.

FORTE M., 2000a, *About virtual archaeology: disorders, cognitive interactions and virtuality*, in Barcelo J., Forte M., Sanders D., 2000 (eds.), Virtual reality in archaeology, Oxford, BAR Publishing (BAR International Series S 843), 247-263.

FORTE M., 2001, *Archaeology and territorial "Musealisation": the virtual-cognitive reconstruction of archaeological landscapes*, in Virtual Archaeology, Proceedings of the VAST Euroconference, Arezzo, 24-25 November 2000 (ed. by F. Niccolucci), BAR International Series, Oxford, 2002, 29-39.

FORMAN R.T.T., GODRON M., 1986 , *Landscape Ecology*, John Wiley & Sons.

GAFFNEY V., OSTIR K., PODOBNIKAR T. and STANCIC Z. 1996, *Spatial analyses, field survey, territories and mental maps on the Island of Brac*, "Archeologia e Calcolatori" 7, 27-41.

GIBSON, J, 1950, *The perception of the Visual World*, Greenwood Press, Westport CT, 1950.

GIBSON, J., 1979, *The Ecological Approach to Visual Perception*, Lawrence Erlbaum Associates. Hillsdale NJ, 1979.

GILLINGS M., MATTINGLY D., VAN DALEN J. (Eds.), 2000, *The Archaeology of the Mediterranean Landscape*, 3rd edn., (eds.) Gillings, M., Mattingly, D. and Van Dalen, J., Oxford, UK: Oxbow Books Ltd., pp.65-82.

GILLINGS M., WHEATLEY D., 2000, *Vision, perception and GIS: developing enriched approaches to the study of archaeological visibility*, in G. Lock (ed.) 2000, 1-27.

HEIDEGGER M., 1927, *Sein und Zeit*, Max Niemeyer, Tübingen.

HEIDEGGER M., 1996, Beme*rkkungen zu kunst – Plastik – raum*, Erker Verlag, 1996.

HIRSCH, E., 1995, *Landscape: Between Place and Space*, in Hirsch E., O'Hanlon M., 1995, The Anthropology of Landscape. Perspectives on Place and Space, Clarendon Press, Oxford.

INGOLD, T., 2000, *The perception of the Environment. Essays in livelihood, dwelling and skill*, London and New York, Routledge.

KORZYBSKI A., 1941, *Science and Sanity*, Science Press, New York, 1941.

LATALOVA M., RACZKOWSKI W., TSCHAN A. P., 2000, *Perception and viewsheds: are they mutually inclusive?*, in G. Lock (ed.), 2000, 28-48.

LENCLUD G., 1995, *Ethnologie et paysage*, in Paysage au pluriel. Pour une approche ethnologique des paysages, Editions de la Maison des Sciences de l'Homme, Paris.

LEROI-GOURHAN A., 1977, *Il gesto e la parola*, Einaudi, Torino.

LEVI-STRAUSS C., 1966 (1958), *Antropologia strutturale*, Il Saggiatore, Milano.

LEVY P., 1997, *Cyberculture. Rapport au Conseil de l'Europe*, éditions Odile Jacob.

LOCK G. (ed.), 2000, *Beyond the Map. Archaeology and Spatial Technologies*, IOS Press, Amsterdam.

MATURANA H., VARELA F., 1980, *Autopoiesis and cognition.*

The realization of the living, Reidel Publishing Company. Doldrecht, Holland.

MATURANA H., VARELA F., 1987, *The tree of knowledge*, Boston, New Science Library.

MYERS, R.E., 1986, *Pintupi country, Pintupi self: sentiment, place and politics among Western desert Aborigines*, Washington, Smithsonian Institution Press.

NASH G. (Ed.) 1997, *Semiotics of Landscape: Archaeology of Mind*, Oxford.

NAVEH, LIEBERMAN, 1990, *Landscape ecology*, XVIII. 356 pp., Springer. Heidelberg.

NELSON, E.W., 1983, *Make prayers to the raven : a Koyukon view of the northern forest.* Chicago. Ill, University of Chicago Press.

PADOVANI G.1998, *L'oppio e le sue molteplici "virtù". L'abduzione in Comte, Peirce e Bateson*, in "Attraverso Bateson. Ecologia della mente e relazioni sociali."ed. Sergio Manghi, Cortina Editore, 1998, 253-270.

REMOTTI F., 1993, *luoghi e corpi. Antropologia dello spazio, del tempo e del potere. Significati antropologici dello spazio*, il Mulino, Bologna.

RENFREW C., 1982, *Towards an archaeology of mind*, Cambridge, Cambridge University Press.

RENFREW C., 1984, *Approaches to social archaeology*, Edinburgh, Edinburgh University Press.

RENFREW C., ZUBROW E. (eds.) 1994, *The Ancient Mind: elements of cognitive archaeology*, Cambridge UP.

ROBINSON J., ZUBROW E., 1999, *Between Spaces: Interpolation in Archaeology*, in Gillings M., D, Mattingly, J. Van Dale, 1999, 65-84.

SHANKS M., TILLEY C., 1992, *Re-Constructing Archaeology. Theory and Practice*, Routledge, New York.

TILLEY C., 1994, *The phenomenology of landscape*, Berg, London.

WITCHER R., 1999, *GIS and Landscapes of Perception*, in Gillings M., D, Mattingly, J. Van Dale, 13-22.

WOOD D., 1992, *The power of maps*, New York, Guilford Press.

WOOD D., 1993, *The fine line between mapping and mapmaking*, "Cartographica", 30, 50-60.

YI-FU TUAN, 1977, *Space and place. The perspective of experience*, University of Minnesota Press , 1997.

ZERUBAVEL E., 1999, *Social Mindscapes. An invitation to Cognitive Sociology*, Harvard University Press, 1999.

ZUBROW, E.B.W. 1994, *Knowledge representation and archaeology: a cognitive example using GIS*, in Renfrew C., Zubrow E. (eds.), 1994, 107-118.

Remote Sensing, GIS and Virtual Reconstruction of Archaeological Landscapes

Maurizio Forte and Stephen Kay

CNR ITABC,
Via Salaria km 29,300, Monterotondo St. (Rome), Italy
maurizio.forte@mlib.cnr.it, stephen.kay@libero.it

1. The Virtual Reality System

Virtual Reality (VR) is a profoundly visual medium and this emphasises the need for an adequate system to re-create and construct from an interpretation the archaeological landscape. However, prior to choosing the system and means of communication, it is necessary to define the requirements of the VR system. In planning the digital phase of the Aksum Project, several themes were identified as fundamental to the successful completion of the project.

The first area identified as integral to the user's sense of immersion in the virtual world was the interactivity provided, to allow a sense of being absorbed, a fundamental aspect to the study of the cognitive landscapes (Forte 2000). As identified by Wheatley and Earl (2002: 5), a virtual model has an emphasis on solidity, mobility and texture, which are important to the visual impact the model will have on the user. Together with this, sound was also identified as important for the virtual world. This would take the form of background noise, such as running water and wind rather than an abstract voice-over commentary. An integral part of interactivity with a VR model is allowing freedom of movement within the virtual space. The user must be allowed to choose their path through the landscape deciding what to explore. Lastly, interactivity should also allow querying of the objects, sights and sounds in the landscape. Whilst this adds an un-real dimension, it is imperative for the archaeological interpretation that the user understands the decisions made in the construction of the model. For example, a structure such as a palace may have more than one reconstruction and the user should therefore be offered a range of alternatives, as well as the data and references with which the interpretation was created. VR offers this unique opportunity to move away from previous dehumanised forms of visualisation, such as 2-dimensional cartographic plans, to a situation where the viewer may choose their own interpretation.

The second area identified as key to the VR requirements related to the construction of the virtual landscapes. The Aksum Project has, since its conception in 1993, collected a range of digital data which has been collated within a GIS. This data forms the materials from which the VR model is built, thus the issue of compatibility with GIS data formats was central to the system requirements. Further, the construction of the ancient Ethiopian landscape required specific vegetation and textures to be acquired, usually in a graphic file format. Therefore, a requirement of the VR system was the ability to apply specific textures to the 3-dimensional models and landscape as well as to import graphic files. These procedures are described more fully in section 3.

The levels of interactivity sort for the model, specifically the ability to query what is being visualised, led to the requirement of connectivity between the VR system and an underlying database constructed in a manner that can be queried by a user. This would contain the full range of digital data, including information omitted from an individual model, allowing the user to adjust the spatial and temporal parameters of the model. However, the key requirement was that textual and graphical information could be recalled and displayed as an object is queried within the virtual landscape.

The final and most important requirement of the VR project was that a system be implemented which would permit a two-way passage of data. An open structure to the system would allow continuous modification and updating, particularly as an aim of the digital project is not to generate a single virtual model, but to create a process allowing constant adaptation. Therefore a product was required that would allow the continuous generation of differing virtual worlds rather than designing one unique model.

The requirements that were defined for this project were based partly upon VR being a profoundly visual medium and also upon a desire to challenge the sterile environment regularly offered in virtual reconstructions. VR offers the opportunity to move away from conventional graphic representations that are often simplified and abstract (Wheatley and Earl 2002), through the ability to create a changeable environment which can also use the other forms of human senses. Furthermore, whilst allowing user interaction, the system must allow the virtual models to be worked with and on rather than just consumed (Shanks, in Gillings 1999). The overall goal of these requirements was to create a system allowing both absorption and interactivity in order to promote an understanding and questioning of the interpretations and the archaeology of ancient Aksum.

2. Virtual Reality Methods

The VR research that has been conducted within this project forms only a part of a much broader study which ranges from excavation to satellite imagery classification . Each of these fields have been unified through the use of a GIS system, which offers "...a powerful set of tools for collecting, storing, retrieving at will, transforming and displaying spatial data from the real world for a particular set of purposes". (Burrough and McDonnell 1996) Through the application of this system, it was necessary to use a VR system which would maintain the spatial context of the data. A range of different techniques were assessed for the project, including Bubbleworld technology, VRML and OpenGL. In investigating rock art in South Africa, Gidlow successfully used Bubbleworlds, as "The panoramas provide a spatial context, in that it becomes possible to visualise where the art has been placed within the landscape." (Gidlow 2002: 19) However, whilst Bubbleworlds have the advantage of being quick to construct and easy to navigate, they are unable to include within them freely navigable 3-dimensional objects, they are essentially describing what is 'real' and tangibly exists. The Aksum Project is a study of an ancient landscape where it is being attempted to reconstruct a landscape that existed in the 3rd and 2nd millennia BC.
The second virtual reality system investigated was the use of VRML, which has seen successful application in landscape studies in several archaeological projects, such as the Negotiating Avebury Project (Pollard and Gillings 1999). The strength of this language is that it may be written in any text editor and is not compiled, rendering it simple to copy and modify in order to create new models. However, as with Bubbleworlds, it is not possible to directly import GIS data, a core requirement of the Aksum Project. Therefore, in order to fulfil the objectives laid down for the Aksum VR, a 3D terrain builder was identified as the most suitable product with which to achieve the projects goals.

In January 2000, a program of scientific cooperation was established between ITABC and the U.S. company Terrex for the use of their software for application in archaeological reconstructions. The choice of using the software package Terra Vista was specifically guided by its use of the OpenGL working environment and its close connection with GIS software packages. Standard file formats used by the most widespread GIS software can be readily imported into the terrain builder, maintaining both their spatial location and associated attribute data. Finally, the format that the terrain is exported in, an OpenGL flight file (FLT), is a widely supported industry standard which can be easily modified and updated in other software programs.

The preparation and acquisition of the data in particular digital formats, whilst a time consuming part of the project (as recently noted by Wheatley and Gillings 2002), ensured that the passage to the subsequent phases was smooth. The choice of software with which to create the virtual reconstruction dictated that three phases would be necessary to complete the simulation: firstly the input of the sources, secondly the creation of the model and lastly the generation, including defining the parameters, of the landscape. The first step therefore was to insert all the desired data into the 3D terrain database program Terra Vista. The software rapidly constructs digital landscapes through its optimisation for drawing, as such an application must be rapidly updated so the viewer sees one continuous uninterrupted landscape. A further advantage of using Terra Vista is the degree of automation of the program, coupled with a accessible GUI based on a Windows style interface.

In summary, the VR system chosen for the Aksum Project offered the best flexibility for producing models and allowed the inclusion of raw data from the many fields of the projects research. VR offers the opportunity of revolutionizing the way in which we engage with and re-present the past (Goodrick and Gillings 2000: 41) and the sense of immersion that it offers is greatly enhanced through the quality and accuracy of the virtual world which is offered. The ability of Terra Vista to offer an infinite number of interpretations was also a key factor to its application. Further, through maintaining the spatial context of the data it becomes easier to visualise the data where it had been in the landscape.

3. Documentation of a Virtual Reality Database

The preliminary stage in the creation of a virtual landscape is the documentation of a database of information from which can be drawn the desired elements. The subsequent step then requires decisions to be made as to which data to include and which presents the most accurate reconstruction of the landscape in question (see figure 1). A brief discussion is therefore useful of some of the data types which were chosen for the models that were created in the course of this research.

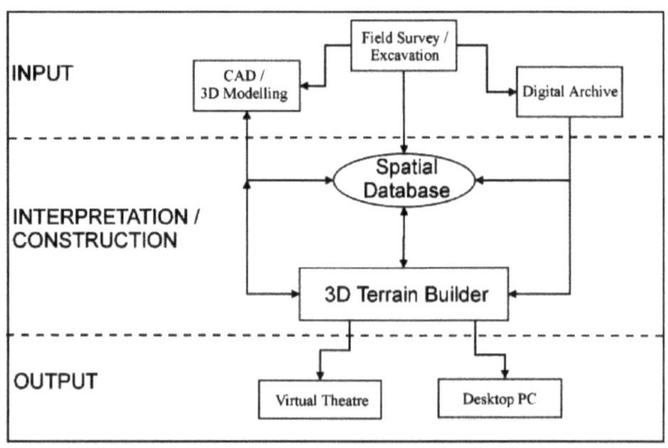

Figure 1. Terra Vista Project schema

A fundamental element within the creation of a virtual landscape is the accuracy of the terrain reproduced, created in Terra Vista through the use of a DEM. Two sets of data were available to create the basis of the model: Data of a GPS survey conducted over several seasons in a localised area on the top of Bieta Giorgis hill in the area of the excavations; Data covering a much wider region around Aksum where contour data was obtained through digitising cartography at 10 meter intervals. Whilst both data sets were imported into the terrain database, due to their differences in scale and accuracy, they were used separately to create localised virtual models on the hill of Bieta Giorgis and for the complete hinterland of Aksum. Terra Vista has the facility to enable datasets required to generate a model and to disable those files not required.

The second category of data which Terra Vista permits the inclusion of is that of geospecific imagery. This term is used to describe georectified aerial photographs or satellite images that may be used to simulate the ground texture. The application of imagery requires that a virtual model have a maximum ceiling and floor imposed in order for the textures to have clarity. Further, areas of modern development seen in the photographs should be removed if the context is that of generating an ancient archaeological landscape. If a walk-through flight were required, it would be necessary to disable the imagery as the texture and instead apply textures captured through photographs taken of the terrain and soil types. The images are imported into Terra Vista as Geo-TIFF files with associated world files (TFW) with all image processing completed prior to importing into the terrain database. It is worth noting that the inclusion of a geospecific image, whilst enhancing the reality of a model, greatly increases the final file size of the virtual world file.

The following stage in the construction of a virtual model is the insertion of vegetation. This will vary greatly for the type of model required, and whilst Terra Vista is provided with a wide choice of generic vegetation types, a majority were unsuitable for application in the Aksum Project. Analysis of the pollen record has shown that in areas that now grow Ecliptic trees (not native to this area) there had previously been Acacia as well as a number of Sycamore, the latter being found at the foot of the hills. Therefore, through cross analysis of satellite imagery and the pollen records it has been possible to reconstruct the type of vegetation present in the ancient landscape. The terrain building software Terra Vista offers a number of methods with which to build the vegetation coverage. The first mode requires the individual insertion of plants through the creation of a point file, either within a GIS package or Terra Vista. Each point is then ascribed a code which is unique for each specie of vegetation. It is further possible to specify the individual height, span and width of each plant therefore creating some diversity. This form of illustration for vegetation is particularly useful when creating walk-through flights as the viewer is much closer to the vegetation and is able to see individual plants and their textures. The drawback of this form of representation is the greater amount of time and processing that is required to build the vegetation, as well as the time required to individually create each plant. An alternative method to simulate vegetation in Terra Vista is through the use of billboards. These are created through closed polygons which define an area of vegetation, such as a woodland. Then a choice of image is made which represents a view of the side of a piece of woodland, which is repeated around the edge of the polygon. This image is set at a chosen height and a canopy placed over the top of the polygon, thus simulating the effect of a closed woodland (as illustrated in figure 2). The advantage of defining vegetation in this manner is the low cost this has on the performance of the flight in runtime. The drawback of this system is the viewer must be kept at a certain distance and altitude from the defined area as the effect when viewed at close proximity is unrealistic. Whilst it is possible to include both representation methods in one flight, it is apparent that one should be used for high fly-through models and the other when the viewers interaction on the surface is greater.

Figure 2. Choice of data types: Point and Billboard.

To create the vegetation to populate the Aksum terrain database, it was necessary to import the correct species into the Terra Vista project. Through analysis of the pollen record, several plant species have been identified that existed in the landscape of ancient Aksum. The first step was to obtain images of these plants, both from the projects digital archive and other sources. Once acquired in digital format (.TIFF), using a graphics program the image was converted into a format whereby only selected parts were visible and the rest transparent. This process is illustrated in figure 3. The first image shows the initial photograph of a tree. The second and third images illustrate the definition of the area and creating of an alpha channel. The fourth image shows the tree as seen in the virtual model. Through manual selection, it is possible to define an area of a tree from within an image, enable this to be seen and the rest of the image is placed within the alpha channel of the image. When imported into Terra Vista and then visualised in a flight file (the output format of the program), the chosen area of the image is seen whilst the rest of the image is transparent. This process was followed for all the vegetation that was visualised in the low flight files generated in the project. Virtual models generated for viewing at a higher altitude used the billboard process and included geospecific imagery.

The focus until this point has been the generation of the natural landscape, correctly detailing topography and vegetation. The next stage is the addition of the habitation and other human activities that took place within the territory. Whilst the landscape was undoubtedly changed by agricultural practices and other human interaction, it is the addition of the 3-D models that begins to transform the landscape from something abstract to that which begins to represent and recreate the ancient landscape. The 3-D resources provided by Terra Vista to generate a landscape are focused towards modern North America, so it has been necessary for the project to establish its own library in order to simulate the type of structures witnessed in the archaeological record.

Figure 3. Process for creating transparent images.

Through the research of the Aksum Project and scholars extensive studies of the territory, in particular by Phillipson (1997), it has been possible to individuate the type of monuments that existed around Aksum. The area is characterised by a number of stele fields, both royal and of lower status. Despite the removal throughout history of these impressive obelisks, a number still remain, both standing and fallen. A majority have been carefully studied, and from these records the dimensions and stone type were extracted and 3-D representations made within a CAD system. Two different forms of stele existed at Aksum, the royal monument which is usually carved, carefully worked stone which may have stood to a maximum height of 20 meters. The other, the lower status stele consist of rough, uneven stone which were simply moved and then placed in an upright position[1]. For the virtual models that were generated, four different forms of royal stele were selected and nine diverse forms of lower status stele. These simpler stele required more diversity as each stone should be different in shape, size and sometimes textures as they represent un-worked stone.

The procedure used by Terra Vista to import individual vegetation, that of creating a point file with an associated unique code, is also used for the more complex 3D models. The models should be imported into Terra Vista as unique flight files (FLT), which may be created through a number of different software packages. The process followed in this research was that of creating the models within AutoCAD, exporting the reconstruction to 3D Studio MAX, applying textures in this program and then exporting, through use of an add-on (PolyTrans), as an OpenGL file. At the current stage of the research program, three different sites have been studied in order to generate a reconstruction. The structure of the work has been to establish the ground plan of the major structures through study of both literary sources and ruins as now seen in the field. From these plans, walls are extruded up to the believed height of the structure. This methodology was followed for the palaces of Enda Sem'on, Mika'el and Ta'kha Maryam which all lay within close proximity to the ancient town of Aksum. A similar process was also followed in order to recreate the type of habitation, drawing heavily on the ground plans seen in the excavations. Investigations have shown that there is much continuity seen between the type of structures witnessed in the archaeological record and those seen in the current landscape.

The final data type that may be entered in to Terra Vista is that of textures, required in order to accurately simulate the ancient landscape. Whilst Terra Vista is provide with a large database of textures, ranging from water types, to soils and roads, the uniqueness of the reconstruction in this project required that numerous texture types be captured in the field for application in the virtual terrain. These images were imported as RGB files, the propriety format of Openflight and added into the Terra Vista library. Whilst applying the correct imagery greatly enhances the simulation and hence immersion of a user, a particular problem was experienced that remains irresolvable at the current stage. Whilst Terra Vista allows the application of different textures in specifically defined areas, the texture is continuously repeated within this area. Whilst it is possible to adjust the size of a image, it must however remain focused. Therefore, textures must be sort that limit the visual impact of repetition, through having no particular feature within the image (such as a large stone in a soil texture). Once all the chosen data types have been entered into the terrain database, and the chosen files enabled, it is possible to pass to the final stage in the terrain generation, that of defining the parameters of the model.

4. Terrain Generation: Definition of Parameters

The previous sections have discussed reasons for the choice of VR system and the data types which were used to construct the terrain database. The final step to create a virtual world, as illustrated in figure 4, is the definition of the parameters of the terrain to be generated, through an interviewing process of the data. Terra Vista allows the modification of all aspects of the terrain, as well as optimising the model for the system upon which it will be

[1] A fuller discussion and drawings of these monuments may be found in Phillipson (1997).

run. This section offers a brief summary of some of the key areas where decisions concerning the parameters of the model must be made which will determine the visual impact of the model. The need to define these parameters prior to the construction of the model is necessary for any VR system, as the terrain builder must optimise the model for a specific use. For example, the choice of textures will vary greatly depending upon the chosen of viewing height for the model.

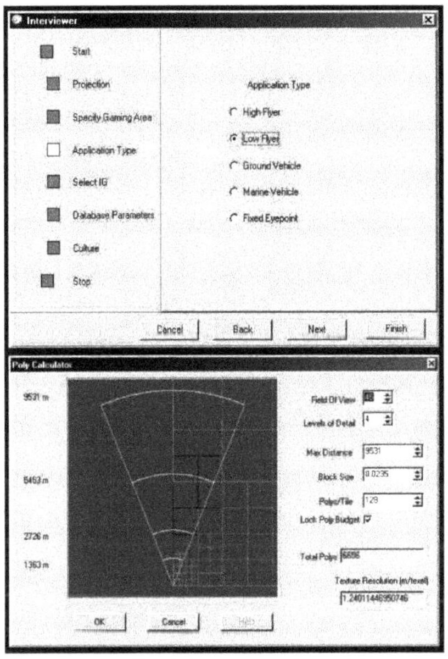

Figure 4. Terra Vista landscape parameters wizard

The procedure as followed in Terra Vista firstly requires the definition of the extent of the landscape model. Factors to consider are the quality and extent of the surface terrain model (DEM), the area of interest and the final file size of the VR model. Closely related to the definition of area is the height of viewing the terrain database. Terra Vista offers five options: a high-fly through (commonly referred to as 'birds-eye'), low fly-through, a walk-through, a marine vehicle and fixed eye-point model. Whilst it is possible to change the viewing height at any point when viewing the terrain, the choice at this stage optimises the database through type of textures and amounts of vegetation. This choice of height for the model is also closely related in Terra Vista to the decision of level-of-detail (LOD) for the model. This important aspect of the program has a considerable effect on the final visual impact of the model. Each LOD represents the terrain at a different polygon density: less detailed for when it is far from the viewer, more detailed when it is closer (illustrated in figure 5). Whether the terrain is displayed in a Virtual Theatre or on a notebook computer, the hardware on which it is rendered will determine some of the parameters for the terrain generation. The number of polygons that can be drawn per second along with the desired frame rate will determine how many polygons can be drawn per frame.

The decision of the number of LOD's (these range from 1 to 6 in Terra Vista) is therefore closely related to the end application for the model. The Aksum Project required models to be optimised for two different platforms, a Virtual Theatre and for visualization on desktop computers. The need to consider the end application of the virtual model is specifically related to the file size of the model. Whilst many factors account for this, specifically the number of datasets used within the model will effect the fluidity of the model.

The last parameter necessary to define is the format of the data output. Terra Vista is provided with two choices, Terra Page and Openflight, both of which are transferable to other software but the second of which is the more widely supported industry standard.

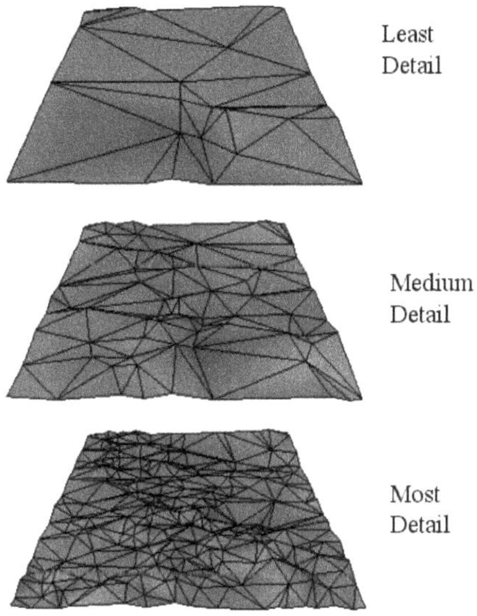

Figure 5. Representation of 3 different levels of detail (LOD) used for visualisation by Terra Vista.

The OpenGL working environment provides high quality graphics on inexpensive PCs through the use of accelerated graphics cards. It is the premier environment for developing portable, interactive 2D and 3D graphics applications through its incorporation of a broad set of rendering, texture mappings, special effects, and other powerful visualisation functions. A further advantage important to this project is that Openflight database is relatively simple to modify and update, which was a necessary step once the data were passed through the Terra Vista program. The subsequent period of post-processing of the data was conducted in MultiGen Paradigm's Creator software, a program particularly suited to working with the Openflight format. Whilst models were optimised for run-time viewing before they were entered into the Terra Vista database, Creator offers the opportunity to manipulate the terrain at the highest level of detail. As the creation of the final terrain Openflight file was an automated function, some errors that visually distort the reality of the terrain were experienced and removal of these was only possible through Creator. For example, the vegetation was inserted into the walk-through terrain as individual points which occasionally coincided with other cultural data, such as a plant appearing on the surface of a river. Through Creator, these can be simply removed or repositioned elsewhere.

The project has sought to develop models for two different areas of application. An aim of the project is to experiment with and develop the viewers perception of a landscape, best achieved through enhancing immersion within a virtual reconstruction. Therefore, thanks to scientific collaboration with the CINECA Institute (Bologna, Italy), virtual models were developed in Terra Vista and Creator to be viewed using the Virtual Theatre facilities of the institute. However, prior to viewing in this form of medium, modifications are necessary to the models and further, the opportunity is given to add atmospheric, temporal and sound effects. These effects were achieved through the use of MultiGen Paradigms Vega software, which was also used to add the stereographic effect thus allowing projection through three Barco Projectors and the viewers to be able to have a fuller sense of immersion through wearing stereoscopic glasses.

Figure 6. Virtual landscape with 3D model of Ta'kha Maryam palace in foreground.

Figure 7. The royal stele field lying close to the reconstructed palaces of Enda Sem'on and Mika'el.

Figure 8. Reconstruction of the ancient Aksumnite landscape and a village.

Figure 9. Reconstruction of a compound with livestock (inserted as directional billboards).

The modifications made to the Openflight dataset in Vega are unfortunately unable to be read by standard desktop computers running software for visualising 3-dimensional Openflight models[2]. Therefore models were generated in Terra Vista, exported in Openflight format to Creator for modification and then saved in Virtual Tree (VT) format to allow simple distribution. A sample selection is offered in figures 6 – 9 of views of a landscape that was generated for use on a PC platform. It is important to emphasise that these images were captured as colour screen-shots whilst navigating the landscape, they do not represent still shots produced through the software.

5. The Virtual Terrain

The aim of the first stage of the Aksum digital project was to develop a methodology and system to be implemented for the creation of virtual landscapes based upon data collected by the entire project. A considerable part of the conducted research has therefore been creating the data archives and testing different procedures. In particular, a process was sort whereby landscapes could be rapidly generated and modified, as well as being transferable and realistic. Two processes were developed for the creation of virtual models for a Virtual Theatre and for use on desktop computers.

The very nature of VR dictates that the most appropriate way to assess the results is through experiencing the landscape that was created. Therefore, the most satisfactory results so far witnessed were those seen in the Virtual Theatre of the CINECA Institute. This structure, through the combination of surround sound and stereoscopic vision, gave the viewer a sense of immersion as they could freely navigate the terrain, although they were unable in this situation to query the underlying database. The theatre gave an important sense of space within the landscape, through its scale and complexity. Textures were also important in drawing attention to the solid surfaces, as texture is an important primal component in the way we perceive the world (Gibson, in Wheatley and Earl 2002). The use of the OpenGL standard for generating the VR meant that the viewing of the model was satisfactory for vision, as passage was smooth and unbroken.

[2] The freeware Audition (CG2) was used by this project.

An aspect of this research has been to utilise portable medium for the data so as the results can be visualised on any hardware. Whilst the VR viewing software is freely available, the specific problem experienced was that of the users immersion in the landscape. In the same manner as it is uncomfortable to read long passages of text from a monitor (Lockyear 1996), so it is that a viewer has a lesser sense of immersion when looking at a virtual model through a monitor. However, through the use of PC's it has been possible to explore and develop, through the use of C++ programming, a process of combining the virtual terrain with a spatial database.

A key aspect in the successful completion of the project thus far has been the use of the terrain building software Terra Vista. The ability to maintain the spatial component of the data has meant that the passage to VR has been a natural extension of the GIS. In future research it is intended to more fully explore this interactivity through applying forms of spatial analysis in a virtual environment, instead of limiting them in their current form to the conventional 2D representation. In particular, as seen from the results of recent research conducted around Stonehenge (Exon et al. 2002) VR may have a particular application in the assessment of visibility analyses.

The 3D reconstruction of the ancient Aksumnite landscape through virtual reality has provided a useful contribution to the development of the project and to the construction of cognitive models. This project represents an application of a new tool for understanding the past, through the integration of spatial techniques (GIS) and visual information dynamic systems (VR). The encouraging results achieved so far, in particular the highly detailed georeferenced terrain created by Terra Vista mean that this avenue of research will continue to be explored.

Acknowledgements

The EU funded grant made through the MINOS program at CINECA (for period October 2001) to Stephen Kay.

The Staff of the CINECA VIS.IT lab for their assistance to the research.

Cinzia Perlingieri for advice concerning the excavation and field data collected.

References

BURROUGH P. A., MCDONNELL R.A., 1998 *Principles of geographical information systems*, Oxford, Oxford University Press.

EXON S., GAFFNEY, V., WOODWARD, A., YORSTON, R., 2002 *Stonehenge Landscapes: Journeys through real-and-imagined worlds*, Oxford, Archaeopress.

GIDLOW J., 2002 Rock art and 'Bubble worlds', in Wheatley, D., Earl, G. and Poppy, S. (eds.), *Contemporary themes in archaeological computing*, University of Southampton Department of Archaeology Monograph no.3, Oxford, Oxbow Books.

GILLINGS M., 1999 Engaging place: exploring the potential of VR in experiential landscape studies, in Dingwall, L., Exon, S., Gaffney, V., Lafflin, S. and van Leusen, M., *Computer applications and quantitative methods in archaeology: proceedings of the 25th anniversary conference, University of Birmingham April 1997*, Oxford: BAR Publishing. British Archaeological Reports S750.

POLLARD J., GILLINGS, M., 1998 Romancing the stones: towards a virtual and elemental Avebury, *Archaeological Dialogues 5(2).*

GOODRICK G., GILLINGS M., 2000 Constructs, simulations and hyperreal worlds: the role of Virtual Reality (VR) in archaeological research, in Lock, G. and Brown, K. (eds.), *On the Theory and Practice of Archaeological Computing*, Oxford, Oxbow Books.

FORTE M., 2000 About virtual archaeology: disorders, cognitive interactions and virtuality, in Barcelo J., Forte M., Sanders D. (eds.), *Virtual Reality in Archaeology*, Oxford, BAR Publishing (BAR International Series S 843).

LOCKYEAR K., 1996 Computer aided publication in practice, in Kammermans, K. and Fenneman, K. (eds.), *Interfacing the past: Computer applications and quantitative methods in archaeology 1995.* Leiden: Analecta Prehistoria Leidensia 28.

PHILLIPSON, D. W., 1997 *The Monuments of Aksum*. Addis Ababa University Press in collaboration with The British Institute in Eastern Africa.

WHEATLEY, D., Earl, G., 2002 Virtual reconstruction and the interpretative process: a case-study from Avebury, in Wheatley, D., Earl, G. and Poppy, S. (eds.), *Contemporary themes in archaeological computing*, University of Southampton Department of Archaeology Monograph no.3, Oxford, Oxbow Books.

WHEATLEY D., GILLINGS M., 2002 *Spatial Technology and Archaeology. The archaeological applications of GIS*, London, Taylor and Francis.

Digital Technologies and Prehistoric Landscapes in the American Southwest

John Kantner and Ronald Hobgood

Department of Anthropology and Geography, Georgia State University
University Plaza, Atlanta, GA 30303-3083
kantner@gsu.edu

Abstract. Archaeologists investigating prehistoric societies of the Southwestern United States gravitate towards two different explanations for the configuration of the built environment. One set of scholars tends to see features such as roadways and masonry towers as facilitating regional interaction by decreasing travel costs and increasing communication. Other researchers regard many of these features as inseparable parts of a political and ideological landscape that had little to do with maintenance of a regional "system." This chapter demonstrates how digital technologies such as virtual reality and Geographic Information Systems can help archaeologists determine how cultural landscapes were perceived by their creators.

1 Introduction

The Chaco Anasazi cultural tradition emerged in the northern U.S. Southwest around A.D. 900 and evolved over approximately 250 years before fading away (Figure 1). In general, the tradition has been loosely defined by the development of an unprecedented architectural style that first appeared in the late A.D. 800s. The best-known feature of this new style is the "great house," a monumental masonry structure that required an exceptional labor investment, especially since the materials were not locally available; tens of thousands of beams had to be hand carried at least 40 km across the semi-arid desert. Supplementing the great house form were "great kivas," enormous subterranean ritual structures whose construction would have also required a great deal of labor.

These features first appear in the isolated and desolate Chaco Canyon in what is now northwestern New Mexico. For many scholars, that this location saw the development of monumental architecture is something of a mystery, for this canyon in the center of the expansive San Juan Basin is one of the most inhospitable places in the northern Southwest. The canyon is characterized by unpredictable and scanty rainfall, a short growing season due to its high altitude, and virtually no unique natural resources. Despite these challenges, Chaco Canyon appears to have been the center of the entire Chaco Anasazi cultural tradition (e.g., Stuart 2000; Lekson 1999).

Roadways built during the Chaco era extend away from the canyon towards the wetter, higher-elevation, and comparatively resource-rich areas on the edge of the San Juan Basin (Figure 1).

Situated in these outlying areas are dozens of prehistoric communities that constructed smaller versions of great houses and great kivas not long after they first appeared in Chaco Canyon. Materials from these outlying areas are found in appreciable quantities in the central canyon, with a few locations providing large volumes of ceramics and lithics (Toll 2001; Cameron and Toll 2001; Mathien, 2001). These distant communities are accordingly referred to as "Chacoan outliers", or, more objectively, "great house communities"(Kantner and Mahoney 2000).

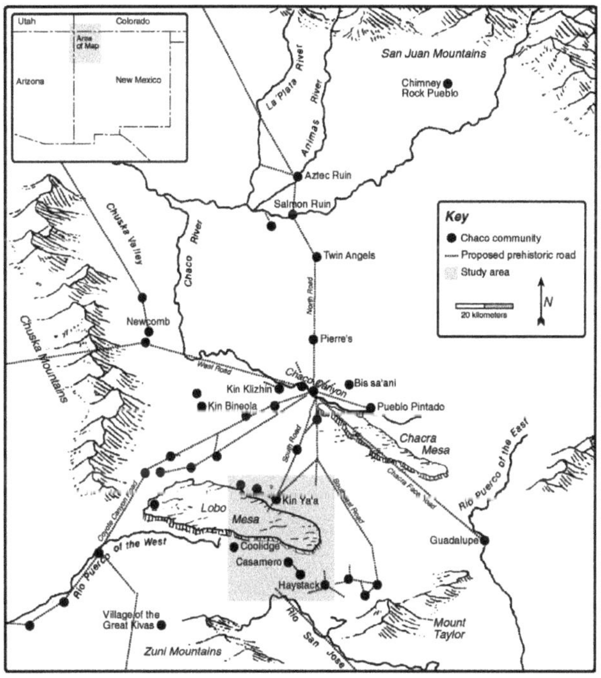

Figure 1: The Chaco Anasazi tradition was centered in Chaco Canyon in northwest New Mexico, but its influence spread over the entire northern Southwest. Based on the remains of numerous roadways found throughout the area, some scholars propose that an expansive road network once existed

Based on the apparent regional primacy of Chaco Canyon, many archaeologists have considered it to have been the center of an expansive political and economic system. Some models argue that the canyon served as a collection and redistribution point that allowed regional participants to contend with variable climatic conditions and regional environmental diversity (e.g., Stuart 2000). According to this view, communities would alternatively send their surpluses to the central repositories in Chaco Canyon or request supplementary foods when needed; roadways facilitated these transactions. Alternative models see Chaco as an imperialistic polity that forcibly appropriated communities and thereby expanded its borders; roadways accordingly served as conduits for military forces, while masonry towers provided defensive cover (Wilcox 1993, 1999). Still other archaeologists contend that the canyon was a pilgrimage center, with the roads envisioned as thoroughfares for both pilgrims and the resources needed for the canyon to host large ceremonial events (e.g., Malville and Malville 2001). One proposal is that a centralized religious tradition based in Chaco Canyon incorporated new communities through a process of missionization (e.g., Kane 1993).

2 Investigating Great House Communities

Many archaeologists have attempted to evaluate these models for the origin of the Chaco Anasazi and the function of the central canyon. Until the last two decades, research was primarily focused on Chaco Canyon itself (for a history of research, see Vivian 1990). More recently, a few scholars have begun to examine the great house communities located outside of the main canyon (see contributions in Kantner and Mahoney 2000). Since most of the models make predictions for how the outlying areas functioned within a larger regional system, one way to evaluate them is to examine how great house communities interacted with one another. Were outlying communities part of an integrated economic system, or were they simply funneling resources towards the center at Chaco Canyon? Is there evidence of military outposts in the outlying communities? Were the communities established as colonies, or were the outlying great houses built as part of a missionization process?

The answers to these questions can be found through intensive investigation of various outlying regions. The research reported here has focused on a 2,500 km2 study area found approximately 40 km south of Chaco Canyon (Figure 2). The center of this study area is Hosta Butte, a prominent geological feature that can be seen from long distances away; it is clearly visible from the largest great houses in Chaco Canyon itself. Hosta Butte sits on a plateau known as Lobo Mesa that divides the study area in half, with the San Juan Basin on the north and the Red Mesa Valley on the south. Prehistoric communities are found on both sides of Lobo Mesa, and all of them saw the construction of great houses in the 10th and 11th centuries, not long after the appearance of this architectural form in Chaco Canyon to the north. Associated with these structures are an array of features often associated with the central canyon, including great kivas, segments of roads, and distinctive pottery styles associated with the Chaco Anasazi.

Several different lines of investigation have focused on the development of the great house communities in the study area and their interactions with one another (Kantner 1996, 1997, 1999; Kantner et al. 2000).

This chapter considers two of the more intriguing analyses, both of which rely on the application of digital technologies, especially Geographic Information Systems (GIS) and three-dimensional (3D) modeling. First, an analysis of the Chaco Anasazi roadways as manifested in

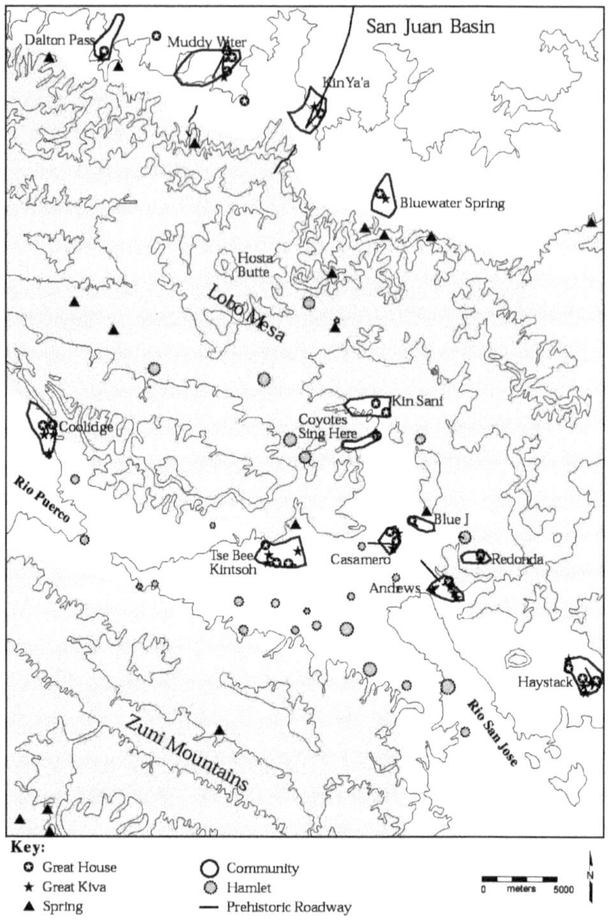

Figure 2: The 2,500-km2 study area containing numerous communities, smaller hamlets, and road segments is centered on the prominent geological feature of Hosta Butte.

the study area is discussed. This work relies on the simulation of cost-pathways using GIS software. Next, the function of unique "tower kivas" found in great house communities is considered through 3D modeling and GIS-enhanced viewshed analysis.

3 Chaco Anasazi Roadways

Our understanding of the famous Chacoan road "system" has changed considerably over the past few decades (Vivian 1997a, 1997b; Snygg and Windes 1998). Based on preliminary investigations, many of which were aerial surveys, the identified roadways were believed to be part of an extensive and well-integrated network (Figure 1). Initial assessments concluded that the roads formed a dendritic network radiating from Chaco Canyon to secondary centers, which in turn built roads connecting to smaller communities scattered over the northern Southwest (Ebert and Hitchcock 1977). These views had a profound impact on our understanding of the Chaco Anasazi tradition, for this reconstruction of the roadways was consistent with explanations that saw Chaco Canyon as a regional redistributive center.

Unfortunately, our vision of the actual archaeological record for the roadways has turned out to be somewhat flawed (Figure 3). Ground-based attempts to confirm the roads identified from the air have discovered that at least some are actually historic tracks (Nials et al. 1987; Nials 1983). In fact, very few prehistoric segments extend beyond one or two kilometers, with the exception of the North and South Roads emanating from Chaco Canyon (Roney 1992). The unresolved question, then, is whether the short segments that remain today were part of the once suspected regional network or whether they may have served some other function yet to be identified.

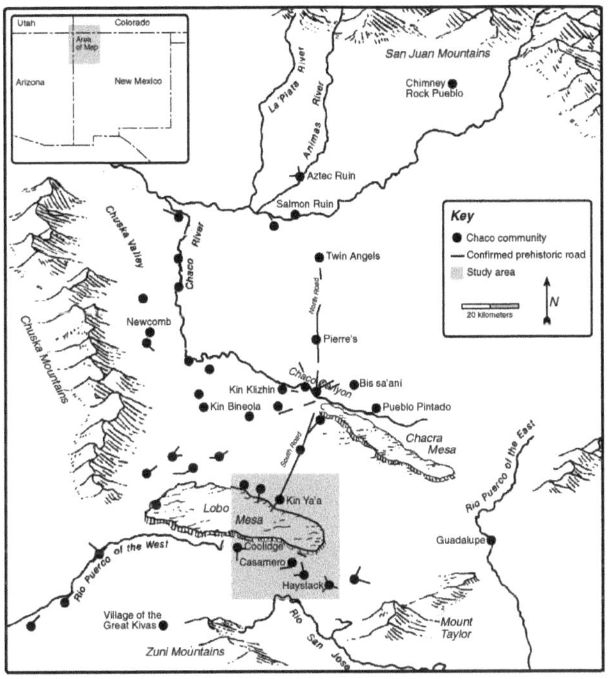

Figure 3: The distribution of roadway segments that have been confirmed through ground investigation reveals that the Chaco Anasazi great house communities were not as well connected as original assessments suggested (e.g., Figure 1).

With these questions in mind, a GIS-enhanced cost-path analysis was conducted on the road segments identified in the Lobo Mesa study area. The goal was to predict where roads should be if they were built to minimize travel effort between any set of points, which were chosen according to the expectations of general models for explaining the evolution of the Chaco Anasazi. The results of these simulated "roadways" could then be compared with the actual road segments to evaluate what goals may have guided the roads' original builders.

To accomplish this analysis, U.S. Geological Service 7.5-minute Digital Elevation Models (DEM) were assembled for the entire 2,500-km2 study area using Arc/Info GIS software produced by ESRI. This created a digital lattice of elevation measurements, with each point 30 m apart. From this base map, a "cost surface" was generated using a "hiking algorithm" created by geographer Waldo Tobler from empirical observations of the movement of Swiss soldiers over variable terrain (Tobler 1993). The resulting cost surface was a lattice of cells, each with a value indicating the cost in time to move across it. Next, least-cost pathways minimizing travel time between any group of points were created; using this technique, entire networks could be built based on the expectations of the hypothesized functions of the roadways (Figure 4). Finally, the actual road segments were overlaid onto the hypothetical networks to determine which provided the closest fit to the actual archaeological record.

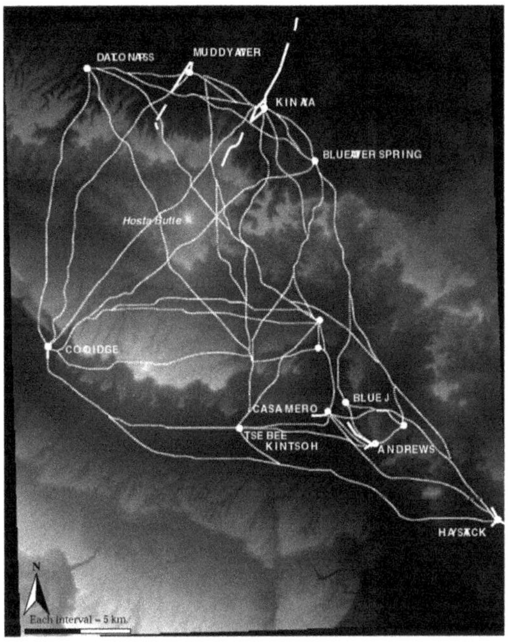

Figure 4: The simulated network of paths is built on the assumption that people moving between communities would have chosen routes that took the least travel time.

One of the first hypotheses to be evaluated was whether the road segments in the study area were the remnants of a comprehensive network that once integrated communities into an economic system. Figure 4 illustrates how such a system would look if its builders wished to construct the most efficient routes for connecting all of the communities together. When compared with the actual prehistoric roadways, however, the match is very poor. For example, road segments identified in the Muddy Water community do not even remotely line up with this hypothetical cost-path network (Figure 5). In fact, none of the remaining roads line up with any part of the integrated network.

What then were the prehistoric roadways used for? Comparison with other cost-paths based on different hypothesized functions suggests two likely functions for the Chaco Anasazi road segments in the study area. First, the majority of actual roadways line up with simulated cost-paths that minimize travel effort from the communities to Hosta Butte itself. In Kin Ya'a, for example, the road segments provide a good fit with a cost-path between the community and the prominent butte (Figure 6). This is especially interesting since Hosta Butte continues to be an important feature to indigenous people living in the region today (Van Valkenburgh 1941). A few other road segments, however, correspond most closely with hypothesized cost-paths that link the great house complexes in community centers with nearby habitation clusters. The Andrews community includes a road that fits this scenario (Figure 7).

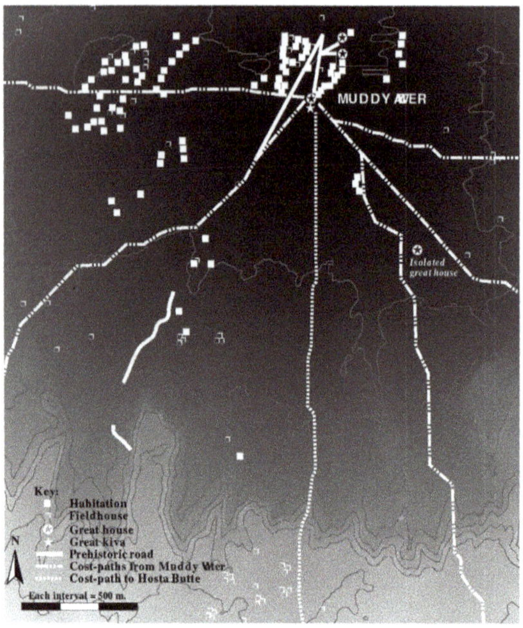

Figure 5: The actual road segments found in the Muddy Water community do not match the simulated cost-efficient routes between communities.

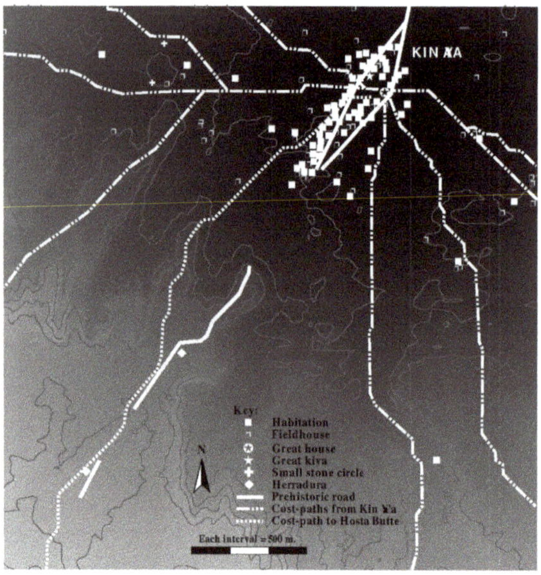

Figure 6: A reasonably close match is seen between the prehistoric roadways in Kin Ya'a and the simulated route leading to the prominent feature known as Hosta Butte.

The results of this analysis demonstrate that the road segments appearing in the Lobo Mesa study area do not match models that see the outlying great house communities as having been closely integrated with one another. Especially clear is that no dendritic redistributive system was supported by the Chaco Anasazi roadways. Instead, these features apparently served the ceremonial great house complexes, linking these monumental structures both with prominent features on the landscape and with nearby settlement clusters (see also Kantner 1997; Till 2001).

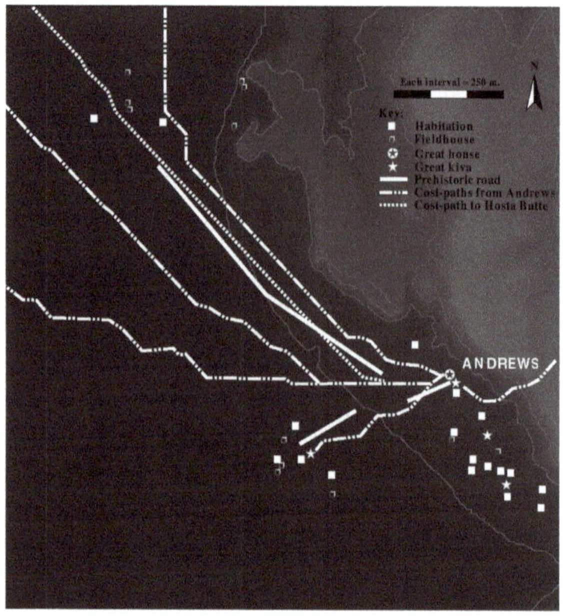

Figure 7: Prehistoric road segments in Andrews correspond both with a simulated route leading to Hosta Butte as well as with a route connecting the great house complex in the center of Andrews with a nearby cluster of houses.

4 Chaco Anasazi Tower Kivas

Tower kivas are unusual features that are associated with Chacoan great houses. Instead of the typical subterranean kiva that is ubiquitous across the northern Southwest, the tower kiva is incorporated into the great house structure and elevated a few stories above ground (Figure 8). Their function is still debated, with proposals that they may have been part of a long-distance communication network that facilitated interaction across the entire Chaco Anasazi region. Other suggestions are that they may have been watchtowers designed to keep track of potential threats to the community or the great house itself. Another possibility is that they functioned not unlike church towers or minarets, both as a symbolic link with the supernatural world but also as a widely visible extension of the ceremonial infrastructure.

Figure 8: A three-dimensional computerized reconstruction of the Kin Klizhin great house, including the square tower kiva that projects above the rest of the structure.

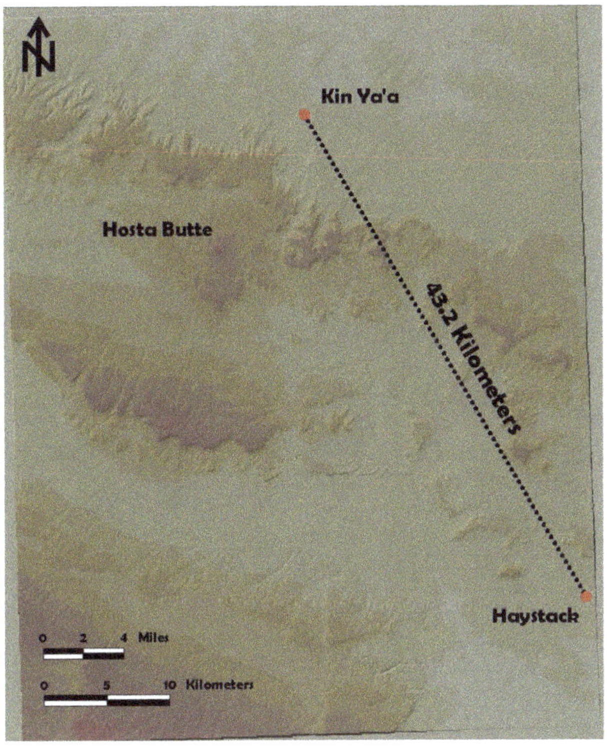

Figure 9: In the study area, two tower kivas have been identified on opposite sides of Lobo Mesa in the great house communities of Kin Ya'a and Haystack.

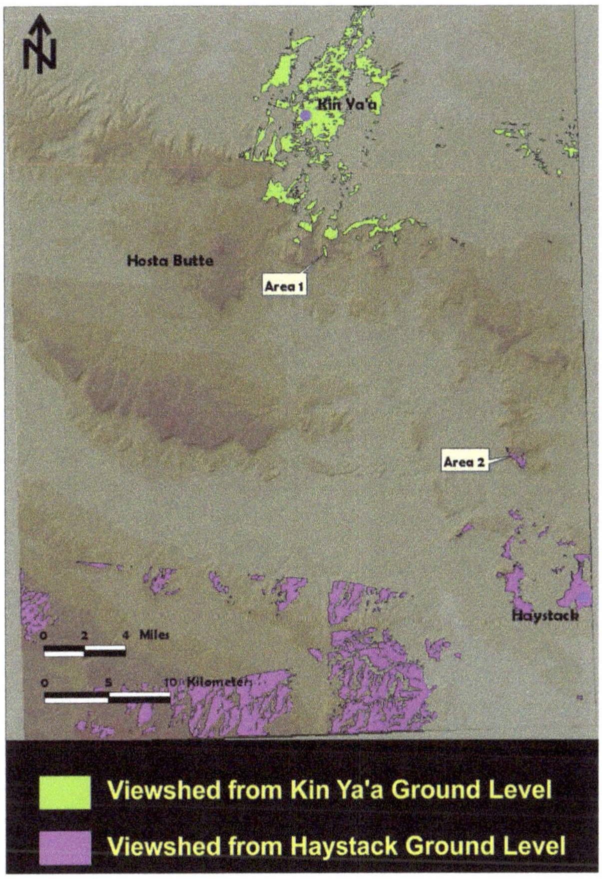

Figure 10: The viewshed analysis indicates that the two communities of Kin Ya'a and Haystack are clearly not intervisible from the ground.

Two tower kivas are found in the Lobo Mesa study area, one in the community of Kin Ya'a and one in Haystack (Figure 9). To assist in evaluation of tower kiva function, viewshed analysis was conducted on these two tower kivas using the same elevation lattice employed for the roadway investigation (e.g., van Leusen 1999, Madry and Rakos 1996; Lake et al. 1998). Using ESRI's Arcview GIS software, the viewable landscapes from the top of the tower kivas were compared with those seen from the ground surface to determine if there were any advantages to the higher elevation of these unique features. Because the tower kivas are no longer intact, this technique was the most reasonable means to evaluate the different hypotheses for tower kiva function.

The first hypothesis to evaluate was whether or not the Kin Ya'a and Haystack tower kivas were part of a regional communication network. Since the two communities are found on opposite sides of Lobo Mesa, they are clearly not intervisible from the ground (Figure 10). The viewshed analysis, however, demonstrated that the two areas were also not intervisible from the tower kivas, even if someone were to stand on their roofs (Figure 11). An alternative hypothesis that was considered was if an intermediate point on Lobo Mesa may have served as a "relay station" allowing visual communication between the tower kivas. To evaluate this idea, areas where the viewsheds from each tower kiva overlapped were identified and targeted for field investigation (Areas A and B in Figure 11). A preliminary assessment of existing survey records indicated that no archaeological sites are found in these potential relay points. However, perhaps more importantly, it was also noted that these same points were also visible from the ground; in other words, building the tower kivas to facilitate communication would have been superfluous since the necessary relay points for any cross-mesa communication were already visible from the ground. The evidence indicates that the tower kivas in the study area were not built as part of a larger regional communication network.

One hypothesis receiving tentative support from the viewshed analysis is that the tower kivas were erected to increase visibility of the local area. The viewshed analysis indicated that, even though the tower kivas did not facilitate long-distance views, the local area visible from the top of these kivas almost doubled in size compared with what could be seen from the ground (Figure 11). This can be interpreted in two ways: either the tower kiva allowed occupants of the great house to see the entire surrounding community, or, alternatively, the tower allowed all of the habitations in the community to see the great house. The investment in this new architectural form therefore may have been intended to increase the intensity of local interaction between great houses and nearby habitation clusters.

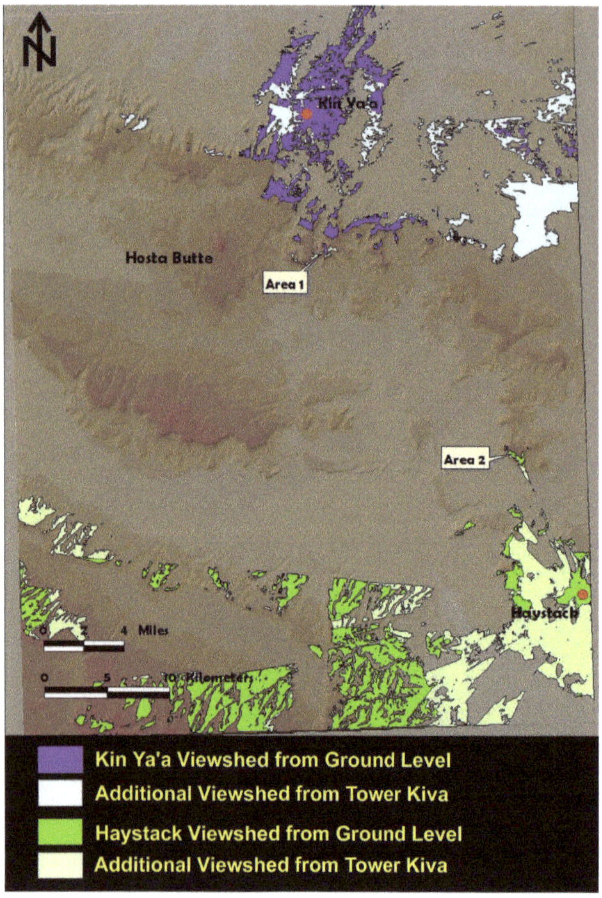

Figure 11: The elevated positions of the tower kivas could not have facilitated intervisibility between Kin Ya'a and Haystack. Two areas were identified on Lobo Mesa as points where the two viewsheds overlapped, but an attempt to identify "relay stations" at these points was unsuccessful. In any case, those same points were visible from the ground, indicating that the tower kivas would have not helped in any way to facilitate visual communication between the communities.

5 Concluding Discussion

The results of the roadway analysis and the tower kiva analysis suggest that neither of these Chaco Anasazi features were designed to facilitate region-wide integration, at least within the Lobo Mesa study area. Road segments and tower kivas instead appear to have been constructed to enhance integration at a much smaller scale, focused at the community level. While this certainly does not mean that no regional interaction occurred among the numerous outlying communities, so far little investment in regional infrastructure has been identified in the study area. These results are consistent with other investigations that demonstrate little movement of pottery across this part of the Chaco world (Kantner et al. 2000; Van Dyke 1997).

These conclusions may appear to contradict the evidence for at least two region-wide roadways—the North Road and the South Road—that emanate from Chaco Canyon (Figures 1 and 3). These two roads appear to have been built at a scale far exceeding that of most Chaco-era roadways, for indeed they extend tens of kilometers across the San Juan Basin. Initial interpretations of these two roads were that they connected the canyon with outlying great house communities in more resource-rich areas. However, closer analysis reveals that these unusual roadways articulate with geological features—Kutz Canyon in the north and Hosta Butte in the south (Marshall 1997). This is intriguingly consistent with the results of the cost-path analysis of roads in the Lobo Mesa study area, but the North and South Roads are simply at a much greater scale, similar to the larger scale of great houses found in Chaco Canyon compared with the more diminutive versions found in outlying communities. These distant communities were apparently trying to do the same thing as Chaco Canyon, but they were only able to do it at a much smaller scale (see also Van Dyke 1999).

Although great house communities may have been emulating the infrastructure seen in Chaco Canyon, the manifestation of roadways and tower kivas suggest that their primary concerns were local. These features were essentially extensions of the great houses found in each community, with the roadways radiating out from the great houses, and the tower kivas extending the vertical profile of the structure. This interpretation leads to two suggestions. First, these features may provide some insight into the sociopolitical context in which great houses appeared and were expanded, especially since the builders of the great house complexes apparently wanted to ensure that the physical manifestation of this ceremonial architecture encompassed as much three-dimensional space as possible. Second, the fidelity of Chacoan features in great houses has been highlighted in other research as a means for determining the circumstances by which these structures were built (Van Dyke 1999, 2000; Hurst, 2000). The variable manifestations and scale of tower kivas and roadways is therefore an additional source for evaluating the intensity of ties between outlying areas and Chaco Canyon.

References

CAMERON, C. M. and H. W. TOLL. 2001. *Deciphering The Organization Of Production In Chaco Canyon.* American Antiquity 66: 5–13.

EBERT, J. I. and R. K. HITCHCOCK. 1977. *Locational Modelling In The Analysis Of The Prehistoric Roadway System At And Around Chaco Canyon, New Mexico.* In: T. R. Lyons And R. K. Hitchcock, Eds. Aerial Remote Sensing Techniques In Archeology. Albuquerque, New Mexico: National Park Service And University Of New Mexico. 169–207.

HURST, W. 2000. *Chaco Outlier Or "Wannabe?" Comments And Observations On A Provincial, Chacoesque Great House At Edge Of The Cedars Ruin, Utah.* In: J. Kantner And N. M. Mahoney, Eds. Great House Communities Across The Chacoan Landscape. Tucson: University Of Arizona Press.

KANE, A. E. 1993. *Settlement Analogues For Chimney Rock: Models Of 11th And 12th Century Northern Anasazi Society.* In: J. M. Malville And G. Matlock, Eds. The Chimney Rock Archaeological Symposium. Fort Collins, Colorado: Rocky Mountain Forest And Range Experiment Station, U. S.

Department Of Agriculture. 43–60.

KANTNER, J. 1996. *Political Competition Among The Chaco Anasazi Of The American Southwest.* Journal Of Anthropological Archaeology 15: 41–105.

KANTNER, J. 1997. *Ancient Roads, Modern Mapping: Evaluating Prehistoric Chaco Anasazi Roadways Using Gis Technology.* Expedition Magazine 39: 49–62.

KANTNER, J. 1999. *The Influence Of Self-Interested Behavior On Sociopolitical Change: The Evolution Of The Chaco Anasazi In The Prehistoric* American Southwest. Thesis (Phd). University Of California At Santa Barbara.

KANTNER, J., N. BAWER, J. LADWIG, J. PERLITZ, S. HATA, and D. GREVE. 2000 *Interaction Between Great House Communities: An Elemental Analysis Of Cibolan Ceramics.* In: J. Kantner And N. M. Mahoney, Eds. Chacoan Great House Communities Across The Anasazi Landscape. Tucson: University Of Arizona Press. 130–146.

KANTNER, J. And N. M. MAHOANEY (Eds.). 2000. *Great House Communities Across The Chacoan Landscape.* Tucson: University Of Arizona Press.

LAKE, M. W., P. E. WOODMAN, and S. J. MAITHEN. 1998. *Tailoring Gis Software For Archaeological Applications: An Example Concerning Viewshed Analysis.* Journal Of Archaeological Science 25: 27–38.

LEKSON, S. H. 1999. *The Chaco Meridian: Centers Of Political Power In The Ancient Southwest.* Walnut Creek, California: Altamira Press.

MADRY, S. L. H. and L. RAKOS 1996. *Line-Of-Sight And Cost-Surface Techniques For Regional Research In The Arroux River Valley.* In: H. D. G. Maschner, Ed. New Methods, Old Problems: Geographic Information Systems In Modern Archaeological Research. Carbondale: Southern Illinois University. 104–126

MALVILLE, J. M. and N. J. MALVILLE. 2001. *Pilgrimage And Periodical Festivals As Processes Of Social Integration* In Chaco Canyon. Kiva 66: 327–344.

MARSHALL, M. P. 1997. The *Chacoan Roads: A Cosmological Interpretation.* In: B. H. Morrow And V. B. Price. Eds. Anasazi Architecture And American Design. Albuquerque: University Of New Mexico Press. 62–74.

MATHIEN, F. J. 2001. *The Organization Of Turquoise Production And Consumption By The Prehistoric Chacoans.* American Antiquity 66: 103–118.

NIALS, F. L. 1983. *Physical Characteristics Of Chacoan Roads. In: C. Kincaid, Ed. Chaco Roads Project Phase I: A Reappraisal Of Prehistoric Roads In The San Juan Basin. Albuquerque*, New Mexico: Department Of The Interior, Bureau Of Land Management. 6-1–6-51.

NIALS, F., J. STEIN, and J. RONEY. 1987. C*hacoan Roads In The Southern Periphery: Results Of Phase Ii Of The Blm Chaco Roads Project.* Santa Fe: Bureau Of Land Management, New Mexico State Office.

RONEY, J. R. 1992. *Prehistoric Roads And Regional Integration In The Chacoan System.* In: D. E. Doyel, Ed. Anasazi Regional Organization And The Chaco System. Albuquerque, New Mexico: Maxwell Museum Of Anthropology. 123-131.

SNYGG, J. And T. WINDES. 1998. *Long, Wide Roads And Great Kiva Roofs.* Kiva 64: 7–25.

STUART, D. E. 2000. *Anasazi America.* Albuquerque: University Of New Mexico.

TILL, J. D. 2001. *Chacoan Roads And Road-Associated Sites In The Lower San Juan Region: Assessing The Role Of Chacoan Influences In The Northwestern Periphery.* Thesis (Ma). Department Of Anthropology, University Of Colorado.

TOBLER, W. 1993. *Three Presentations On Geographical Analysis And Modeling.* Santa Barbara: University Of California At Santa Barbara (Technical Report, 93-1).

TOLL, H. W. 2001. *Making And Breaking Pots In The Chaco World.* American Antiquity 66: 56–78.

VAN DYKE, R. M. 1997. *Tracking The Trachyte Boundary: A Southern Perspective On Exchange And Interaction Among Chacoan Communities. 62nd Annual Meeting Of The Society For American Archaeology*, Nashville, Tennessee, 1997.

VAN DYKE, R. M. 1999. *The Chaco Connection: Evaluating Bonito-Style Architecture In Outlier Communities.* Journal Of Anthropological Archaeology 18: 471–506.

VAN DYKE, R. M. 2000. *A Chacoan Landscape: The View From The Red Mesa Valley.* In: J. Kantner And N. Mahoney, Eds. Chacoan Great House Communities Across The Anasazi Landscape. Tucson: University Of Arizona Press.

VAN LEUSEN, M. 1999. *Viewshed And Cost Surface Analysis Using Gis. In: J. A. Barcelo, I. Briz, And A. Vila, Eds. New Techniques For Old Times.* CAA98. Computer Applications And Quantitative Methods In Archaeology. Oxford: BAR. 215–223.

VAN VALKENBURGH, R. F. 1941. *Diné Bikéyah. Window Rock,* Arizona: U.S. Department Of Interior, Office Of Indian Affairs, Navajo Service.

VIVIAN, R. G. 1990. *The Chacoan Prehistory Of The San Juan Basin.* San Diego: Academic Press.

VIVIAN, R. G. 1997a. *Chacoan Roads: Function.* Kiva 63: 35–68.

VIVIAN, R. G. 1997b. Chacoan Roads: Morphology. Kiva 63: 7–34.

WILCOX, D. R. 1993. The Evolution Of The Chacoan Polity. In: J. M. Malville And G. Matlock, Eds. The Chimney Rock Archaeological Symposium. Fort Collins, Colorado: U. S. Department Of Agriculture, Forest Service. 76–90.

WILCOX, D. R. 1999. A Peregrine View Of Macroregional Systems In The North American Southwest, A.D. 750–1250. In: J. E. Neitzel, Ed. Great Towns And Regional Polities. Albuquerque: University Of New Mexico Press. 115–141.

NASA archaeological research: a remote sensing approach

Marco J. Giardino[1] Troy E. Frisbie[1] Michael R. Thomas[1]

[1] NASA Stennis Space Center,
Mississippi in the Earth Science Applications Directorat
Mike.Thomas@jpl.nasa.gov

Abstract. Since the 1970s, Stennis Space Center in Mississippi has developed and implemented a wide variety of remote sensing applications to archaeology. Using a multitude of passive and active sensor systems, Stennis and other NASA centers have identified Indian roads in the southwestern US, buried cities in the Near Eastern desert, structures in the jungles of Thailand and Native American earthworks in the flood plain of Mississippi. As NASA's Lead Center for Earth Science Applications, Stennis is initiating new remote sensing applications that provide decision-makers from State, Local and Tribal governments the necessary geospatial information to identify, delineate and mitigate cultural resources. Using multispectral, hyperspectral, RADAR and LIDAR systems, Stennis scientists and technicians collect data from a variety of environments. Digital images are processed to identify high probability locations for archaeological sites. The spatial and spectral information relevant to site location and size is often field tested by our Federal and State partners, including the U.S. Army, Corps of Engineers, the Minerals Management Service, USGS, and National Park Service. These agencies, together with the State Historic Preservation Officers from numerous states, work with Stennis to determine whether remote sensing offers a more efficient and effective alternative to the discovery and mitigation of cultural resources. Following successful deployment of these types of geospatial applications, the Applications Directorate at Stennis works with commercial partners to provide systems that support the routine, operational use of these new technologies and methods. True to its overall mission, NASA provides remote sensing R&D that improves the study and the preservation of the world's cultural resources.

1 Introduction

NASA's Stennis Space Center, located in Hancock County, Mississippi, is one of the leading institutions for the application of remote sensing technology and research to archaeology. For nearly 30 years, Stennis has pioneered remote sensing research that benefit of the archaeological community. Since the early 1970s, initially under the leadership of Dr. Tom Sever, researchers at Stennis developed a variety of remote sensing techniques for studying archaeological sites. Stennis' scientists were among the first to utilize the thermal portions of the electro-magnetic spectrum to discover prehistoric pathways in the Southwest United States. Stennis provided some of the first satellite and multispectral airborne data to enhance the mapping of the Poverty Point site in Louisiana. Additional work followed in Central America, in Mississippi and in Europe. Advanced classification algorithms and registration techniques provided archaeologist with valuable tools for discovering and delineating a variety of sites located in various ecosystems. The Stennis Space Center hosted a conference in 1990 that provided a through review of the many applications of remote sensing to anthropology, and particularly archaeology (Behrens and Sever 1991).

During the last decade, the Stennis Space Center has continued the development and application of remote sensing to archaeological research and application topics. Progress has been achieved in the area of geophysical prospecting, data fusion, multispectral and hyperspectral data analysis and geo-rectification of imagery with historic maps. This work was greatly enhanced by the recent funding from Congress to implement the NASA's Scientific Data Purchase Program.

2 NASA Scientific Data Buy Program

The Scientific Data Purchase (SDP) is a demonstration program developed in response to former President Bill Clinton's Space Policy, directing NASA to purchase remote sensing data from the private sector. Initiated in fiscal year 1997, the SDP was funded under the Earth Science Enterprise (ESE) Program to provide scientific data to the ESE science community. The $50 million program was an opportunity to advance global-systems research, to strengthen the U.S. economy through development of remote sensing technologies, as well as, to test a new way for NASA doing business. The program was again funded in FY01 for $20 million to continue this mission.

The NASA Earth Science Applications (ESA) Directorate at the John C. Stennis Space Center is managing the SDP program. The Directorate's mission is to enhance U.S. economic competitiveness through development of remote sensing technologies. The spaceborne, airborne, and in-situ commercial remote sensing data being made available through the SDP program were identified as data sets that are needed and have high value to science, including archaeology. Participating vendors such as EarthSat Corp., Positive Systems, Earthwatch /Digital Globe, Intermap Technologies and Space Imaging have offered scientists a wide variety of products including Digital Elevation Models, RADAR Mosiacs, various LandSat Mosiacs and 1 meter high resolution imagery. To date, over 600 NASA affiliated scientist and researchers have benefited from this program. The response from the science

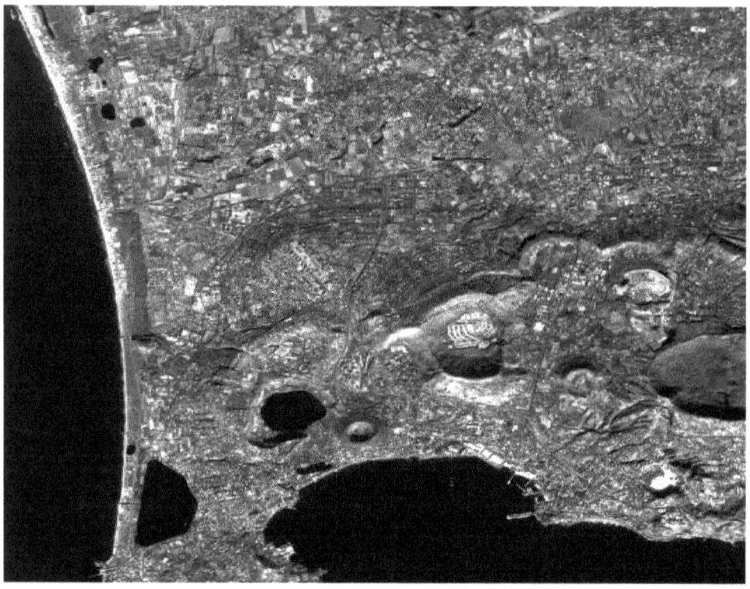

Figure 1. IKONOS image over the site of Cumae (Campania) Italy. Courtesy of Space Imaging, Inc.

community has been extremely positive with users rating the SDP program an average of 8.5, on a scale of 10, based on factors such as "Quality of Data" and their "experience with the SDP process". In all, the SDP program has provided a quality product that may have otherwise been unaffordable and unattainable to the general science community.

NASA, together with its commercial partners, is working to expand the resources available to the ESE science community in its quest for knowledge about the Earth and its changing environment.

Specifically, the SDP has provided for archaeologists a series of high-spatial resolution, multispectral data sets over numerous important archaeological and paleoanthopological sites. NASA's university partners in the SDP include Boston University's archaeology project at Aksum in Ethiopia, UCLA and the American Academy in Rome studying the remains of Cumae in Campania (Figure 1) and Rutgers University working in Koobi Fora, Kenya. NASA's Goddard Space Flight Center partnered with Stennis in the use of SDP products to augment research being conducted at ancient Troy in Asia Minor. Currently, high spatial resolution imagery from the SDP is being collected to documenting a variety of archaeological sites currently on the World Heritage register. Although the NASA SDP program ends in October 2002, the data will continue to be available through commercial sources.

3 Co-registering Historic Maps and Modern Digital Data

Armed with high-spatial resolution multispectral data from orbiting platforms, NASA researchers have pursued the co-registration of historic plats and maps to the digital data. This method, besides providing more precise locations of historic sites and features, produces very accurate site maps, particularly when the data are registered to them from active sensors like LIDARs and RADARs. Digital Elevation Models (DEMs) that are developed with the use of active sensors are utilized in combination with X and Y coordinates derived from high spatial resolution imagery to produce accurate 3-D renderings of sites and the surrounding environments. When applied to large regions, the co-registration of maps to satellite imagery allows development and testing of predictive site location models.

A project where this modeling method is being applied presently is in support of the archaeological research associated with the Lewis and Clark Corps of Exploration. In anticipation of the Corps's bicentennial, commencing in 2004, the Stennis Space Center entered into an agreement with the Columbia Gorge Discovery Center and Museum in Oregon. NASA researchers partnered with Dr. Ken Karsmizki of the Discovery Center in 2001. Karsmizki has been searching for Lewis and Clark sites for several years. He provided NASA with copies of the original Clark maps and with a prioritized list of research objectives. Remote sensing scientists at Stennis Space Center then enhanced satellite imagery (initially LANDSAT and AVHRR) with elevation models to project the historic Lewis and Clark maps onto the current topography (Figure 2). William Clark produced the vast majority of the maps collected during the Corps of Discovery Expedition (1804-1806). When examining the route that Lewis and Clark traveled to the Pacific Ocean, it became clear that co-registration of modern imagery with the historic maps was difficult due, in part, to the significant changes that have occurred in the courses of the rivers over the last 200 years.

To facilitate the georectification of the two data sets, modern and historic, Stennis' scientists developed elevation models over the entire route using Digital Elevation Models and RADAR data collected from a variety of airborne and orbiting platforms. Since Clark often denoted the location of bluffs on his charts, the elevation models served to refine the prediction of site locations. Using this approach, NASA has developed predictive models of site locations for significant Lewis and Clark camp sites such as Fort Clatsop and Fort Mandan. These data were provided to Ken Karsmizki who will examine the accuracy of these predictions during the upcoming 2002 field season. Co-registration of modern images with his historic maps, including the renderings of bluffs and other elevations, narrows the search for related historic localities, thus saving time, lessening subsoil disturbance and avoiding excessive transit through private lands.

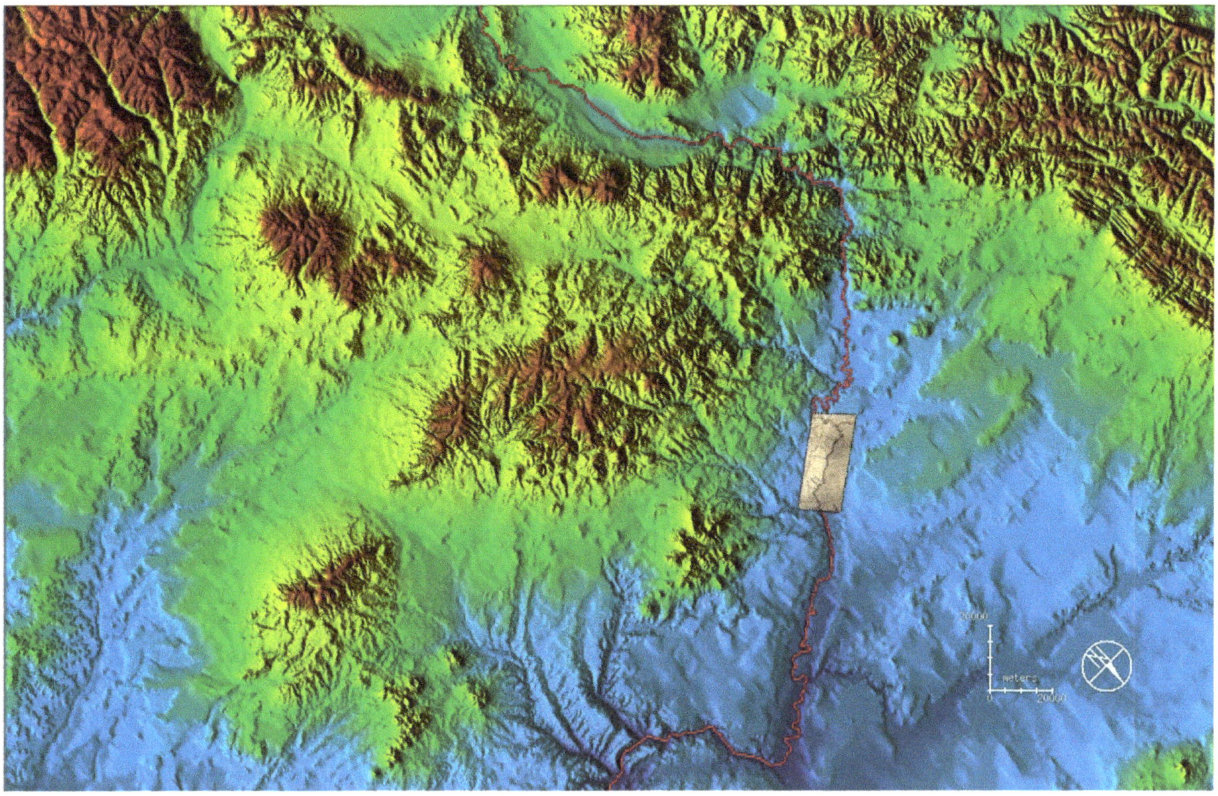

Figure 2. Clark map registered to LANDSAT image draped with DEM.

NASA's development of more accurate co-registration algorithms is being refined at the site of Gainesville, Mississippi. This once thriving town was the county seat of Hancock County during the mid-19th century. Gainesville was one of five historic towns that became part of the NASA Stennis Space Center Buffer Zone and Fee area in 1962 with the advent of the Saturn rocket program that launched the Apollo spacecrafts to the Moon. All the towns, which by the time of NASA's arrival were sparsely populated, were removed for safety reasons prior to the inception of rocket testing.

Gainesville, now nominated to the National Register of Historic Places, is being studied and preserved through use of a wide range of remote sensing techniques. In essence, NASA is developing the Gainesville site as an archaeological validation and verification (V&V) site for the application of a variety of remote sensing data collection and interpretation methods. Results from this research should have broad methodological applicability to a variety of historic sites. Data collected to date over Gainesville have included high spatial resolution multispectral data from two orbiting platforms owned and operated by U.S. companies: the IKONOS, a product of Space Imaging, Inc. and Quick Bird manufactured by Digital Globe, Inc. Both satellites capture image data at spatial resolutions varying from 4 meters in the multispectral to less than a meter in the panchromatic bands. Under ideal conditions, the standard archaeological test unit (normally two square meters) can be visible from orbit.

These satellite data are merged with other digital images acquired from airborne platforms, including thermal data collected by the NASA's ATLAS sensor, a fifteen band research tool flown on a LearJet. Further enhancing the remote sensing research at Gainesville is the inclusion of a variety of more traditional geophysical archaeological prospecting techniques including ground penetrating radar, magnetometry and resistivity.

Finally, LIDAR and airborne RADAR data, together with hyperspectral data derived from airborne platforms are fused to the other data sets to provide a comprehensive view of the surface and subsurface remains that constitute the Gainesville site. Overlying a grid over an image correlates all data sets. The remotely sensed data is then co-registered with historic plats and maps produced by Gainesville's surveyors at during the mid-19th century (Figure 3). The grid, aligned and denoted by UTM coordinates, is replicated in the field. All the data sets, including historic maps, are then fused using the UTM grid as the baseline. In this way, historic roads and building are more easily correlated with remotely sensed signals and anomalies. Scientific excavations provide the final validation and verification of remotely sensed signals.

4 Applications to Cultural Resource Management

In another project, NASA has partnered with the U.S. Army, Corps of Engineers in New Orleans to advance the use of remote sensing as an efficient and effective tool in

Figure 3, Gainesville showing location of survey transects and historic houses overlain on IKONOS image. Image courtesy of Space Imaging, Inc

identifying and mitigating cultural resources. The project area is located in the wetlands of Southeastern Louisiana, an area where traditional surveying techniques are complicated by the difficult terrain. The Corps ofEngineers is responsible for complying with the statutes specified under the National Environmental Protection Act and the National Historic Preservation Act as they relate to cultural resources. The joint project utilizes both LANDSAT and high-resolution hyperspectral and multispectral data from an airborne platform to develop predictive models of site location within the project area. Site files located at the State Archaeologists' Office in Baton Rouge were reviewed and those with relatively accurate location information were entered into a GIS. This enabled the location of sites on LANDSAT imagery. The study area was then flown with a airborne scanning radiometer at a spatial resolution of 4 meters to refine site locations. The data was analyzed to extract spectral response curves for a variety of sites including shell middens and earthen mounds and the results were used as the basis for supervised classifications to identify potential sites not yet discovered and entered in the state's site files. The Corps of Engineers will test the predictive site model this year.

Although this approach is not entirely new (see for example Limp, Frederick, 1993 for a summary of similar studies), it is being augmented by the use of hyperspectral data. Hyperspectral instruments collect data in much narrower bands (usually 10 nm wide versus the 10 micrometers that are common for multispectral sensors) thereby providing new information on the physical properties of land features, particularly vegetation and soils. It is hoped that through the use of hyperspectral data, site location and boundaries in the Louisiana marshes can be refined, specifically though the analysis of plant phenomenology.

Hyperspectral sensors increase the potential for identifying plant species, as well as plant vigor (or lack thereof). Research indicates that variations in plant health can be detected in vegetation reflectance curves in specific regions of the red and near infrared regions of the electromagnetic spectrum (EMS). It is possible that plants growing on organically enriched midden soils grow more vigorously than those on "sterile" soil.

In the marshes of Southeast Louisiana large Rangia shell middens are covered with marsh grass and other vegetation. Hyperspectral imagery collected over several shell middens and covered by a single specie of plant (i.e., Spartina sp.) is being assessed to determine whether plants found on these sites differ from those growing off the site by either growing more vigorously due to the increased organic content, or not growing as well due to the hard substrate. It may be possible to associate spectral variability in a homogenous plant stands like Spartina with the nature of the substrate. If these conditions can be resolved spectrally, than large regions can be examined using classification algorithms and digital processing.

Even in imagery with high-spatial resolution, most pixels will actually be a mixture of several different materials on the ground. For example, grasses, such as Juncus or Spartina, dominate a typical marsh pixel. However, the pixel may also contain water, soil, and perhaps other plant species. This means that the spectral values of any particular pixel are really composite values for each of the materials present in the pixel. It is important to realize that a pixel's values are simply the average of the reflectance of that pixel's constituents, weighted by the relative abundance of each material in the pixel. This fact has an exciting consequence; if the reflectance curves for the imaged materials are known, then the pixel's reflectance can be mathematically decomposed into the reflectance of its constituents. This provides a direct estimate of the abundance of each material on the ground (Jones and Giardino 1997).

5 Software and hardware

Facilitating the application of remote sensing digital image analysis to archaeology are improvements in both hardware and commercially available software. During the last five years, computers capable of processing and storing the very large data sets that normally result from processing the data collected during remote sensing missions (commonly, hundreds of megabytes to a few Gigabytes) have become widely available and relatively affordable. The development of image processing software such as ERDAS' Imagine, RSI's ENVI and ESRI's ArcView provide the commercial tools necessary to conduct even the most quantitative remote sensing analysis. While the necessary equipment to properly conduct remote sensing analysis in archaeology averaged over $20,000 just a few years ago, today a fully operational setup costs less than $5000. Similarly, the cost and accessibility to relevant data sets have improved considerably as new commercial vendors have entered a market that was once serviced almost exclusively by Federal agencies. This last outcome is perhaps the greatest contribution of the SDP program and greatly enhances the capability of archaeologists throughout the world to use digital data collected remotely to conduct research.

Conclusions

Remote sensing technology and image analysis are currently undergoing a profound shift in emphasis from broad classification to the detection and identification of the condition of specific materials, both organic and inorganic. In the last few years, remote sensing platforms have grown increasingly capable and sophisticated while programs like NASA's SDP have made more of these types of data accessible to archaeologists. Sensors currently in use, or nearing deployment, offer significantly finer spatial and spectral resolutions than were previously available. Paired with new techniques of image analysis, this technology may make the direct detection of archaeological sites a realistic goal (Jones and Giardino 1997). The increased accuracy of the methods should results in efficient and effective surveys of large regions. Predictive models of site locations can be developed and tested, often without extensive and expensive field campaigns. Advanced image registration and data fusion will narrow the search for archeological sites, saving time and resources.

Archaeologists have depended always on the accuracy of their surveys to study and preserve archaeological sites. In this way, these scientists are predisposed to accept new and more thorough methods for discovering and delineating sites, both prehistoric and historic. The merging of varied data, including historic documents, is becoming increasingly available to archaeologists, even without extensive technical training in software development or digital data processing and analysis. There are exciting times on the horizon for the use of remote sensing to further archaeological research.

References

BEHRENS, CLIFFORD AND THOMAS SEVER 1991, *Applications of Space-Age Technology in Anthropology. Conference Proceedings,* November 28, 1990. NASA Science and Technology Laboratory, John C. Stennis Space Center, Mississippi.

JONES, PATRICK AND MARCO GIARDINO, 1997, *Remote Advances in Remote Sensing Techniques.* Presentation at the Southeastern Archaeological Conference, abstract published in SEAC Bulletin 40, page 34.

LIMP, W. FREDERICK, 1993, *Multispectral Digital Imagery.* In Jay Johnson (editor) The Development of Southeastern Archaeology. Tuscaloosa and London: University of Alabama Press, 184-206.

Genetic Programming, and Traditional Statistics: towards Interpretation of Ancient Landscape and Social Simulation

Andrea Manzo[1], Cinzia Perlingieri[1]

[1]Istituto Universitario Orientale
P.zza S. Domenico Maggiore, 12 - 80134 Naples, Italy
ybeyene@iuo.it

Abstract. Reconstructing cultural processes with traditional and traditional computer based methods is often not enough to explain changes that occur and determine the processes themselves. Mainly when we deal with internal factors of change, we need to establish which factors do influence and modify the other ones. The application of Genetic Programming, a correlation system already tested in the field of biomedicine and genetics, to our archaeological data is an attempt and a preliminary approach to a new experimental method that represents an interesting application to archaeology. It consists of automatic calculations to combine and cross together different classes of data previously grouped by statistical methods. The input for this system are formalized databases, the desirable output consists of correlations linking different variables expressed in mathematical form. In the case of materials from Bieta Giyorgis, Genetic Programming made evident how the changes in relationships among variables produces innovation and changing in a diachronic perspective (e.g. the link between the success of a ceramic type and a certain environmental situation; the conditions favoring the success of certain types of artifacts). Genetic Programming succeeded in pointing out the changing of relationships among variables producing innovation in Aksumite ceramic production from Bieta Giyorgis. Moreover, the application of this procedure to the data-base of the sites recorded on the top of Bieta Giyorgis might throw new light on the reasons for site location, on their rise, and decline in this area.

1. Theoretical Framework

Statistic analysis are a topic apparently unrelated to the reconstruction of landscape. We think that this diversity is just apparent. Indeed, our goal is not the reconstruction of a general theorical landscape, like in a video-game but the real ancient landscape in a specific area at a specific time.

In a general ecosystemic approach widely accepted in archaeology showing all possible interactions, the study of artefacts and ecofacts points to the relationships among natural and anthropic variables which actually take place, in which way, and with which intensity they did (see e.g. Butzer 1982). The deep links between landscape archaeology and regional scale archaeology, and their common focus on human/environment dialectic were already stressed. The study of archaeological remains and their relationships with natural features can make a general neutral and uniform space varied, made of places very different from one another by their natural characteristics and by human activities taking place there. The effectiveness of this ecological approach as theoretical framework for landscape analysis was already suggested (Aldenderfer 1996, 7; Wise 2000) and it does not exclude but complement other cognitive approaches. As rightly suggested by Christopher Fisher and Tina Thurston (1999, 630): "landscape in this sense is a board, inclusive, holistic concept created intentionally to include humans, their anthropogenetic ecosystem and the manner in which these landscapes are conceptualised, experienced and symbolized". In other words, in the reconstructions of an ancient landscape, space should be conceived a container where the social system is structured, and interacts with the natural resources.

To sum up, the archaeological remains (artefacts and ecofacts), i.e. the data processed by statistics, are the traces left by the systems of internal and external (i.e. with anthropic and natural environment) relations characteristic to each human group. In building up our reconstruction of landscape and three-dimensional models (see Foccillo, Manzo, Perlingieri, and Perlingieri this volume) the study of artefacts through statistics make again what ancient man made in the real ancient space, i.e. organizes and codifies it.

The statistic study of artefacts contributes both to the reconstruction of the sincronic geometric organization of ancient landscape and of the diacronic development of the landscape through time, it will help to create the context and also to put the text in it, to reconstruct the paleoecological relationships that take place and their changes through time, it embrace complex anthropic activities, and, at the same time, allows us to perceive their changes and the changes of their context.

Statistic analysis of archaeological data can interact with the reconstruction of landscape in two main ways and in two different moments:

⇒ when the reconstruction of landscape is elaborated: statistic methods and procedures performed on the available data can contribute to point out associations among the components constituting the landscape.

⇒ when the landscape is interpreted: the organization and structure of the landscape are by-product of relationships among concrete and specific components that statistics can point out providing elements for interpretation. The attributes and patterns of relationships suggested by statistic analysis can integrate the ones suggested by the direct perception of the reconstructed landscape (see e.g. Perkins 2000). Nevertheless, it should be stressed that statistic analysis can point out patterns of relationships which sometimes are not immediately evident to a simple perception (see e.g. Woodman 2000).

Statistics are a strong link between the reconstructed landscape and the data on which it is based, avoiding the danger of building up wonderful computer graphics applications without any element to substantiate them (as rightly pointed out by Daly, Franchetti, and Okkonen 2000, 97) or without any useful impact in terms of knowledge and solving research problems (Forte 2000, 251; Forte and Beltrami 2000, 276, 286-287).

As we will see (R. Fattovich in this volume) the reconstructed ancient landscape can be conceived also as a theatre to perform simulations (but being a part of the simulation by itself, i.e. changing through time together with the relationships between environment and social actors and following general environmental trends). It was already suggested that simulations can help to develop diachronic historical forms of spatial analysis (Aldenderfer 1996, 13-14; Wheatley 2000, 129) which is greatly needed (see Zubrow 1990, 67-68). This aspect is even more emphasized in our case study in which the use of statistics is associated with a strong ecological approach, as ecology has produced a body of theory well suited to the study of people and their relationships with their surroundings through time (Wise 2000, 142). The potential of this kind of approach and its possible applications is not yet fully understood and will be explored in the next decades (Doran 1996; Forte and Beltrami 2000, 278, 290-291).

Statistic analysis can support the building up of a social simulation when:

⇒ the model environment ("simulated landscape", see Gilbert and Troitzsch 1999, 11-12, 167, Auer and Norris 2001) of the social simulation and the agent attributes are set up (Gilbert and Troitzsch 1999, 159-164, Auer and Norris 2001). Actually, the simulation environment and the agents are constructed on the basis of concrete data of a certain area, i.e. archaeological data processed by means of statistic;

⇒ agent and agent-environment interaction patterns (starting values) and behaviour (decision rules) are set up (see Gilbert and Troitzsch 1999, 164-168, Auer and Norris 2001), i.e. when rules expressing patterns and relationships among components (streams, soils, geological formations, man-made structures, sites etc.) of the social simulation are fixed. Actually, if the environment and the agents are constructed on the basis of concrete data of a certain area, and they are parts of a system, then the rules underlying the structure of this system can be inferred. These rules can be used to "move" the social actors. When the reconstruction of landscape is translated into computer simulation programs, the rules pointed out by statistics would be formalised making possible a concrete and not abstract simulation, as they reflect (they are) the behaviour (action and reaction) of the "inhabitants" of the landscape. The "inhabitants" of the landscape are (at different scale) sites, man-made landscape features and buildings, components of buildings, objects, i.e. the actions of man and aggregations of men.

Moreover, the comparison of the situation(s) depicted at the end of simulation runs with the statistic (and territorial) models built up from real data can be used as a system of validation and evaluation of the models proposed by the simulation (see Sallach 2001). On the other hand, this comparison can help us to evaluate and integrate to some extent the statistic (and territorial) models, as they inherit uncertainty and gaps natural for some (almost all) classes of data (see Renfrew and Bahn 1991, for artefacts 41-60 and, at a different scale, for sites 64-68).

To support both the reconstruction of ancient landscape and the social simulation, we think that a further step in the application of statistics to archaeology is greatly needed.

Statistics are used in archaeology to extract information about specific and circumscribed questions and "to give precision to the statements that we make all the time" (Morris 1992, 24) without a true impact on the research methods and strategies. These methods are never used in the phase of interpretation when we use all the information to give a general reconstruction of the way of live of ancient people. There is no real direct impact or a very limited impact of computer applications and quantitative methods in the correlation of different classes of data and in the interpretative phase (as stressed by Doran 1988, 239; D'Andrea and Niccolucci 2000, 15-16). We used statistics not only in the first stages of data analysis, but also in the further pre-interpretation stage (see De Falco, Iazzetta, Manzo, Perlingieri, and Tarantino *in press*). In agreement with our ecosystemic approach, we assumed that there were relationships among the different types of data and we formalised databases in which each entry consists of a context and the variables (attributes of each entry) consist of the associated assemblage of artefacts and ecofacts.

Correlations among different variables are pointed out by means of runs of Genetic Programming (GP) and Principal Component Analysis (PCA) (De Falco, Iazzetta, Manzo, Perlingieri, and Tarantino *in press*). This method can be applied at different scale in a synchronic perspective, making evident spatial relationships among variables (e.g. settlement types and environmental context, settlements and traditional track network), or they can make evident how the changing of relationships among variables produces innovation and changing in a diachronic perspective (e.g. the link between the success of a ceramic type and a certain environmental situation; the conditions favouring the success of certain types of artefacts).

The interpretation of these relationships and links among different classes of variables can be evident by itself or not evident at all. In any case, the scholar has to intervene to suggest an interpretation of the detected links.

From the observation and interpretation of these relationships rules can be extracted (e.g. the relationship between geographic setting and human groups can give rules on the type and organization of economic exploitation of space; the relationship between certain ceramic shapes and paleobothanical and environmental variables can produce rules on the use, circulation and production of goods, energy and information).

2. First case study: database of archaeological assemblages

A database of archaeological contexts excavated at Bieta Giyorgis and the linked assemblage was processed by means of GP to detect relationships among artefacts, the subsistence system, and the environment and their changes trough time (De Falco, Iazzetta, Manzo, Perlingieri, and Tarantino *in press*).
The problems studied by means of GP were:

1. Presence of typical Early Aksumite decoration,
2. Presence of Eragostis tef ,
3. Presence of Grey Ware,
4. Presence of Red Ware,

The negative and positive relationships pointed out by these rules can be summarized as follows:

1. **Decoration Early Aksumite** is directly linked with the presence of **non-arboreal pollen taxa, shrubs and herbs/grasses** and **pollens of highland trees**, the presence of imported **glass**, and **closed ceramic cups**.

2. **Teff** and **Grey Ware** are variables linked with an higher presence of **bovines** than of caprovines , moreover **Grey Ware** is linked with the presence of **jars**.

3. **Red Ware** is linked with the presence of certain amounts of imported **glass**.

The same database was processed also by means of PCA and, on the basis of the rotated component matrix[1], we could confirm the links pointed put by the GP and several other relationships were made evident:

1. Link between the presence of **Hordeum** and the presence of **Triticum**;

2. Relationship linking **non-arboreal pollen taxa**, pollens of **small trees**, and pollens of **highland trees**;

3. Minor link between ceramic **open cups**, **jars**, and **ceramics imported from Sudan**;

4. Link between **ceramics imported from the Mediterranean** and **ceramic beakers and cups**, with minor links with **Decoration Early Aksumite**, **teff**, and **jars**.

The lack of GP rules similar to these PCA components is explained by the fact that in GP the operator has to pose problems to the system and we did not pose the problems of the presence of Hordeum, Triticum, of the different pollen assemblages, of ceramic open cups, of ceramic materials imported from the Mediterranean, and of ceramic beakers and cups. Also these relationships pointed out by PCA can be considered valid, although we plan to verify them with dedicated runs of GP.

Some of the variables of our database have also a chronological value, as Red Ware pottery, Grey Ware pottery, and Early Aksumite Decoration are good chronological markers in the Aksumite cultural sequence (see Fattovich, Bard, Petrassi, and Pisano 2000, 69-72). Some of the relationships pointed out by GP and PCA can be related to specific periods of Aksumite history, and explain the cultural changes. Thus, when the reconstruction of landscape in certain phase of the Aksumite culture is elaborated, these systems can point out associations among the components constituting the landscape, as they did in the Virtual Model of an Aksumite elite residential structure we elaborated (see Foccillo, Manzo, Perlingieri, and Perlingieri in this volume) and the organization among these components.

In a social simulation perspective, our analyses contributed also to the set up of the simulated landscape, of the agent attributes, and of the agent and agent-environment interaction patterns (starting values) and behaviour (decision rules).

Actually, from our results we can infer that:

1. The production of Red Ware pottery, at the beginning of the Aksumite culture (Proto-Aksumite period 4th-2nd century BC) is related to the production of (Red Ware) basins (used in funerary offerings), to the presence of ceramic materials imported from Nubian-Sudanese area in a process of adoption, modification and sharing of funerary practices with the Nubian-Sudanese area suggested also on the basis of typological and cultural analogies (see also Fattovich, Manzo, and Bard 1998).

2. The diffusion of Early Aksumite decoration took place from the 1st century BC and is related with the presence of glass vessels whose fluted surfaces were imitated by the grooved and impressed decorations typical of this pottery (see also Manzo nd.a).

3. The diffusion of the Grey Ware pottery in Late and Post-Aksumite times (from 6th century AD) seems linked to the cultivation of *Eragostis tef* and an higher presence of bovines than of caprovines. It can be suggested that these economic changes had a direct impact on the alimentary habits and, thus, on the tableware, and pottery production. Dealing with the link between a crop system based on teff and the higher presence of bovines, it should be noted that still today the teff is systematically

[1] Extraction method: Principal Component Analysis. Rotation method: Varimax Kaiser Normalization

used as fodder for bovines[2].

The strong link between *Triticum durum* and *aestivum* and *Hordeum volgare*, with very few and weak other links suggests that these variables are almost completely independent from the cultural developments of the studied phases (4th century BC- 14th century AD). Most likely, their cultivation was already established in the 4th century BC (Barnett 1999) and continued afeterwards, although its importance was affected by the progressive adoption of a new system based on teff.

The weak relationships between the associations of pollens and the other cultural and economic variables seem to suggest that the general framework of vegetal coverage was already established in the studied area at the beginning of the period covered by our data, i.e. in the 4th century BC (Bard, Coltorti, DiBlasi, Dramis, and Fattovich 2000).

Generally speaking, it seems that the following general behavioural patterns which can be formalised as rules in a social simulation can be inferred from this database:

1. The adoption and elaboration of external stimulus play a central role in the changes of the Aksumite culture (as in the discussed cases of the red ware pottery and typical Early Aksumite decoration).

2. The changes of economic variables are almost unrelated to the cultural ones (except for the case of teff and paleozoology linked to the spread of Grey Ware pottery).

3. The environmental variables are almost unrelated with the cultural (except for the case of some pollinic assemblages and the pottery with Early Aksumite decoration) and economic ones.

3. Second case study: database of sites

In this database each entry corresponds to an archaeological site[3]. Main aim of this analysis was to point out patterns and rules of the settlement system and their possible changes through time. The variables of the database are attributes of the sites and are related to their **location**, to the presence of **streams**, the **dimensions**, **artefacts density**, **structures** visibile on the surface, possible ancient **use**, present **vegetation cover** and **pedology**, **dating** and **length of the occupation**.

Since the organisation of the database, it was clear that the variables related to land use, pedology, and vegetation cover represent the present situation which can be very different from the past one. These variables were retained because they can suggest remarks on the survey strategy and the factors favouring and/or disfavouring the visibility of sites on Bieta Giyorgis.

On the basis of the rotated component matrix[4], we distinguished the following factors affecting the data pool:

1. Component linking present **land use**, **vegetation**, and **soil type**;

2. Component linking **length of the occupation**, **dimensions**, **density of the artefacts**, and the **structures** visibile on the surface;

3. Component linking the location of the sites in different **sectors** of the hill and their **topographic** location;

4. Component linking the possible ancient **use** of the sites, the presence of **streams**, and the **structures** visible on the surface;

5. Component showing just minor links among **dating**, and possible **ancient use** of the sites.

Actually, from these results we remark that:

1. The reasons of the links among present land use, soil type, and vegetation cover pointed out by the first component are immediately evident and could be expected ;

2. The second component points out that dimensions of the sites, the density of artefacts on the surface, and the kind of structures are related to the length of occupation of the sites. The link between structures and dimensions shows the presence of more impressive structures on the biggest sites;

3. The third component points out that there is a link between the location of the sites in a sector of the surveyed area and their topographic location;

4. The fourth component points out a link between the presence of streams in the area of the sites and the type of the site, with a minor link with the kind of structures recorded at the sites. As it could be expected, the typology (and probable use) of each site is strictly related to the type of structures visible on its surface amd to the availability of water;

5. The fifth component points out the apparent lack of significant links involving the variable "dating".

To sum up, all these remarks make sense and could be expected from the beginning except for the last one. **It might suggest that the general parameters of the settlement pattern were constant through the studied period (7th century BC-13th century AD)**. This is noteworthy mainly with regards to all the variables like the topographic location of the sites and the presence of streams.

---

[2] C.A. D'Andrea, pers. com.

[3] A complete report on the analysis performed of this database is in progress and will be published separately.

[4] Extraction method: Principal Component Analysis. Rotation method: Varimax Kaiser Normalization.

Also the type of soil is apparently unrelated to the other attributes of the sites. Nevertheless, it does not mean that soil quality was not important for ancient people as it is for modern Tigreans. Most likely, it means that the present soil conditions (the only ones we can take into account and record in the database) are different from the past ones[5] (which might have affected the settlement pattern).
However, it seems that for the whole period covered by our data the sites continued to be located in the same topographic positions, and this explains also the high ratio of multi-phase sites in the surveyed area. This remark on the continuity is qualitative and it does not mean that the number of sites, their total dimensions (most likely to be related to demography and/or social factors), and, as a consequence, the average dimensions were constant trough the whole studied period.

4. Final Remarks

These results can be used to implement the reconstruction of ancient landscape and the behaviour of its inhabitants and the understanding of the relations among its components. Moreover, the remarks suggested by our analysis can be translated into rules for possible social simulations at different scales (where the social agent can represent a single actor, a social group, a site, an human group, etc.).

But are these rules true?

The question should be posed because they are strongly affected by intrinsic limits of all archaeological studies[6]:

1. The intrinsic gaps and relativness affecting all classes of archaeological data, as we never have a complete collection of what was used and left (see Renfrew and Bahn 1991, for artefacts 41-60 and, at a different scale, for sites 64-68). In the case of our first database, the data entry was affected by the problem posed by paleozoological, palinological, and paleobotanical variables which can be definied for a very limited number of archaeological contexts. Of course, similar problems[7] of availability of data were recorded also for the database of the surveyed sites.

2. The problem of representation in the database of archaeological data characterized not only by their intrinsic uncertainty but also by problems of classification, attribution, measurement, and formalization (Crescioli, D'Andrea, and Niccolucci 2000; Niccolucci, D'Andrea, and Crescioli 2001).

[5] As confirmed also by Thomas Schmid.

[6] Not to speak about all the problems and possibility of errors in the excavation, sampling, labeling, recording and further phases of data processing.

[7] Not to speak about the usual problems posed by data collection in a survey e.g. in calculating the dimensions of a site and how (and if) they reflect the dimensions of the ancient site and if and how they are (or were) affected by post-depositional factors. Of course, the same problems are posed also by the concentration of artefacts and all the other variables.

Moreover, there is a more general limit common to all the sciences:

3. The fact that statistic analysis can point out chance link and relationships among variable. Indeed, this possibility is less probable and we can be more confident in our results as much as we expand the number of the entries in our databases.

In the future, the multidisciplinary archaeological research project at Bieta Giyorgis will make available more data making the results of our statistic analysis more reliable.

4. The fact that theoretically the factors and variables to be taken into account are countless. In our case (like in all other cases), the classes of evidence and factors are laid down by the experimenter, and, of course, if an entire class of phenomena is omitted, phenomena of that type cannot be described in the analysis and in the further steps of study in which the outputs of the statistic analysis will be used, like in possible social simulations (see Doran 1996, 1235).

A good example of this fact is what we noted trying to make some final remarks on the factors affecting the location of the sites, i.e., according to our results, the distance from streams and the topographic location, which apparently did not change trough time. Nevertheless, most likely, these remarks are affected by the fact that in our database very important data on the geological setting of the sites were not recorded, as they were not yet available when we performed our preliminary analysis.

Dealing with the first database, we plan to include new variables related to the ideological system (e.g. the use of the cross as a decorative pattern in the pottery), in order to understand better how the ideological system interacted with the other variables and affected the cultural changes. Actually, it seems that ideological aspects had a primary importance in the Ethiopian cultural history. In the future we plan to extend these databases with other cases and other cultural, economic and environmental variables, but also after the inclusion of these new variables, how many classes of phenomena whose importance (ore existence) is not so evident are we going to omit?

Because of these limits, it should be always stressed that we propose remarks on the basis of our databases. Our present final remarks, like all archaeological (and scientific) conclusions should be regarded as provisional. Indeed, a work of this type reveals a multitude of variable parameters for the social simulation and reveals just how complex is the dependence upon them (Doran 1996, 1243). In any case, the impossibility to include all (or enough) classes of data and to represent in a more appropriate way the complexity of realty is an important limit of rules and remarks extracted by this kind of statistical applications and it should always be taken into account also when they are used in the further steps of the research.

References

DOERR, M. and KALOMOIRAKIS, D., 2001. *A Metastructure for Thesauri in Archaeology*. In Stančič, Z. and Veljanovski, T. (eds.), Computer Archaeology for Understanding the Past CAA2000, Proceedings of the 28th Conference, Ljubljana, April 2000, Oxford BAR Publishing (BAR International Series 931).

ALDENDERFER M.S., 1982. *Methods of cluster validation for archaeology*, World Archaeology, 14, 1, 61-72.

ALDENDERFER M.S., 1996. *Introduction*, M. Aldenderfer and H.D.G. Maschner, Anthropology, Space, and Geographic Information Systems, New York and Oxford, 3-18.

AUER K and T. NORRIS, 2001. *"ArrierosAlife" a Multi-Agent Approach Simulating the Evolution of a Social System: Modelling the Emergence of Social Networks with "Ascape"*, Journal of Artificial Societies and Social Simulation, 4, 1, www.soc.surrey.ac.uk/JASSS/4/1/6.html

BARD. K.A., M. COLTORTI, M. DIBLASI, F. DRAMIS & R. FATTOVICH, 2000. *The Environmental History of Tigray (Northern Ethiopia) during the Holocene: A Preliminary Outline*, The African Archaeological Review, 17 (2): 65-86

BARNETT T., 1999. *The Emergence of Food Production in Ethiopia*, Cambridge Monographs in African Archaeology, 45, BAR International Series, 763, Oxford.

BUTZER K.W., 1982. *Archaeology as human ecology*, Cambridge.

CRESCIOLI M., A. D'ANDREA and F. NICCOLUCCI, 2000. *A GIS-based analysis of the etruscan cemetery of Pontecagnano using fuzzy logic*, G.R. Lock, Beyond the Map: Archaeology and Spatial Technologies, Amsterdam, 157-179.

D'ANDREA A. and F. NICCOLUCCI, 2000. *L'archeologia computazionale in Italia: orientamenti, metodi e prospettive*, A. D'Andrea e F. Niccolucci, Atti del I Workshop Nazionale di Archeologia Computazionale, Archeologia e Calcolatori, 11, 13-31.

DALY P., M. FRANCHETTI, and J. OKKONEN, 2000. *GIS and Early Åland: Spatial analysis in an archipelago of south-western Finland*, K. Lockyear, T.J.T. Sly, and V. Mihăilescu-Bîrliba, CAA 96. Computer Applications and Quantitative Methods in Archaeology, BAR International Series, 845, Oxford, 91-99.

DE FALCO I., A. IAZZETTA, A. MANZO, C. PERLINGIERI and E. TARANTINO, in press. *Genetic programming in archaeology. Extracting rules from excavation data: the case of Bieta Giyorgis*, Archeologia e Calcolatori.

DORAN J., 1988. *Expert systems and archaeology: what lies ahead?*, L.N. Ruggles and S.P.Q. Rahtz ed., Computer and Quantitative Methods in Archaeology *1987*, *BAR International Series*, 393, Oxford, 237-241.

DORAN J., 1996. *Artificial societies and cognitive archaeology*, III Convegno internazionale di Archeologia e Informatica, Archeologia e Calcolatori, 7, 1231-1245.

FATTOVICH R., K.A. BARD, L. PETRASSI and V. PISANO, 2000. *The Aksum Archaeological Area: a Preliminary Assessment*, Naples.

FATTOVICH R., A. MANZO and K.A. BARD, 1998. *Meroe and Aksum: New Elements of Comparison*, Archéologie du Nil Moyen, 8, 1998, 43-53.

FISHER C.T. and T.L. THURSTON, 1999. *Special section. Dynamic landscapes and socio-political process: the topography of anthropogenic environments in global perspective*, Antiquity, 73, 630-631.

FORTE M., 2000. *About Virtual Archaeology: Disorders, Cognitive Interactions and Virtuality*, J. Barceló, M. Forte, and D.H. Sanders, Virtual Reality in Archaeology, BAR International Series, 843, Oxford, 247-259.

FORTE M. and R. BELTRAMI, 2000. *A proposito di Virtual Archaeology: disordini, interazioni cognitive e virtualità*, A. D'Andrea e F. Niccolucci, Atti del I Workshop Nazionale di Archeologia Computazionale, Archeologia e Calcolatori, 11, 273-300.

GILBERT N. and K. G. TROITZSCH, 1999. *Simulation for the Social Scientist*. Buckingam: Oxford University Press., 159-164

GIRDEN E.R., 1996. *Evaluating research articles from start finish*, Thousand Oaks.

GOLDBERG D., 1989. *Genetic Algorhitms in Search. Optimization and Machine Learning*, Reading Mass.

HOLLAND J., 1975. *Adaptation in Natural and Artificial Systems*, Ann Arbor.

JOLLIFFE I.T., 1986. *Principal Component Analysis*, Springer Verlag.

KOZA J., 1992. *Genetic Programming: on programming Computers by means of Natural Selection and Genetics*, Cambridge Mass.

MCHUGH F., 1999. *Theoretical and Quantitative Approaches to the Study of Mortuary Practice*, BAR International Series, 785, Oxford.

MANZO A, nda. *Skeuomorphism in Aksumite pottery? Remarks on the origins and meaning of some Aksumite ceramic types*, paper submitted for publication to *Aethiopica*.

MORRIS I., 1992. *Death-ritual and Social Structure in Classical Antiquity*, Cambridge.

NICCOLUCCI F., A. D'ANDREA, and M. CRESCIOLI, 2001, *Archaeological Applications of Fuzzy Databases, CAA 96.* Computer Applications and Quantitative Methods in Archaeology, BAR International Series, 931, Oxford, 107-115.

PERKINS P., 2000. *A GIS investigation of site location and landscape relationships in the Albenga Valley, Tuscany*, K. Lockyear, T.J.T. Sly, and V. Mihăilescu-Bîrliba, CAA 96. Computer Applications and Quantitative Methods in Archaeology, BAR International Series, 845, Oxford, 133-140.

RENFREW C. and P. BAHN, 1991. *Archaeology. Theories, Methods, and Practice*, London.

SALLACH D.L., 2001. "*Review* of *Dynamics in Human Primate Societies: Agent-Based Modeling of Social and Spatial Processes*, edited by T.A. Kholer and G.J. Gumerman, Oxford University Press 2000", Journal of Artificial Societies and Social Simulation. jasss.soc.surrey.ac.uk/4/3/reviews/kohler.html

WHEATLEY D., 2000. *Spatial technology and archaeological theory revisited*, K. Lockyear, T.J.T. Sly, and V. Mihăilescu-Bîrliba, CAA 96. Computer Applications and Quantitative Methods in Archaeology, BAR International Series, 845, Oxford, 123-131.

WISE A.L., 2000. *Building theory into GIS-based landscape analysis*, K. Lockyear, T.J.T. Sly, and V. Mihăilescu-Bîrliba, CAA 96. Computer Applications and Quantitative Methods in Archaeology, BAR International Series, 845, Oxford, 141-147.

WOODMAN P.E., 2000. *Beyond significant patterning, towards past intentions: the location of Orcadian chambered tombs*, C. Bucl, V. Cummings, C. Henley, S. Mills, and S. Trick, U.K. Chapter of Computer Applications and Quantitative Methods in Archaeology, BAR International Series, 845, Oxford, 91-105.

ZUBROW E.B.W., 1989. *Contemplating space: a commentary on theory*, K.M.S. Allen, S. Green, and E.B.W. Zubrow, Interpreting Space: GIS and Archaeology, London, 307-318.

Preliminary recognition and analysis of archaeological mounds in the lower Sourou Valley (Burkina Faso)

Paolo Mozzi[1], Aldino Bondesan[1], Armando De Guio[2], Francesco Ferrarese[1], Giovanna Pizziolo[3]

[1] Dipartimento di Geografia, Università di Padova,
via del Santo 26, 35123 Padova (Italy)
[2] Dipartimento di Scienze dell'Antichità, Università di Padova,
Piazza Capitaniato 7, 35139 Padova (Italy)
[3] Dipartimento di Scienze dell'Antichità, Università di Firenze,
Piazza Brunelleschi 3-4, 50121 Firenze (Italy)

Abstract. The Souoru valley in NW Burkina Faso presents a high concentration of archaeological mounds, distributed along the Sourou floodplain. Their chronology is poorly known, as no specific archaeological investigations have up to now been carried out: by comparison to nearby, better known areas (Volta valley), ages should probably span between the VIII and XIII century AD. During the present research, still at a preliminary stage but with already planned near-future developments, these sites have been studied through an integrated approach of ground survey and GIS analysis (elaboration of DEM, geothematic layers, georeferenced aerial photographs). More than twenty mounds, in an area of ca. 1000 Km2 of thorny savanna, have been detected and described in terms of morphometry and position in the physical landscape, the latter object of detailed palaeoenvironmental and geomorphological investigations. A first reference framework has been set concerning the landscape archaeology of this area, a corridor between the Volta basin to the south and the Niger basin to the north.

1 Introduction

The Sourou river is a tributary of the Mou-Houn river (*Volta Noire*) (Fig. 1); the northernmost portion of its catchment basin lies in Mali, while the lower reach, approximately between latitude 13°30' N and 12°40' N, belongs to Burkina Faso. This southern segment of the Sourou valley represents a case study for a project currently under way at the Department of Geography of the University of Padua, directed by P. Faggi, regarding the use of water and related environmental, social and economical problems in the "hydraulic territorialisation" of dry Africa (Faggi, 2000a; Faggi, 2000b; Bethemont *et al.*, 2002). Within this framework, several field trips have been carried out in the lower Sourou valley since 1999. It was during the campaigns of 2000 that it became evident that important archaeological settlements, up to now almost unknown and poorly studied, are present in the Sourou valley (Bondesan & Mozzi, 2000). The definition of ancient landscapes is relevant for achieving a thorough picture of the relationships between man and environment in the region. Archaeological investigations have thus been inserted within the agenda of the research project. The main settlements are generally in the form of large anthropogenic mounds, some of which were surveyed during the field trip of November 2000. An integrated approach of ground survey and GIS elaborations for the detection and preliminary, quantitative evaluation of these archaeological features in the whole lower Sourou valley, together with a first appraisal of the characteristics of the archaeological landscape, are presented and discussed in this paper.

2 The environment

North-western Burkina Faso is characterized by a large plateau ("*plateau Mossi*") of elevations ranging between 250 and 350 m, cut in crystalline rocks. Only few, isolated hills, capped by ferrallitic crusts ("laterite"), break the otherwise monotonous, gently rolling landscape. Part of the Sourou hydrographic basin extends on the north-western portions of this plateau, while the lower Sourou river contours its westernmost fringes.

A low-gradient pediment links the plateau elevations to the Sourou valley floor, representing the eastern flank of the valley. Remnants of the pediment surface are also found nearby the Sourou floodplain, partly covered by alluvial mud. The seasonally inundated floodplain, consisting of silt and clay, reaches a maximum width of 8 km. It is characterized by many shallow ponds, some hundred metres wide and not more than 2 m deep, locally known as "*ko*", linked one another by a network of secondary channels. A large plain, consisting of fluvio-lacustrine silty clayey sediments, extends for tens of kilometres to the west of the Sourou (*plaine do Gondò*), till the base of the great rock escarpment of the *Falaise de Bandiagarà*. On the pediment, soils are of the tropical ferruginous type; in the alluvial plain brown soils are the most represented, passing to hydromorphic soils in the floodplain.

The climate of the Sourou valley is of the hot dry steppic type (BSh in the Köppen-Geiger classification) (Bondesan & Mozzi, 2000); vegetation is of the thorny savanna type, at the transition between the sud-Sahelian and the Soudanian biogeographical zones (Monod, 1986). Precipitations in the Sourou basin are thus scarce, in the

Fig. 1 The study area (square in the lower frame).

average around 650 mm/year, and concentrated between June and September. A characteristic of the Sourou is that it also receives waters from the Mou-houn. When the large floods of this latter river, which have its sources in the humid regions of SW Burkina Faso, reach the confluence, part of the water enters the Sourou valley. The very low gradient of the lower Sourou floodplain, approximately 0.3‰, allows the water to move upstreams for tens of kilometres, beyond the present-day border with Mali. These remounting floods, and not the low and discontinuous downstream discharges, are responsible for the seasonal flooding of the valley floor. The geomorphic history of the Sourou can account for the existence of such phenomenon. During the Late Quaternary the Mou-Houn was a tributary of the Niger river, and it flowed to the north through the Sourou valley (Leprun, 1969; Guillobez, 1993). Aeolian sand deposits probably blocked this drainage direction during a dry phase of the Pleistocene, possibly in correspondance to the last glaciation. This apparently led, in a first stage, to the formation of a lake occupying the present lower Souoru valley, and then to a reversal in the drainage direction of the palaeo-Mou-Houn, which cut its way southwards assuming the present configuration. Thus, when flooding the Sourou valley, the Mou-Houn is following what used to be its former drainage direction. This natural mechanism is controlled since 1984 by two small dams and an artificial canal which deviate the Mou-Houn directly into the Sourou some kilometres upstream from the confluence. This in order to prevent the drying out of the reservoir during the dry season and allow year-round irrigation of the fields nearby the river. The more important artificial change in the hydraulics of the river consists in forcing a continual presence of water in the former seasonally dry floodplain, while the height of the water is set at approximately the same average maximum of the natural regime (ca. 252 m a.s.l.).

3 The archaeological framework: from previous knowledge to critical "depressions" and horizons of expectations

The major line of archaeological evidence is, first of all, the locally "emerging past" (Whimster, 1989) and "emergent complexity" (Arnold, 1996), which in the Sourou valley, beyond our morphogenetic metaphors, consist of the very mundane occurrence of a great number of "tells". They represent a dramatic and paradigmatic instance of "morphological objects" (cp. De Guio, 1995), easily detectable on a GIS modelling base (cp. further) thanks to their basic geometrical properties (tending to a conical shape) and relative altitude against background (local delta and "eminence"). The actual morphometrics taken from GIS analyses (from 70 to 700 m in diameter, from 1 to 5 m in height, along with inferred assessments on volume: cp. below) have been largely confirmed by our ground-truth analysis carried out in the course of our two-weeks field-survey in November 2000 (cp. further).

These mound features correspond to a palimpsestic over-imposition of dwelling and/or productive sites (cp. below). The mechanics of such site-forming processes are easily discernible, and are still operative today on the large majority of traditional settlements, some of which (roughly 70%) show a further (and potentially diagnostic: cp. further) spatial coincidence with the mounds themselves, involving continuity or post-abandonment re-occupation. In fact the tells mainly derive from the diachronic accumulation and subsequent weathering of mud-brick debris: the raw material is still very commonly used today, due to easiness of access and processing. We have observed a great number of present-day/yesterday surface features, telling us a lot about differential stages of abandonment/post-abandonment (and relative conservation) of mud-brick buildings and huts.

A recurrent crucial feed-back circuit has been observed, which operates on the archaeological record of these palimpsestic settlements in relation to mud-brick processing: the last being a seasonal (dry season), non-gender specific operational chain, draining a substantial amount of daily workload and energetic budget (possibly in the order of 20%, at least in the November time-window under our observation). The clay is often extracted at the near-site level of spatial resolution and quite often at the hill-foot of the tells, thus forming a number of depressions (our "critical depressions"), which are ergonomically exploited as ponds to store water for the mud processing itself. The actual practice is very frequently causing the exposure of "stratigraphic windows" on the underlying proper archaeological record (involving the previous settlement and sometimes intercepting a nearby graveyard: cp. below): the "impact" thus acts as a systematic source of damage to cultural heritage, causing a systematic problem to any Eco-Cultural Resource Management policy, but, at the same time, it offers to the archaeologists a vast amount of zero-cost stratigraphic information, in a fascinating "ethnoarchaeology on archaeology" analytical scenario, very stimulating for our theoretical horizons of expectation.

Although this kind of evidence is quite striking and widespread, there is still a nearly absolute lack of archaeological research in Sourou , so we are operating in a sort of basically "*terra incognita*" frame of reference : the only very preliminary evaluation offered by our main research partner, dr. Koté Lassina (Dept. of Archaeology-Univ. of Ouagadougou) suggests that, on the comparative base of evidence emerging from a recent survey project in the nearby upper Mou-Houn valley, the mounds of the Sourou valley could be mainly referred to a quite dramatic demographic expansion in the time-range between the VIII and XIII century AD, followed by an equally dramatic collapse. Further insights by the same author suggest that we have to expect a clear evidence of size and functional differentiations among sites, pointing to a "complex", hierarchical landscape of power involving local specializations, with peculiar emphasis on metalworking (bronze, iron and gold) and related short to long-distance exchange networks: this is a well established domain of research in the (ethno)archaeology of Burkina Faso, outside the Sourou valley (Cline, 1937; Mauny, 1957; Andah, 1973; Kietega, 1978,1993; Etienne-Nugue, 1982; Kote, 1985; Saontouma, 1987; Grebenhart, 1988; Armbuster, 1991; Millogo, 1993; A.A.V.V., 1996).

4 The November 2000 field survey

The November 2000 archaeological campaign has been just a preliminary approach to a proper field survey (programmed for the 2002 field season) and can be more properly classified as an istance of "reconnaissance" (cp. De Guio, 1995). In fact it involved just three members of our research team (A. Bondesan, A. De Guio, P. Mozzi) for no more than ten days and consisted first of all of a purposive (non probabilistic) research design, aimed mainly at a more formal recognition of our "previous knowledge" (an "*in terra cognita*" scenario): this was basically the list of mounds already known from a soil survey (c.p. further) and from a preceding campaign (February 2000) carried out by two of us (A. Bondesan, P. Mozzi) in the context of a more geographic- oriented theoretical and operational frame of reference (sub-domain: "cooperation and development") (Faggi & Mozzi, 2000). In any case our incidental "haphazard/ opportunistic encounter strategy" in the field, gave us the opportunity to discover a considerable amount of previously unknown evidence, with special regard to:

a) new instances (11) of mounds of the same type already depicted;
b) new sources of geomorphological and (palaeo)environmental evidence (e.g. a trial excavation at the bottom of a "*ko*" near Gouran and detailed inspection of some major "stratigraphic windows" exposed by large transect trenches in the Sourou plain;
c) a wide spectrum of contextual ethnoarchaeological evidences (cp. below), sometimes insisting on the proper archaeological ones (according to the above mentioned "ethnoarchaeology on archaeology" perspective), in a quite exciting positive feedback circuit.

We have been able to individuate, georeference and offer a preliminary evaluation of 24 mounds. At the basis of one of them (Niassan) we discovered extended traces of ironworking (with numerous extraction pits and an impressive amount of slags). In two sites (Di-Sobaso and Lanfiera), we detected the graveyards: at Di a simple inhumation occurs, while Lanfiera presents a very characterized practice-inhumation in dolia, which seems to represent a burial custom spread in a wide space/time range in the sub-sahelian macro-region (cp., e.g., a Burkina Faso case-study from Tougou of "inhumation en jarres horizontales", showing a tradition persisting well into the XIX century AD: cp. A.A.V.V., 1994, p. 37).

It has been possible to make some preliminary drawings and field-notes and take some ecofactual and artefactual samplings (with special reference to the local ceramic sequence, that seems, at a first look, strongly characteristic, and in part continuing into the present-day with key features such as some patterns of impression and plastic decoration on large dolia).

5 The mounds in the GIS

The whole area of the lower Souoru valley is covered by topographic maps at the scale 1:20,000, carrying both general information, such as villages, tracks, paths and hydrography, and 0.5 m contour lines and many elevation spots with centimetric resolution. These maps, derived from aerophotogrammetry, served as the cartographic base for soil maps made in the 1980s by the Soil Survey of Burkina Faso (BUNASOLS). The first step in the construction of the GIS (using the software Idrisi32^{TM}) has been the digitalization of all the topograhic maps, soil maps, and applicative derived maps concerning the aptitude of soils to different cultivations. Information have been distributed in thematic vector layers: elevation spots as vector points; contour line, hydrography, roads, tracks and paths as vector lines; villages, soil units, and the dam on the Sourou as polygons. Digitizing has been carried out with an A0 Calcomp digititizer for the northern portion of the study area, while the southern part was digitized on screen on the previously rasterized and georeferenced maps. Georeferencing root mean square (RMS) errors range between 8 m and 15 m.

1:10,000 glossy paper schemes of irrigation projects were also available, provided by the Authority for the Management of the Sourou valley (AMVS), and have been georeferenced and digitized in vector format. Official maps of the Geographic Institute of Burkina Faso

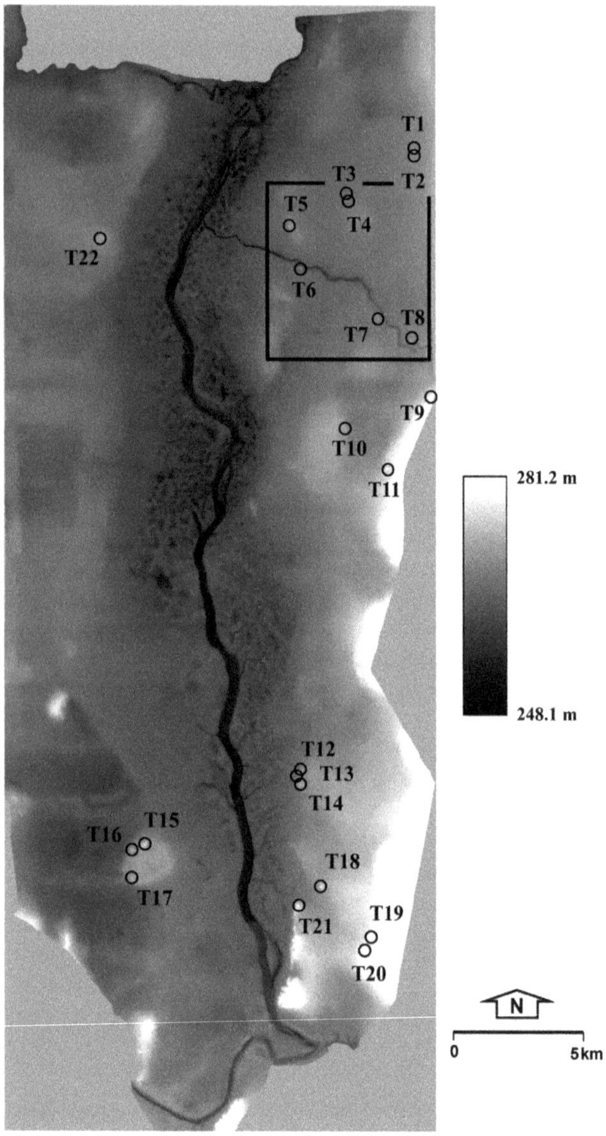

Fig. 2 the Digital Elevation Model of the souoru valley, with localization of the mounds. The square marks the area enlarged in fig. 6 and 7.

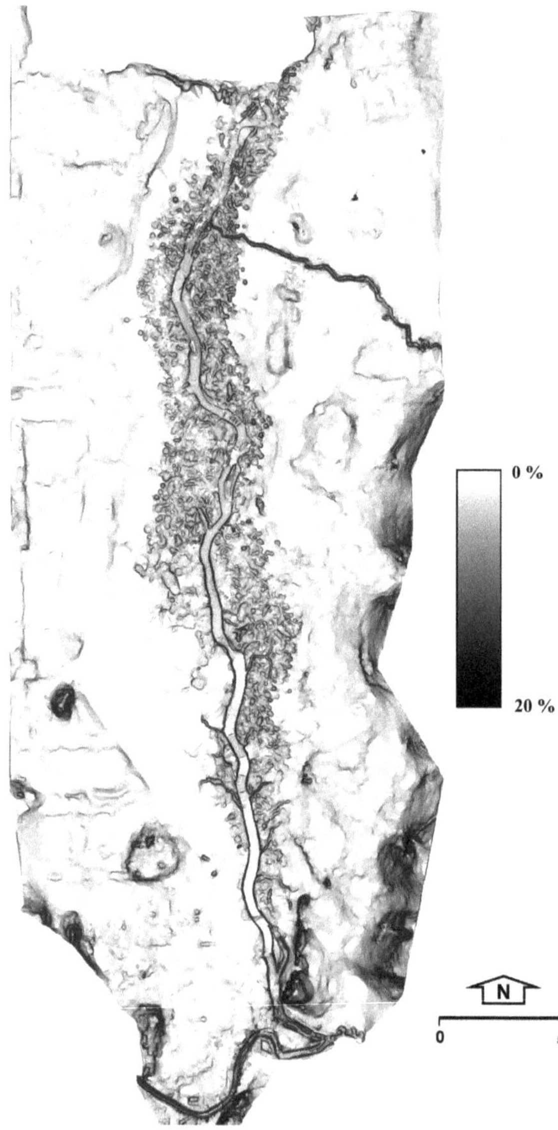

Fig. 3 The slope map of the Souoru valley (DEM elaboration).

(IGB), at scales 1:50,000 and 1:200,000, have been scannerized and georeferenced in raster format. Several black and white panchromatic aerial photographs, at scales 1:10,000, 1:20,000 and 1:50,000 have also been scannerized, georeferenced, and assembled in controlled photo-mosaics.

A triangular irregular network (TIN), using the Delaunay's method for detecting triangles, has been the base to create a raster surface representing elevation values (digital elevation model), with a spatial resolution of 10 metres (Fig. 2).. Triangles have been originated from elevation spots and contour lines vertices, the latter derived every 20 m along the lines. A series of critical points, selected from those automatically obtained by the algorithm, has allowed the determination of the top of ridges and the bottom of depressions. Slope and aspect maps, together with 3D visualisation, have been performed on the DEM.

The detection of the anthropogenic mounds, and the measuring of their morphometric values, have been carried out in the GIS through the analysis and query of DEM (Fig. 2, Fig. 3, Tab. 1) and soil layers (Fig. 4, Fig.5, Tab. 2). In the DEM the mounds stand out as the more marked morphological anomalies in the landscape. The evidence of one or more sub-circular contour line that draw an height gives good clues for the detection of these structures, which also become very evident when orthographic views of the DEM are displayed. Slope analysis (Fig. 3) allows for the automatic sorting out of surfaces with different gradients: mounds are evident in this elaboration as high gradient, sub-circular areas. By interpreting DEM views and elaborations, 22 features were found out in an area of 970.7 Km^2. Several of them have been surveyed in the field, while one corresponds to an area where the soil map indicates "anthropic soil". The correspondence of these independent information (Fig. 6) shows that the interpretation of the anomalies in the DEM as archaeological features is valid. Panchromatic aerial photographs have proved less effective for the detection of the mounds (Fig. 7), as the reflectance values and vegetation patterns of the latter are quite similar to those

Tell	Area (m^2)	Perim (m)	D max (m)	D min (m)	D ave (m)	D ide (m)	Eccentricity	Height (m)	Vol/area Ratio (m)	Vol approx (m^3)
T1	18127	512	205,1	113,3	159,2	152,0	1,07	5,1	3,2	58076
T2	3274	206	68,5	60,7	64,6	64,6	1,02	5,3	2,0	6670
T3	16993	474	166,7	125,0	145,8	147,1	1,03	1,5	0,6	10062
T4	40157	812	336,0	153,9	245,0	226,2	1,14	5,8	0,9	37711
T5	63266	1028	400,7	221,2	310,9	283,9	1,15	3,6	0,8	49404
T6	38008	720	247,8	181,8	214,8	220,0	1,04	2,5	1,7	65687
T7	99680	1155	398,3	319,5	358,9	356,3	1,03	1,8	0,9	88388
T8	40634	759	288,5	180,0	234,2	227,5	1,06	1,8	0,6	23626
T9	14468	438	141,4	127,2	134,3	135,8	1,03	1,3	0,2	2278
T10	12272	405	142,7	116,8	129,7	125,0	1,03	1,5	0,4	4465
T11	42877	769	293,1	196,2	244,6	233,7	1,05	2,7	1,0	42941
T12	69271	988	382,6	253,6	318,1	297,1	1,06	1,3	0,5	36515
T13	41155	732	244,2	227,1	235,7	229,0	1,02	1,9	0,7	29718
T14	122468	1329	505,3	287,8	396,6	395,0	1,07	4,1	1,4	166823
T15	199953	1731	677,1	428,7	552,9	504,7	1,09	4,0	1,5	294896
T16	79412	1052	384,5	288,9	336,7	318,1	1,05	2,6	0,7	54877
T17	76624	1011	340,6	296,5	318,5	312,4	1,03	1,6	0,6	42979
T18	105887	1178	417,1	338,4	377,7	367,3	1,02	4,8	1,6	169360
T19	37128	711	268,8	184,7	226,8	217,5	1,04	0,9	0,5	18935
T20	21755	537	193,0	144,6	168,8	166,5	1,03	0,7	0,4	8664
T21	94635	1403	544,3	292,5	418,4	347,2	1,29	3,5	0,9	84162
T22	298274	2044	797,3	518,5	657,9	616,4	1,06	1,4	0,5	161051
average	69832,6	908,9	338,3	229,9	284,1	253,7	1,1	2,7	1,0	66240,4
st. dev.	68599,8	446,2	177,0	110,4	141,6	106,3	0,1	1,5	0,7	72180,4
min	3273,8	206,3	68,5	60,7	64,6	64,6	1,0	0,7	0,2	2278,0
max	298274,0	2044,0	797,3	518,5	657,9	504,7	1,3	5,8	3,2	294895,8
range	295000,2	1837,7	728,8	457,8	593,3	440,1	0,3	5,0	3,0	292617,8

Tab. 1 – Main morphometric parametres of mounds in the Sourou valley, from DEM elaboration: area, perimeter, maximum diameter (D max), minimum diameter (D min), average diameter (D ave), diameter of the theoretical circle with area equal to the mound (D ide), eccentricity, maximum height, volume/area ratio, volume.

of the surroundings areas (Fig. 8).

DEM-derived morphometric analysis of the mounds (Tab. 1) shows that they are about 70,000 m^2 wide in the average, but can reach 300,000 m^2 (T22). Diameters go from 70 m to more than 700 m. Heigths can vary from less than 1 m to more than 5 m, with an average value of 2.9 m. Although the mounds usually have simple and regular shapes, there are no significant correlations between the different morphometric parameters. Only values of area and volume grow with coupled trends, displaying a quite strong linear regression (area = 6925 + 0.85 * volume; correlation coefficent R = 0.81).

In the northern sector of the valley the mounds concentrate on the eastern side of the Sourou river, having the Debe river, an ephemeral stream tributary of the Sourou, as an approximate axis. Another cluster lies in the southernmost portion of the valley, showing a somehow more pronounced simmetry between the eastern and western banks of the Souoru. Most of the mounds are located on low hillocks, remants of the pediment surface, or on the distal fringes of the pediment itself. Such landforms are strongly weathered, and, in fact, when considering the distribution of mounds in relation to the soilscape, it results that 64% of the mounds lie in areas with ferruginous soils (Tab. 2).

6 Present-day traditional organization of the territory: an analogue for modelling ancient settlement patterns?

In order to define models which could provide some clues for the understanding of the archaeological landscape, the investigation of the present-day, traditional territorial organisation of the study area seems to represent an interesting perspective. "Traditional" are here considered those systems of land use, settlement and socio-economical practices which were developed before the building of the hydraulic infrastructures in the last decades, and that still persist today in parallel to the "modern" making of the territory along the straight lines and right angles of the irrigation canals. The focus is on the understanding of the territorial management and of the way in which environmental resources (i.e. water,

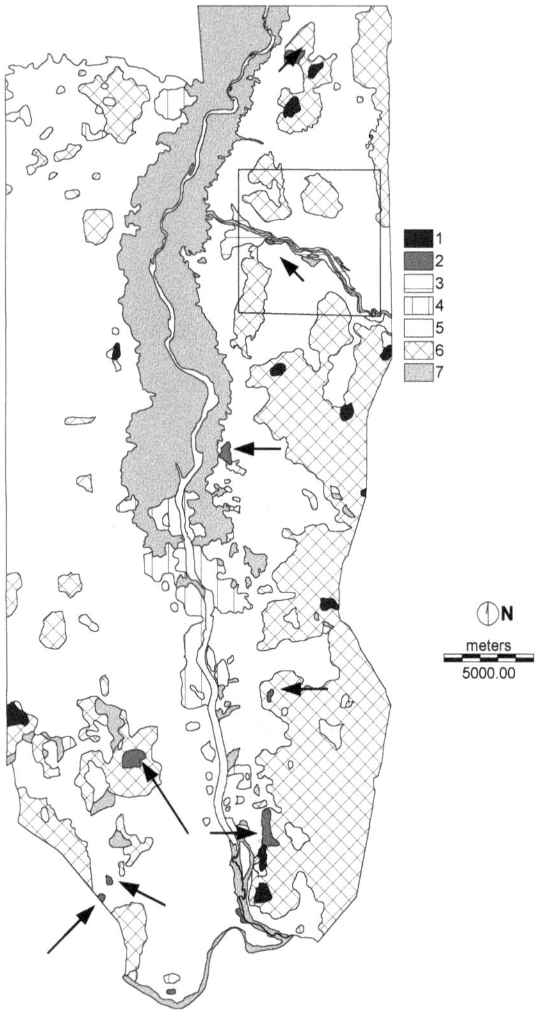

Fig. 4 Simplified soil map of the Souoru valley (reclassification of the «soil» layer in the GIS): 1. lithosols on cemented ferrallitic horizons; 2. anthropic soils (indicated also by arrows); 3. poorly developed soils on recent alluvium; 4. vertisols; 5. brown soils on older fluvial and lacustrine deposits; 6. tropical ferruginous soils, with leached and hard horizons; 7. hydromorphic soils. The square marks the area shown in fig. 6 and 7.

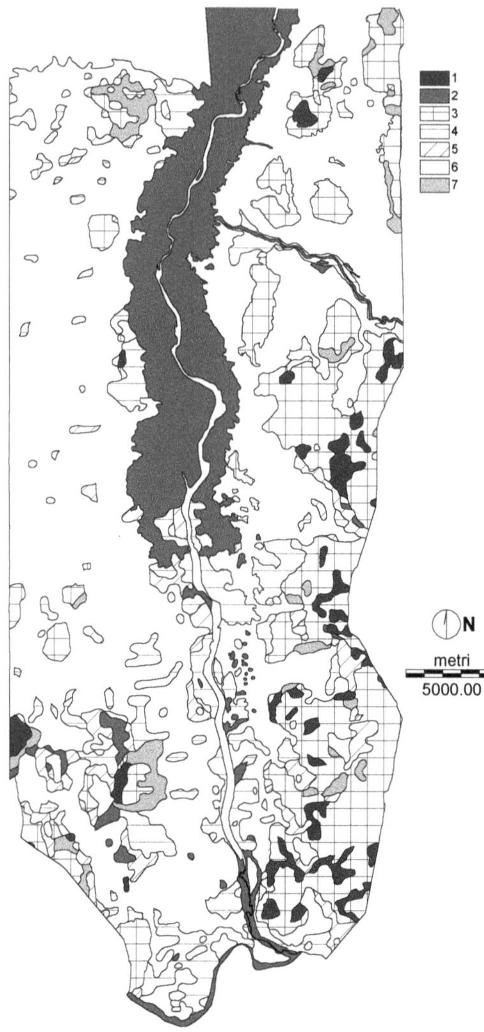

Fig. 5 Map of the suitability of soils to different cultivations (reclassification of the «soil» layer in the GIS): 1. not suitable for maize, horticulture, soya-bean, sorghum, wheat, cotton, rice; 2. not suitable for maize, horticulture, soya-bean, wheat, cotton; low suitability for sorghum; medium suitability for rice; 3. low suitability for maize, horticulture, soya-bean, sorghum, wheat, cotton; not suitable for rice; 4. low suitability for maize, horticulture, soya-bean, sorghum, wheat, cotton; medium suitability for rice; 5. low suitability for maize, horticulture, soya-bean, sorghum, wheat, cotton; medium suitability for sorghum; not suitable for rice; 6. medium suitability for maize, horticulture, soya-bean, sorghum, wheat, cotton, rice; 7. medium suitability for maize, horticulture, soya-bean, sorghum, wheat, cotton; not suitable for rice.

soil, pasture, game and fish, forest) are exploited, as well as on how the cultural landscape is structured. The valley, comprising a wet riverine floodplain environment, and a dry lowland environment, is characterised by a complex interaction between systems of sedentary agriculturists and fishermen, and transhumant stockbreeders, involving the presence of different ethnic groups (Bellot-Couderc & Bellot, 1978; Bertoncin & Pase, 2000; Zoungrana, 2000). Landscape is organized and perceived in various ways by single groups according to their cultural and productive needs. It is, therefore, crucial the identification, and subsequent mapping into the GIS, of the different settlement strategies.

Data to be used derive from various sources. Historical aerial photographs dating before hydraulic transformations are precious data, as they provide information on land use and land management, including pathways and enclosures related (or not) to settlement structures. It is moreover important to reconstruct the demographic and ethnic composition of each village in order to highlight changes in territorial management or relationships between different groups. Such type of data, as well as information on land management, derive from reports carried out by French authorities during the fifties. Another fundamental source is represented by interviews to people belonging to the villages. In particular, it is of great interest to gather information on cultural elements involved in land management. Oral and written information, analysed and then "georeferenced", may be

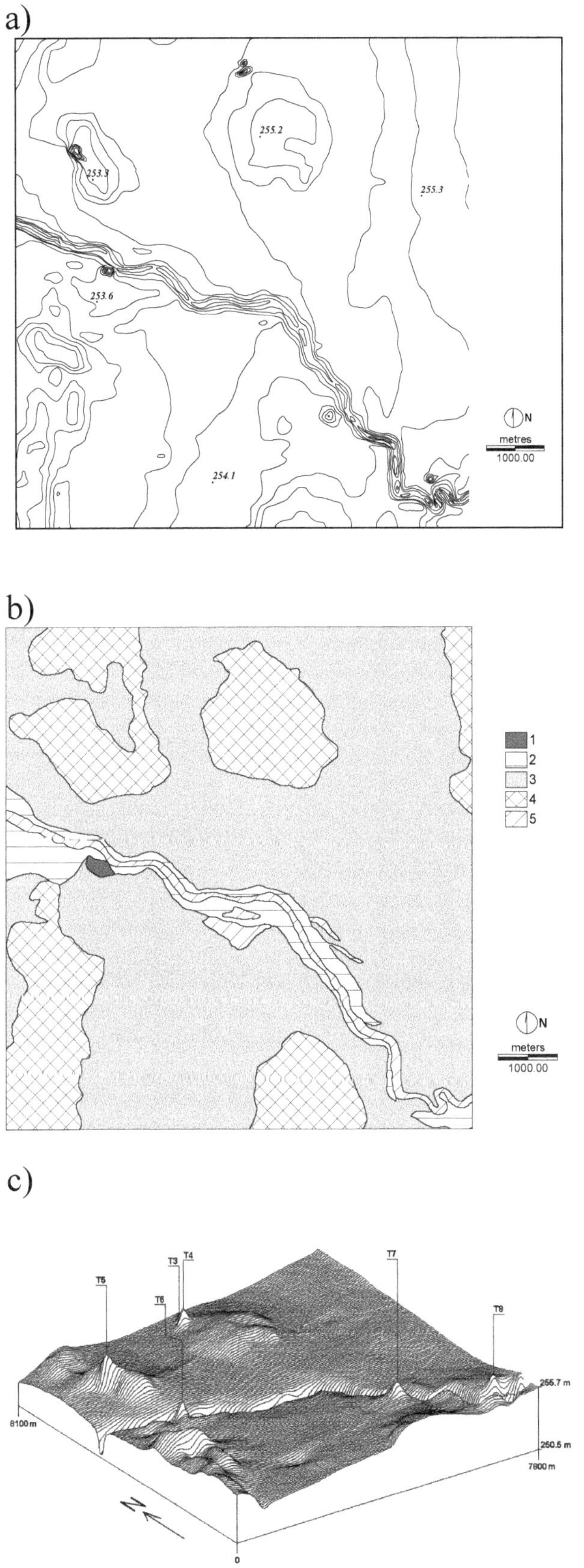

Fig. 6 The Debe area (see Fig. 2 and Fig. 4 for localization): a) elevation map with 0.5 m contour lines; b) soil map: 1. anthropic soils, 2. poorly developed soils, 3. brown soils, 4. tropical ferruginous soils, 5. hydromorphic soils; c) 3D visualization of the DEM with indication of mounds.

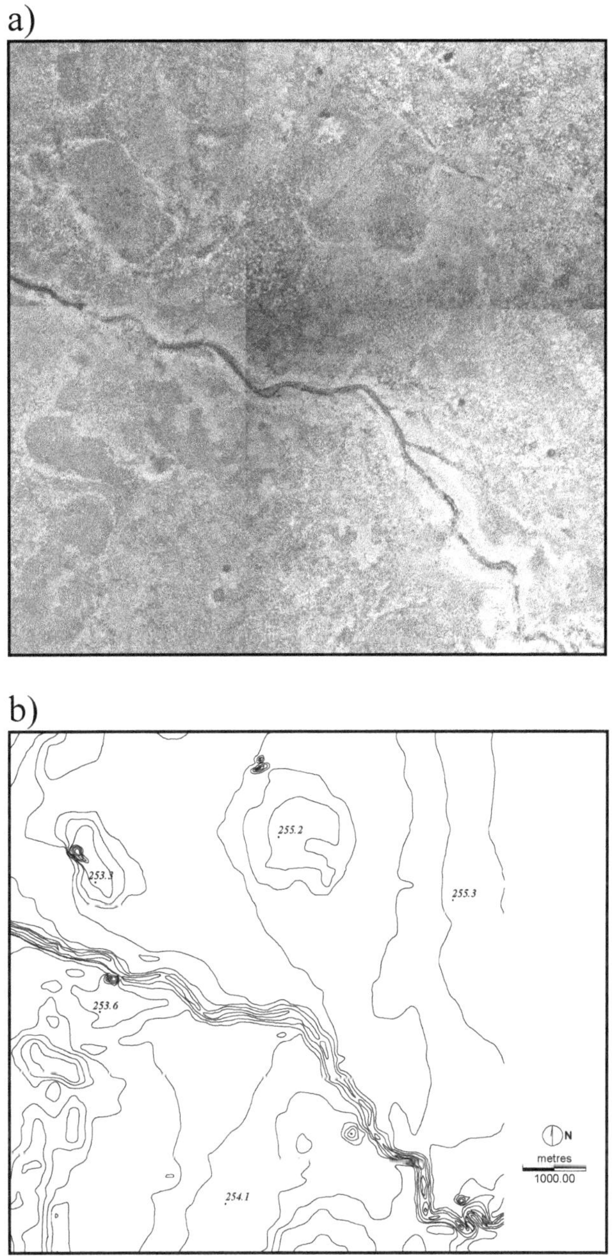

Fig. 7 The Debe area (see Fig. 2 and Fig. 4 for localization): selection of the georeferenced photomosaic (a), with altimetric map for reference (b).

tells	village near tell	village in tell proximity (0.5 - 1 km)	soil types (in French classification)	min dist (m)	nearest neighbour tell
T1	Koromé		Sols ferrugineux tropicaux lessives indures moyennement profond (40 - 60 cm)	102	T2
T2	Koromé		Sols ferrugineux tropicaux lessives indures moyennement profond (40 - 60 cm)	102	T1
T3	Lo		Sols ferrugineux tropicaux lessives indures peu profonds (20 - 40 cm)	165	T4
T4	Lo		Sols ferrugineux tropicaux lessives indures peu profonds (20 - 40 cm)	165	T3
T5	Niassan		Sols ferrugineux tropicaux lessives indures peu profonds (20 - 40 cm)	2039	T6
T6		Débé	Sols mineraux bruts anthropiques modaux	2039	T5
T7	Leri		Sols bruns eutrophes tropicaux ferruginises	2030	T8
T8	Kourani		Sols bruns eutrophes tropicaux vertiques	2030	T7
T9	Yo		Lithosols sur cuirasse ferrugineuse	2849	T8
T10		Gouran	Sols ferrugineux tropicaux lessives indures peu profonds (20 - 40 cm)	2882	T11
T11	Ban		Sols ferrugineux tropicaux lessives indures peu profonds (20 - 40 cm)	2882	T10
T12	Bissan		Sols peu evolues d'apport alluvial hydromorphes	393	T13
T13	Bissan		Sols peu evolues d'apport alluvial hydromorphes	295	T14
T14	Bissan		Sols ferrugineux tropicaux lessives indures peu profonds (20 - 40 cm)	295	T13
T15	Sono		Vertisols a drainage externe nul, vertiques	572	T16
T16		Sono	Sols ferrugineux tropicaux lessives indures superficiels (0 - 20 cm)	572	T15
T17			Sols ferrugineux tropicaux lessives indures moyennement profond (40 - 60 cm)	1377	T16
T18	Nion		Sols ferrugineux tropicaux lessives indures moyennement profond (40 - 60 cm)	1624	T21
T19			Sols ferrugineux tropicaux lessives indures superficiels (0 - 20 cm)	662	T20
T20			Sols ferrugineux tropicaux lessives indures peu profonds (20 - 40 cm)	662	T19
T21			Vertisols a drainage externe nul, hydromorphes	1624	T18
T22	Koubé		Sols ferrugineux tropicaux lessives indures peu profonds (20 - 40 cm)	9416	T5

Tab. 2 Neighbourhood values of the mounds of the Souoru valley, and their position in the soilscape.

Fig. 8 A mound near Bissan, in the T12-T13-T14 cluster (see Fig. 2 for location). The scarps at the foot of the mound are the result of diggings made by local people in order to extract clayey materials for mud bricks (picture by P. Mozzi, February 2000).

Fig. 9 This man is not an archaeologist. Quarrying of anthropic deposits in the same mound shown in Fig. 8, nearby Bissan, for mud bricks production. Such activity brings about serious problems for the conservation of the archaeological record in the Souoru valley (picture by P. Mozzi, February 2000).

stored into the GIS. Field surveys are useful to improve accuracy in the location of features and in singling out landscape markers. Important elements for the landscape reconstruction are obviously represented by environmental and paleoenvironmental data (i.e., landforms, soils, extension of floods in different historical periods, dynamics of the vegetation cover), necessary to provide a correct physical context.

Most of the data are still being acquired and analysed, with the aim of highlighting the role played by the various elements of the system, and of reconstructing the system as a whole. Once the framework will be set, experimentation on neighbourhood relationships and relative influence areas between villages may be performed via GIS, taking into consideration a variety of criteria. A particular attention will be paid to the integration between environmental data, such as soil characteristics, and land use / settlement locations, both to understand the productive management and to look for anomalies and evidence of cultural settings in the landscape (i.e. sacred areas or territorial exchanges between villages). The characteristics of the traditional landscape will then be compared to the archaeological data, in order to define possible relationships. The traditional system will be used to explore continuity and changes in the landscape and to stimulate directions of investigation. Basic questions will be formulated on the continuity of settlement patterns, and answers will be relative to the different scales of analysis.

Conclusions

Archaeological and palaeoenvironmental investigations in the Sourou valley are still at a preliminary stage. Nevertheless, the reference framework has been set, through both reconnaissance field surveys and GIS elaborations of DEM, geothematic layers and aerial photographs. The area presents a considerable variety of ecotypes, spanning from the dry plain and pediment to the floodplain swamps. Due to its Late Quaternary palaeohydrographic evolution, the Sourou river represents a sort of "wet branch" which extends northwards from the Mou-Houn basin towards the Niger inner delta. Ancient settlements, in the form of anthropogenic mounds ("tells"), are numerous in the valley, and apparently concentrate in clusters at different scales. Some general trends seem to be repeated, as for the west side of the Sourou Valley, which has scarce archaeological evidence (and is equally less inhabitated today). Local patterns, such as the larger mounds or the concentration of mounds along the Débé river, may be related to areas endowed with special attraction, intensely frequented in terms of time and/or number of inhabitants, with multiple productive activities. In some cases the spatial distribution of sites seems to be related to elements of the physical environment, such as areas with ferrallitic soils, convex landforms, riverine areas, but the actual patterns are still to be fully defined. In order to unravel the evolution of the archaeological landscape, some major focuses of interest have been pinpointed, which give a problem-orientation to our planned research design for the near future (starting from the incoming September 2002 field-season):

a) settlement systems (space/time patterns and mobility): demographic insurgence, climax and collapse; locational/allocational strategies; local and non-local resource management. This seems to be, among other, an optimal case-study for application of a number of (pre and post-GIS) morpho-genetic detectors, including our "Survival Analysis" borrowed from biomedical sciences (De Guio, 1985);
b) sites visibility/intervisibility;
c) communication and connectivity: network analyses;
d) landscape of power (projection in the space, time and functional dimensions of social complexity);
e) metastability and ecotonality: natural processes and social engineering at the root of systematic imbalance and overexposure to ana/catastrophic events and conjunctures;
f) risk and uncertainty management: strategies for coping with seasonal or otherwise cyclical phenomena in the natural and social domains (space/time geography);
g) local to long-distance trade (especially: gold, iron, salt);
h) ethnicity (social dynamics in the present/past ethno/eco-mosaics);
i) frontiers and boundaries;
l) landscape ecology and mindscape;
m) the impact issue (on archaeological and palaeoenvironmental resources) (Fig. 9);
n) ECRM (Eco-Cultural Resource Management) and Public Archaeology: evaluation, *mise en valeur*, monitoring;
o) "Archaeology and Development": collaboration with ONGs and "associations paysannes" involving studying, recovering and promoting traditional well-balanced (environmentally and socially) practices in the agrarian sphere of interactions.
p) the ethnoarchaeological domain, with special emphasis on the following locally-sensitive topics: 1) water management; 2) mud-brick processing (Fig. 10); 3) fishing (Fig. 11); 4) ceramic processing ("canari"); 5) metallurgy (gold, iron, bronze); 6) food processing; 7) pre/abandonment/post-abandonment in the systemic and (ethno)archaeological records; 8) "sacred landscapes"; 9) ethnicity and ethno-related processes; 10) oral tradition and "*griots*"; 11) cognitive maps and toponymy (multi-layer/multi-ethnical: cfr. our special "TopoGIS" experimental approach: De Guio, 2001).

Fig. 10 Present-day production of mud bricks in the Souoru valley (picture by P. Mozzi, February 2000).

Fig. 11 Fishing in the swamps of the Sourou floodplain, nearby Yaran (picture by P. Mozzi, February 2000).

Acknowledgements

This research has been undertaken within the research project COFIN 1999 *"Development projects, natural resources and social participation dynamics in dry Africa*" (local director P. Faggi) in the national research framework " *Territorial organisation and environmental protection in Africa"* (national director A. Turco), funded by the Italian Ministry of University and Scientific Research.

References

A.A.V.V., 1994, *Vallées du Niger*, Paris, Sépia.

A.A.V.V., 1996, *La préhistoire de l'Afrique de l'Ouest*, Saint-Maur, Sépia.

ANDAH B., 1973. *Archaeological reconnaissance of Upper-Volta*, unpublished PhD thesis, University of California, Berkley.

ARNOLD A. (ed.), 1996, *Emergent complexity. The evolution of intermediate societies*, Ann Arbor, International Monographs in Prehistory.

ARMBUSTER B., 1991, *A study of lost wax casting processes in Mali and Burkina-Faso in Western Africa. Workshops and casting techniques as ethno-archaeological demonstration material*, in Formigli E. (ed.), Antiche officine del Bronzo, Murlo, Nuova Imagine Editrice, pp. 153-159.

BELLOT-COUDERC B. & BELLOT J.M., 1978, *Pour un aménagement du cours moyen de la Volta et de la vallée du Sourou*, Les cahiers d'outre-mer, Bordeaux.

BERTONCIN M. & PASE A., 2000, *Rationalite projectuelle hydraulique dans la Vallée du Sourou, in Faggi P. & Mozzi P. (eds.), La territorialisation hydraulique dans la vallée du Sourou (Burkina Faso): lignes pour la recherche*, "materiali" - Dipartimento di Geografia – Università di Padova, 22, pp. 29-37.

BETHEMONT J., DE MARCHI M., FAGGI P. & ZOUNGRANA T. P., 2002, *La valle del Sourou (Burkina Faso): per una geografia della territorializzazione idraulica in Africa*, Rivista Geografica Italiana, 109, pp. 200-222.

BONDESAN A.& MOZZI P., 2000. *L'environnement physique et les problématiques environnementaux dans la gestion des projets d'irrigation, in Faggi P. & Mozzi P. (eds.), La territorialisation hydraulique dans la vallée du Sourou (Burkina Faso): lignes pour la recherche*, "materiali" - Dipartimento di Geografia – Università di Padova, 22, pp. 39-55.

BUNASOLS, *Carte pédologique de la zone d'aménagement de la Valleé du Sourou, echelle* 1:20.000, 10 feuilles, Bureau National des Sols, Ministère de l'Agriculture et de l'Elevage du Burkina Faso.

CASAMASSA G., DE GUIO A. & VERONESE F. (eds.), 2000, *Paesaggi di potere: problemi e prospettive*, Roma, Quasar.

CLINE W., 1937, *Mining and metallurgy in Negro Africa*, Menhansa.

DE GUIO A., 1985, Archaeological applications of survival analysis, in Voorips A. & Loving S.H. (eds.), To pattern the past, Louvain, P.A.C.T.,11, pp.361-381.

DE GUIO A., 1995, *Surface and subsurface: deep ploughing into complexity*, in HENESEL W., TABACZYNSKI B., URBANCZYK P. (eds.), Theory and practice of archaeological research , II, Warszawa, Institute of Archaeology and Ethnology, Commiteee of pre- and Protohistoric Sciences, Polish Academy of Sciences, pp. 329-414.

DE GUIO A., 2001, *"Superfici di Rischio" e C.I.S.A.S. Se lo conosci, non lo eviti*, in Guermandi M.P. (ed.), Rischio Archeologico: se lo conosci lo eviti, Firenze 2001, All'Insegna del Giglio, pp. 265-306.

ETIENNE-NUGUE J., 1982, *Artisanats traditionnels. Haute-Volta*, Dakar, Institute Culturel Africain.

FAGGI P., 2000a, *Presentation*, in FAGGI P. & MOZZI P. (eds.), La territorialisation hydraulique dans la vallée du Sourou (Burkina Faso): lignes pour la recherche, "materiali" - Dipartimento di Geografia – Università di Padova, 22, pp. 1-6.

FAGGI P., 2000b, *La territorialisation hydraulique du Sourou: le cadre théorique*, in FAGGI P. & MOZZI P. (eds.), La territorialisation hydraulique dans la vallée du Sourou

(Burkina Faso): lignes pour la recherche, "materiali" - Dipartimento di Geografia – Università di Padova, 22, pp. 11-16.

FAGGI P. & MOZZI P. (eds.), 2000, *La territorialisation hydraulique dans la vallée du Sourou (Burkina Faso): lignes pour la recherche*, "materiali" - Dipartimento di Geografia – Università di Padova, 22.

GREBENHART D., 1988, *Les premiers métallurgists en Afrique occidentale,* Paris, Èditions Errance/Les Nouvelles Étitions Africaines.

GUILLOBEZ S., 1993, L*e Mouhoun, un ancient affluent du Niger: exemple de capture en milieu tropical aride*, C. R. Acad. Sci. Paris, t. 317, Serie II, pp. 537-542.

HEBERT E., 1984, *Red gold of Africa*, Wisconsin.

KIETEGA J.B., 1978a, *États des recherches archéologiques en Haute-Volta,* Bamako, IIe Congrè de l'Association Ouest Africaie d'Archéologie, 11-18 dec., pp. I-XXX.

KIETEGA J.B., 1978b, *Le laboratoir d'archéologie de l'Université de Ouagadougou*, "Cahiers du CERLESHS", Université de Ouagadougou, pp. 150-157.

KIETEGA J.B., 1993a, *Le cycle du fer au Burkina Faso, Association* "Découvertes du Burlina", Bécouvertes du Burlina, II, Paris, Ouagadougou, Sépia-A.D.D.B., pp. 73-96.

KIETEGA J.B., 1993b, *Le cycle de l'or au Burkina Faso*, Association "Découvertes du Burlina", Bécouvertes du Burlina, II, Paris, Ouagadougou, Sépia-A.D.D.B., pp. 97-125.

KOTE L., 1985, *Recherches archéologiques au Burkina Faso.* État actuel des connaissances, "Mémoires de Maîtrise", Université Paris X, Nanterre, pp. 176-196.

LEPRUN J-C., 1969, *Évolution géomorphologique de la vallée du Sourou et de ses bordures voltaïques*, Maîtrise de géographie, année universitaire 1968-69, Université de Dakar, pp 49.

MAUNY R. (1957), *État actuel de nos connaissances sur la préhistoire et l'archéologie du Haute-Volte*, "Notes Africaines", 7, janvier, pp. 1-3.

MILLOGO K.A., 1993, *Récherces préhistoriques au Burkina Faso*, "L'Anthropologie", 97, 1, pp. 97-118.

MONOD T., 1986, *The sahel zone north of the Equator*, in Evenary M., Noy-Meir I. & Goodall D.W. (eds.), Ecosystems of the world: hot deserts and arid shrublands, Elsevier, pp. 203-243.

SAONTOUMA I., 1987, *La métallurgie ancienne du fer dans la region de Koumbori (Yatenga-Burkina Faso*), "Mémoires de Maîtrise", Université Ouagadougou, Départment d'Histoire et d'Archéologie, pp. 1-119.

WHIMSTER R., 1989, *The emerging past. Air photography and the buried landscape,* London, Royal Commission on The Historical Monuments of England.

ZOUNGRANA T.P., 2000, *Transfert techonologique et aptitudes paysannes a la gestion d'un périmètre irrigue: Di, dans la vallée du Souoru*, in FAGGI P. & MOZZI P. (eds.), La territorialisation hydraulique dans la vallée du Sourou (Burkina Faso): lignes pour la recherche, "materiali" - Dipartimento di Geografia – Università di Padova, 22, pp. 29-37.

Archaeological Subsurface Site Reconstruction Using Computer Processing of GPR Data

Sheldon S. Sandler[1], Alan Crandall[2]

[1]Research Fellow, Department of Archaeology, Boston University; Technical Advisor, Geo-Centers Inc, Newton Centre, MA; Professor Emeritus, ECE Department, Northeastern University
[2]Senior Engineer, Geo-Centers, Inc, Newton Centre, MA
ssandler@geo-centers.com

Abstract. Many archaeological sites have overburdens or have been reused over wide spans of time resulting in layers of habitation. Cultural objects such as fireplaces or stone monuments may also be present. This presentation is pointed toward producing two and threedimensional presentations of the subsurface stratigraphy using data from a Ground Penetrating Radar (GPR). This type of radar does not give an easily interpretable real space image but rather a radar space image that contains a bewildering array of patterns. In order to simplify the image presentation and make it more useful for 3-D reconstructions, a GPR image is reduced to a single line plot. The line plot shows the presence of a layer or a localized object. Multiple 2-D line plots are subsequently stacked and interpolated to form a 3-D surface reconstruction of the earth. A short description of the computer processing background is given and application is made to GPR data acquired at an 8000-year-old archaeological site in Carver, Massachusetts (USA). Results are given for the original GPR traces, line plot representations, and a 3-D surface plot.

Introduction

Each vertical line on a GPR trace is a gray or color representation of the two way travel time of an electromagnetic (EM) pulse. Because the GPR antenna is moved in the horizontal direction, a 2-D image results. This image or trace roughly represents a slice of the earth. Because waves move out spherically from the antenna a point on the vertical line may come from many locations in the earth that are at the same distance. The GPR also produces signals that are system dependant such as earth coupling effects. Nonetheless it is assumed that the GPR image is a reasonable representation for a slice of the earth underneath the track of the antenna, although it may be difficult for the untrained eye to interpret. Engineers or geophysicists can use their knowledge of electromagnetics and wave propagation to readily understand the various radar patterns that are produced by buried objects or interfaces. For example, buried objects generally produce hyperbolic patterns while interfaces produce lines. Radar traces are also degraded by resonance (SANDLER, S.S. and JONES, D. 1998) and multiple scattering effects that produce a multiplicity of patterns as well as other scatterers in the field of view such as earth voids. The archaeologist is faced with dealing with radar as producing images that are not only hard to interpret but hard to explain in terms of a technology that is mostly unintelligible. For this reason the approach here is to produce a simple radar image that is quite easy to interpret. Briefly put, the 2-D radar image is reconstructed as a line in real space whose vertical position shows the top most features. The line is a reconstruction based on physical principles with logical application to the radar trace. The lines from multiple traces can be stacked and interpolated into a surface plot in three dimensions.

1. Physical Principles of Ground Penetrating Radar (GPR)

In order to understand the interpretation of the radar data for archaeological ground studies a short review of the relevant principles will be given. Along with this, the meaning of the data images will be discussed in light of the physical principles.

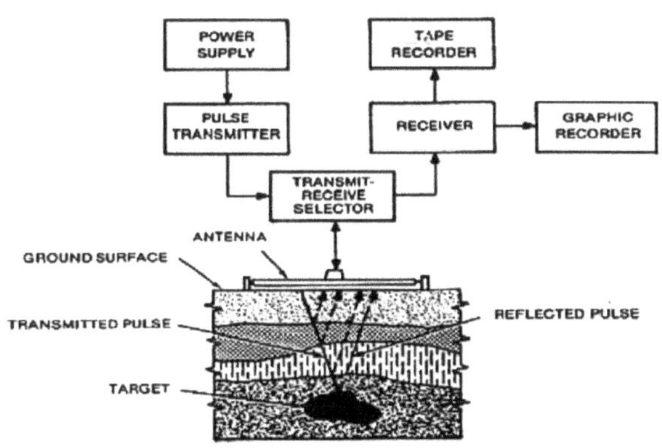

Fig. 1-1 Block diagram of a GPR system .

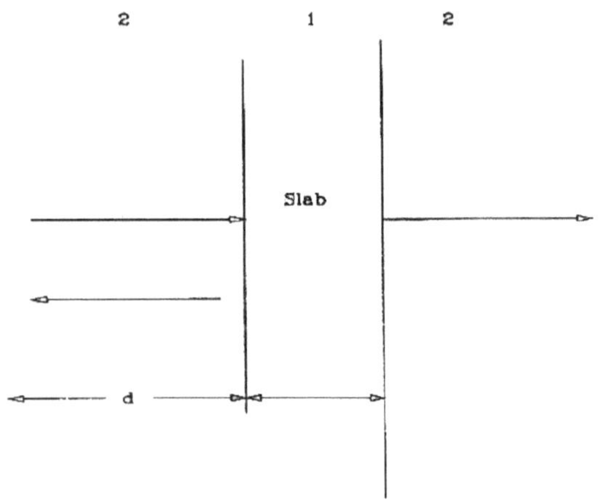

Fig. 1-2 Diagram of a limestone slab in earth.

A block diagram of a typical GPR system is shown in Fig.1-1. The antenna section contains both a transmitting and receiving antenna, which both sends and receives a short time limited pulse. First a pulse is sent into the earth and then the receiver is switched on to listen to pulses, which are reflected from both interfaces and buried targets such as rocks or voids. Interfaces in this study could consist of layers of habitation, which in the course of time are covered by an overburden of sand or soil. A model-based study [1] on reflections from a limestone slab and a limestone sphere in earth (loam) are used as examples for image production. A GPR signal or pulse is incident on the left hand face of a semi-infinite slab. Some of the pulse is reflected from the first interface while the remainder travels through the slab and is reflected by the right hand face. This reflection passes back through the slab and emerges moving to the left.

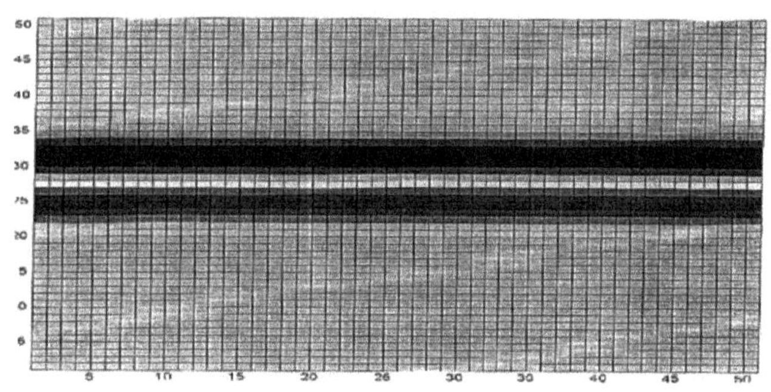

Fig. 1-3 Radar return from the slab of Fig. 1-2.

Fig.1-2 illustrates the process. The resulting image in radar space is shown in Fig.1-3 where the horizontal axis is the distance along the surface of the earth and the vertical axis is time. The first horizontal band represents the pulse reflection from the left hand slab interface and the second band is the signal produced by the pulse that was reflected from the right hand slab interface which travelled to the left through the slab and out to the left. The images are bands because the distance from the air earth interface where the antenna is located does not change as the radar is moved in the horizontal direction. Another example concerns radar reflections from a buried limestone sphere. Here the radar image shown in Fig.1-4 consists of a set of peaked curves with the maximum value at the point where the radar is vertically over the sphere. The light and dark traces are due to internal wave reflections inside the sphere. An actual experimental result (HORTON, K.A., R..M. MOREY, R.H. BEERS, V. JORDAN, S.S. SANDLER, and ISACSON., 1981), shows the validity of the models.

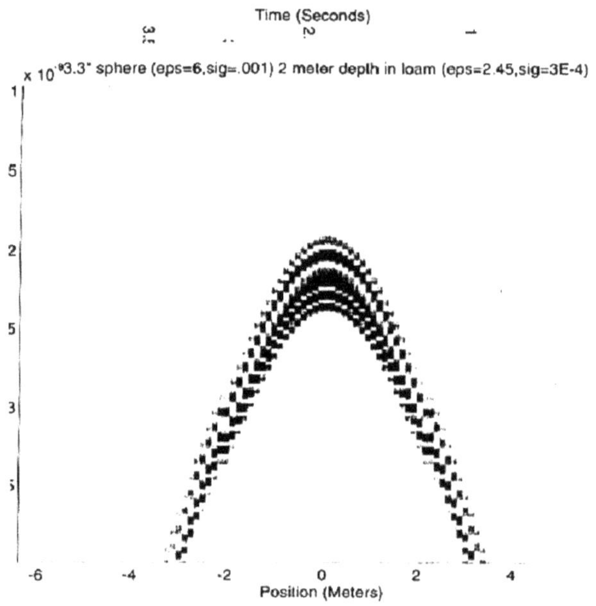

Fig.1-4 Radar return from sphere.

Here a number of metal and plastic barrels were buried in a sand trench and covered over using a bulldozer.

The resulting radar image shows a number of curves similar to those previously found for a buried sphere. In addition there are many horizontal and nearly horizontal traces which are related to those found from our model of a semi-infinite slab. These are due to the layering caused by the bulldozer as it cleared out the sand pit.

2. Reconstruction

Reconstruction involves the production of a 2-D or 3-D image of a scene or object from a limited set of views. The views are taken remotely without direct access to the internal parts of the scene. Remote sensing of the human body, for example, takes place by passing X-ray beams through human tissue. The beams pass through the body and are recorded by a detector. The angle of the beams is then changed and the result again recorded by the detector. Using a computer and well-known numerical algorithms, a slice of the interior of the body is reconstructed. By laterally moving the beams further slices can be obtained and subsequently stacked to produce a three-dimensional image of the interior of the body. This is the principle of the Computer Aided Tomography (CAT) procedure used for finding tumors and defects in many parts of the body. The preceding procedure is illustrated in Fig. 2.1. In Fig. 2.1 the Xray beams are incident on a slice from the left. They pass through the slice and are detected along the projection line P. Subsequently a number of projections are recorded at different angles. These projections are combined through a numerical algorithm to produce the best reconstruction, or image, of the slice interior [4]. The CAT scan procedure works very well because of the large number of views available and the fact that the beams go in straight lines through the body. Things are much more complicated the GPR case for a variety of reasons. A GPR antenna produces radiation which is spread over a wide angle. In practice the radiation is approximately confined to a 3-D conical volume. Response from the radiation is recorded as if it occurrs directly under the GPR, although it could be due to signals anywhere within the 3-D conical sector. Trying to use the GPR images to reconstruct a 2 or 3-D image is a formidable problem. In this paper a simple yet physically meaningful approach to the problem will be taken.

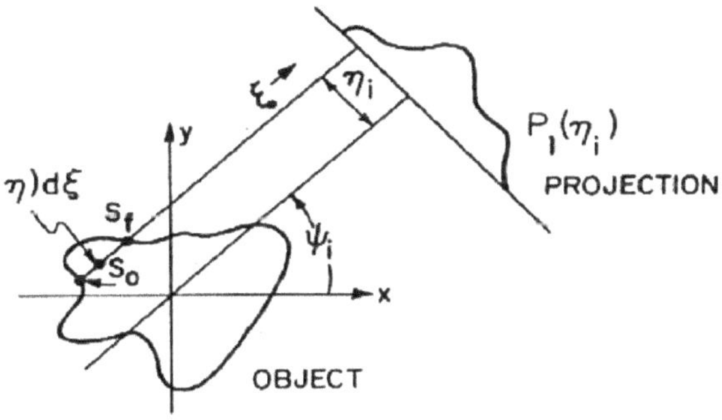

Fig. 2-1 Geometry for a single slice.

3. Time Slices

There are many papers in physical science journals (e.g. IEEE Transactions on Antennas and Propagation, IEEE Transactions on Geoscience and Remote Sensing) on reconstruction. However the most important work on reconstruction using archaeological GPR data is due to Goodman and Conyers [5]. A brief description of their method follows to point out the difference between their time slice and the present approach.

Using data from a plan view of the earth over a subsurface anomaly, vertical time planes are established. Only a few slices are used in the stacking, and the remainder of time slice planes is discarded. Many results of this method are illustrated in their book on radar [5] which clearly show a three dimensional perception of a localized shape like a burial mound. The time slice method is not a reconstruction, but a very intelligent use of the original unprocessed data.

4. Basic Reconstruction Algorithm

The objective of data processing is to clearly delineate the first subsurface layer or target. From an archaeological point of view, it is the first discontinuity found by digging. It could be a habitation level or it could be a stone fireplace above the habitation level. If lower levels were desired, the parameters for the algorithm would be adjusted for the desired result. The physical basis for the algorithm is involved with the energy measure for each radar return. The returned signal, represented by a gray-scaled vertical trace on the radar image, has an amplitude, which is proportional to the electric field reflected from a subsurface discontinuity such as a rock or floor. The energy density in the signal is proportional to the square of the electric field of amplitude. If each squared amplitude point in the vertical trace is summed, the total energy for the trace results. Note that zero energy corresponds to the top of the trace (i.e. the earth air interface) and maximum, or 100 per cent energy corresponds to the bottom of the trace (i.e. maximum depth). The new reconstruction algorithm fixes an energy level that will draw a line for the first discontinuity level in the earth.

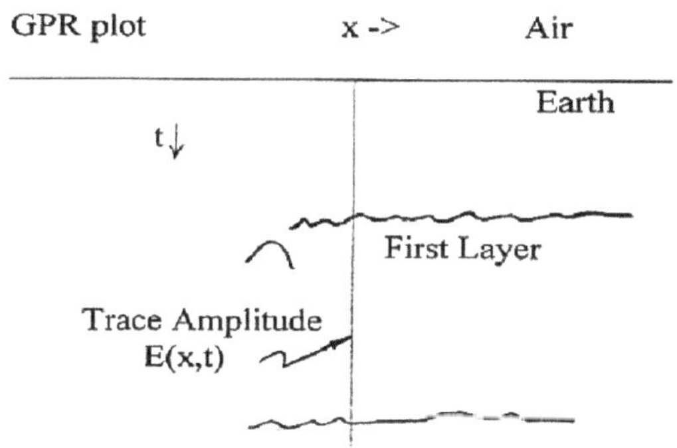

Sum Squared Averaged

$$SSA = \underset{(time)}{Sum} [E(x,t)]^2 \ / \text{Averaged Value}$$

Set Value of SSA for first layer or target (rho)

rho * SSA

Fig. 4-1 Basic reconstruction algorithm and GPR plot.

Fig. 4-1 shows a simplified subsurface radar trace along with the mathematical form of the reconstruction algorithm. A quantity called the SSA (sum squared averaged) is seen to be proportional to the total energy in the trace, which is proportional to the electric field squared in the trace. In order to pick out the first layer or target a value of the constant rho is set, which when multiplied by the SSA will pick the energy value to construct a line for mapping the discontinuity.

5. Archaeological Site

The site chosen for surveying is in the Annasnappet Archaeological District in North Carver Massachusetts, which is close to Cape Cod. More closely it is near the present day end of State Route 44, about 1000 feet from High Street near Annasnappet Pond. It is an 8000-year-old site with long time occupancy dating from the late Paleo-Indian/ archaic eras. Dianna Doucette, a Ph.D. student in the Department of Anthropology at Harvard University, is working the site. Her grid lines were used for the present study. The survey was taken on a 5m by 5m grid where the long traverses were taken the West-East direction for a length of 30m. Three parallel long traverses were taken with the 5m points marked on the magnetic storage tape. The GPR was a GSSI Model SIR 10 and the 500 MHz antenna was used. Although radar returns could be found at depths of 100 feet, the controller was set to sound only the first two meters in the earth.

Fig. 5.1 shows the heavily forested terrain where the survey was taken. A tape for the survey path is shown just right of the center of the picture. Fig. 5.2 is a pit excavated by Dianna Doucette, who is also in the picture. A fireplace made of stones was found in a corner of the pit and was picked up by the radar along with the pit itself.

Fig. 5-1 Woods where survey was taken.

Fig. 5-2. Excavated pit with fireplace.

6. Presentation of the Data and the Reconstruction

A section of the original unprocessed GPR data is shown in Fig.6.1 for a horizontal individual block length of about 24 feet. The original traverse lengths were approximately 120 feet. The three horizontal segments correspond to the three parallel traverses. Even without processing this data set clearly shows the pit boundary and the buried rock fireplace with a hyperbolic type pattern along with other types of anomalies. To vastly improve the clarity of the image the data was processed using background subtraction.

The background subtracted consisted of the first vertical trace for each of the vertical segments in Fig. 6.1.

The background subtracted subsurface image is shown in Fig. 6.2 along with the line reconstruction of the first layer or anomaly. Note that the line reconstruction "reaches" down to indicate the first buried object or layer. The algorithm is very robust and finds vertical structures that are not readily apparent to the eye.

Using a MATLAB program described in the next section, the three line reconstructions are stacked to produce a 3-D surface image that is shown in Fig.6.3. Linear interpolation was used to fill in the surface from line to line.

The excavated pit is the first line segment at width equals one and note that the pit extends from approximately 35 to 45 on the front horizontal scale. The fireplace is near 40 on the horizontal scale. A shaded 3-D image is shown in Fig. 6.4 with a reversal in viewing angle from Fig.6.3. Although the 3-D reconstruction is shown for only a small area the methodology is easily extended to much larger surfaces.

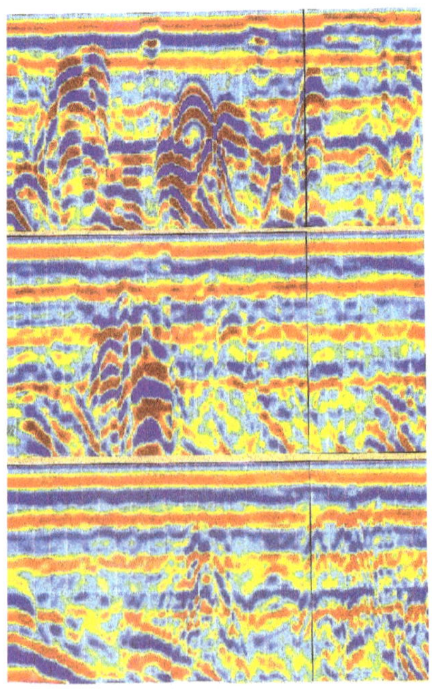

Fig. 6-1 Original unprocessed data.

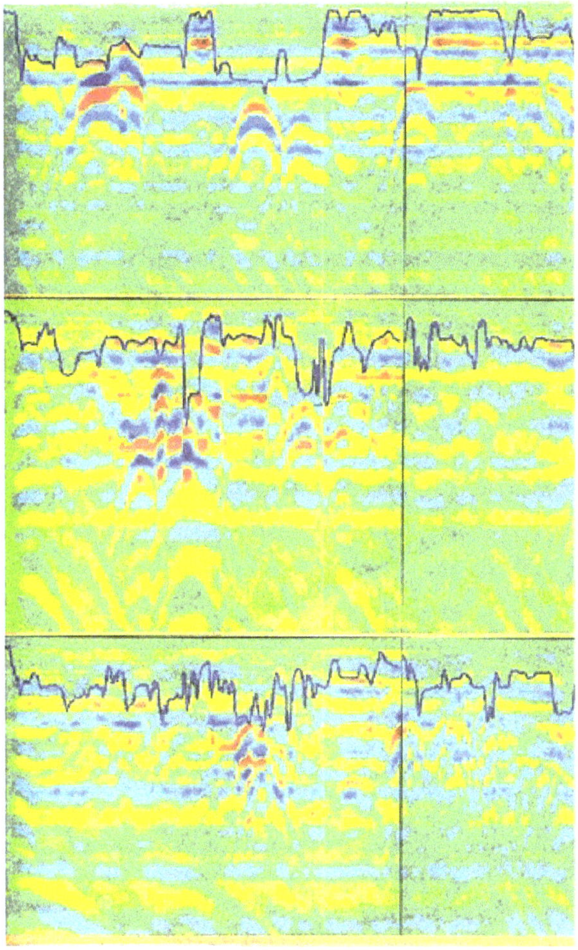

Fig. 6-2. Background subtracted image.

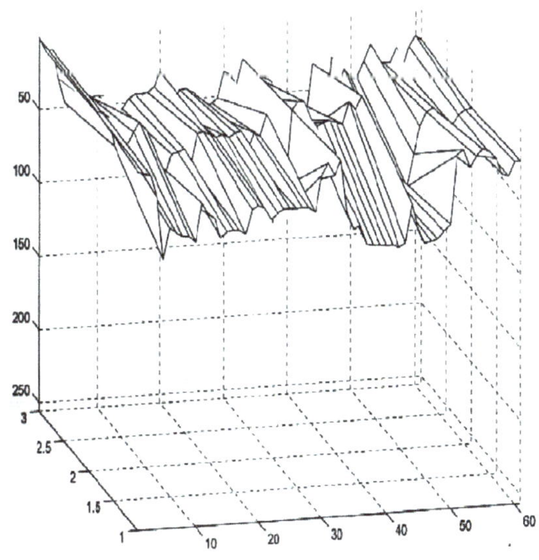

Fig. 6-3 3-D surface image reconstruction.

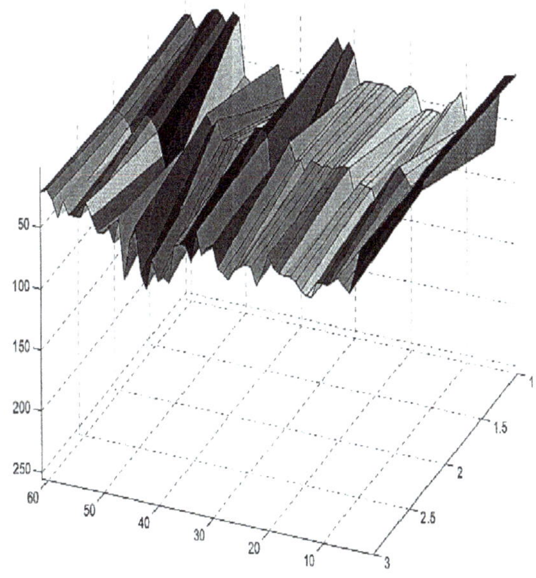

Fig. 6-4 Shaded image.

7. Steps in the Computer Processing

Processing on the reconstruction was undertaken using MATLAB for the mathematical manipulation and Paint Shop Pro for the presentation of the original and background data. The 2-D and 3-D images were calls in MATLAB, which used the line reconstructions as inputs. First the data is transferred to the PC using a GSSI supplied SirLink program. It is then converted to 16 bit numbers using SIR2RADW. The ASCII files are then loaded into MATLAB and resized to the longest file. The background image is produced and the gain adjusted to reduce the effect of deeper ringing. The steps to find the line reconstruction are preformed as explained in Section 4. The MATLAB programs are called to present the final 3-D images.

Acknowledgement

The authors are thankful to Geo-Centers, Inc for supplying the GPR for this study. Thanks are also due to Dianna Doucette of Harvard for helping with the field measurements and helping to correlate the results with her investigations. A further paper is planned with her to go over the detailed correlation of the archaeological results with the GPR results. Thanks also go to my wife Barbara for preparation of the manuscript.

References

CONYERS, L.B. AND D. GOODMAN.1997. *Ground penetrating Radar, an Introduction for Archaeologists,* Ch. 5. Alta Mira Press, Div. Of Sage Publications. Walnut Creek, CA.

HORTON, K.A., R..M. MOREY, R.H. BEERS, V. JORDAN, S.S. SANDLER, AND ISACSON 1981. *An evaluation of ground penetrating radar. Geo-Centers*, Inc., Newton Center, MA. (GC-TR81-171)

SANDLER, S.S. 1975. *Picture Processing and Reconstruction.* Ch. 2. Lexington Books Div. Of D.C.Heath, Lexington, MA.

SANDLER, S.S. AND JONES, D. 1998. *A closer look at GPR data interpretation for archaeological applications*. Intl. Conference on Remote Sensing in Archaeology, Boston University, Boston.

Remote Sensing and the Location of the Ancient Tigris

Elizabeth C. Stone

Stony Brook University, Department of Anthropology,
Stony Brook, NY 11794-4364, 516/632-7627
estone@sunysb.edu

Abstract. When Robert McC. Adams conducted his seminal research into the settlement patterns of the Lower Mesopotamian Plain, modern imagery was largely unobtainable, and thus he assumed that the Tigris River played only a minor role in the ancient Mesopotamian settlement and history. The availability of new imagery, especially SPOT and Corona images, have changed this picture. We used SPOT imagery to identify the branch of the Tigris that passed near Mashkan-shapir and, more recently Gasche and Cole have combined DEMs from SPOT imagery with textual data to reconstruct the second millennium B.C. river channels in the northern portion of the plain. In this paper, I will use a combination of SPOT, Corona and Aster Terra data to attempt a reconstruction of the role of the Tigris in the late third and early second millennia Mesopotamian history and settlement.

In this paper I will explore the role of the Tigris in ancient Mesopotamian settlement and history using satellite imagery to identify ancient watercourses and surface survey data to determine contemporary sites. Once the lines of the ancient river have been determined, excavated data can be used to refine our chronologies and textual sources to identify ancient routes–but these are steps beyond the scope of this paper. The main focus of this project, ancient Babylonia–the land bounded by Baghdad in the north, the Gulf in the south, the Tigris in the east and the Euphrates in the west–is an ideal place for this kind of work. The southern Mesopotamian plain is one of the world's great deltas, a huge flat plain created by the outflow from the Tigris and Euphrates rivers. In antiquity the rivers flowed through what was, until the post-Gulf war agricultural expansion, a central desert, a place where traces of ancient watercourses are readily seen in aerial imagery. Moreover, much of this area has been subjected to archaeological survey, and excavations at many of the major sites have yielded cuneiform tablets, some of which include geographical information.

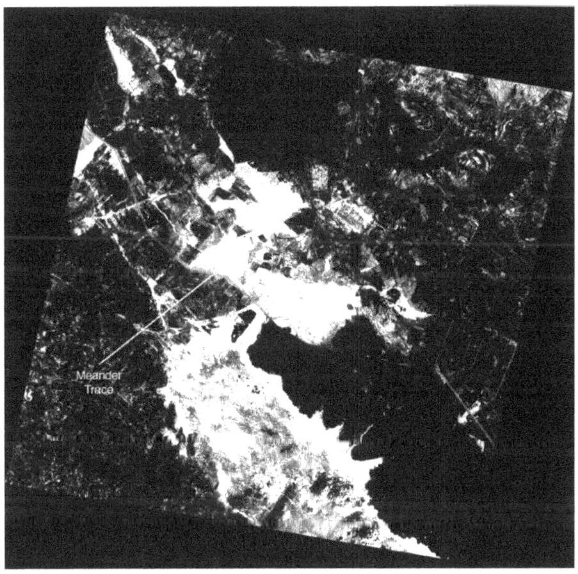

Figure 1. 1988 Spot Panchromatic Image of Central Babylonia with ancient watercourse visible.

Until 15 years ago, the Tigris river was thought to have played little role in Mesopotamian settlement and agriculture between the fourth and the late first millennia B.C. Since that time, however, a number of different lines of evidence have come together to suggest that this interpretation can no longer be held.

The first indication that the Tigris might have played a more important role than had been thought heretofore came as a result of our survey project at the site of Tell Abu Duwari. Finds from the first season at this large early second millennium urban center suggested its identification with the ancient Mashkan-shapir, a city which was known to have been located on the Tigris on the basis of cuneiform records. Today, however, the site is located some 40 km to the west of the modern Tigris, well within the area then thought to have been watered by branches of the Euphrates. To resolve this apparent conflict, we asked the Spot Image Corporation to take a Panchromatic view of the area (Fig. 1).

They took the image in May 1988, at one of the rare times when the spring rains were sufficient to make the desert bloom. Presumably as a result of differential plant growth, this image provided remarkable detail of subtle differences within the desert such that both the site itself and the large curve of its associated river are visible even when the whole image is examined–and with still more clarity (Fig. 2). The watercourse seen in this image is too large to have been a part of the Euphrates system, and was therefore immediately interpreted as the Tigris (Stone and Zimansky 1995). It could be traced throughout the image, and was associated with some of the early second millennium sites known in that area.

The Tigris was brought further into prominence by the collaborative work of Gasche and Cole (1998) who used Spot DEMs to locate the ancient levees in the northern part of Babylonia and textual data to identify their dates and ancient names. In their reconstruction of the early

Figure 2. Blow up of the 1988 Spot Panchromatic Image of Central Babylonia showing ancient Tigris channel and the site of Mashkan-shapir.

second millennium B.C. watercourses, they see the Tigris as flowing in two main branches in northern Babylonia–and their map of the location of these can be directly linked with the ancient river traces that we had already mapped on the basis of the Spot image of the Mashkan-shapir area. This demonstrated the importance of the Tigris in both northern and central Babylonia in the early second millennium on the basis of both textual and archaeological evidence.

Until now, information on the role played by the Tigris in southern Babylonia has been based solely on textual data. As much as a decade ago, Heimpel (1990) argued that the Tigris must have flowed through the southeastern part of Babylonia at least in the late third millennium B.C., and more recently Piotr Steinkeller (2001) has enlarged on this theme, arguing that the watercourse once thought of as the easternmost branch of the Euphrates–a branch flowing through the ancient city of Adab--was in fact the Tigris. Steinkeller has published maps of what he believes was the course of the ancient Tigris, but these only partially correspond with the traces of the watercourses that we see on the ground.

The task that I have set myself for this paper is first to determine where the Tigris was likely to have been within late third and early second millennia Babylonia and to see how it related to the various channels of the Euphrates. First, however, a discussion is in order on the ways in which remote sensing techniques can be used to identify these ancient river traces.

The southern Mesopotamian floodplain is one of the largest deltas in the world, as shown in Fig. 3, a DEM with the grey scale stretched between sea level and 100 feet in elevation. As such it is extremely unstable, with the locations of deserts, marshes and irrigation agriculture shifting continuously. As a result, the plain has been sculpted through a complex process of both alluviation and deflation. This means that a single feature, such as an ancient Tigris bed, may not leave the same traces in all parts of the alluvium, and that imagery taken at different times will show quite different features. For this project we relied most heavily on Corona imagery. Two series are available, one taken in the spring but with a few popcorn clouds in the central desert and a second series taken in August where the skies are clear but moving dust obscures the surface in some areas. We also have five Spot panchromatic images, taken under the ideal conditions of May 1988 when the desert had bloomed, and some recent Aster Terra data. We have not tried Thematic Mapper data because of its tendency to be oversaturated in desert areas.

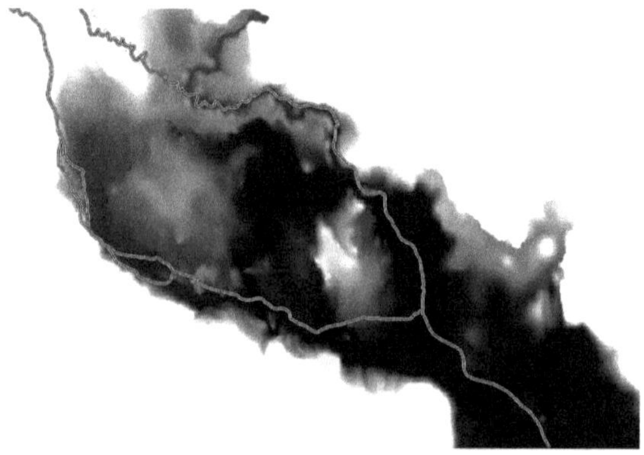

Figure 3. 30 Arc-Second Digital Elevation Model of Southern Iraq with the modern Tigris and Euphrates river courses superimposed. Grey scale from Sea Level (black) to 100 feet or more in elevation (white).

The levees on which the Tigris once flowed can still be seen, especially in the northern and eastern part of the plain, and especially within heavily irrigated areas. Here linear traces of uncultivated land within large scale irrigation systems reflect the slight differences in elevation involved.

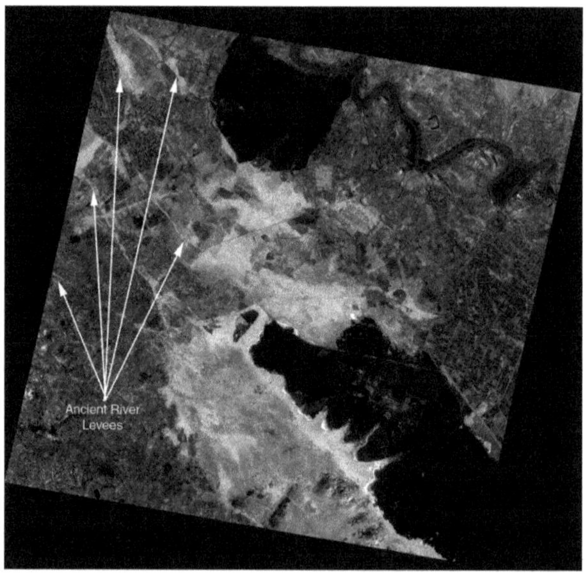

Figure 4. 1988 Spot Panchromatic Image of Central Babylonia showing ancient river levees, seen as linear uncultivated areas within the irrigated districts.

It appears that these levees are less well preserved further south and east since they are visible as linear areas of uncultivated land within irrigated areas in the 1968

Corona data (Fig. 4), but can hardly be seen in recent Spot data (Fig.5).

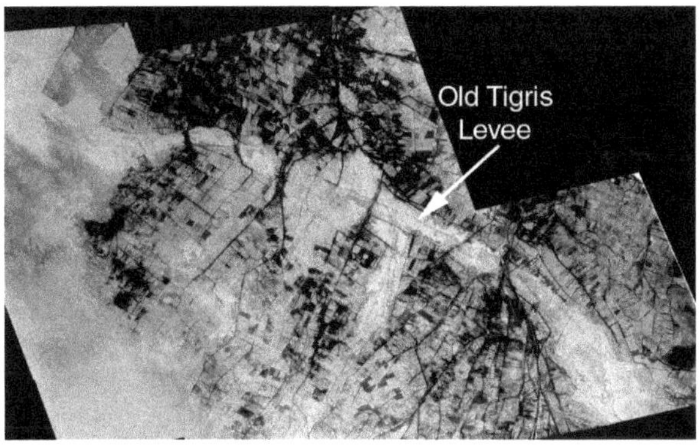

Figure 5 Detail from 1968 Corona Imagery of an old river levee, shown as a uncultivated line within the irrigated zone.

Nevertheless, traces of the ancient river bed are visible within the desert areas, and even, sometimes in lightly cultivated zones. In some places the ancient bed is represented by discolorations within the desert which show up in both the Corona and the Spot data (Fig. 6). These are probably the result of increased access by plants to subsurface water. This phenomenon has been clearly documented for small canals and there is no reason to suppose that a similar process might not be involved for large rivers. In these areas recent–post Gulf War--irrigation projects are not impeded by the traces of this watercourse, suggesting that the combination of alluviation and deflation has eliminated the elevation differences that once existed between the river levees and the rest of the plain. These traces lack the meanders that characterize the modern rivers and may indicate areas where artificial means were used to direct the flow.

Figure 6. The same area as shown in Fig. 5 from a recent Spot Panchromatic Image showing how the old river levee has now been almost overwhelmed by irrigation channels.

In other parts of the desert a more natural, meandering river is indicated by the presence of meander scars etched into the desert surface (Fig. 7), which can also be traced beneath shallow marsh deposits (Fig. 8). Here, although the meander scars themselves are indicated by slight differences in elevation, the ancient river system itself is no longer characterized by raised levees. Thus different types of imagery, taken at different times, are more or less effective in identifying these ancient river traces in the various parts of the plain.

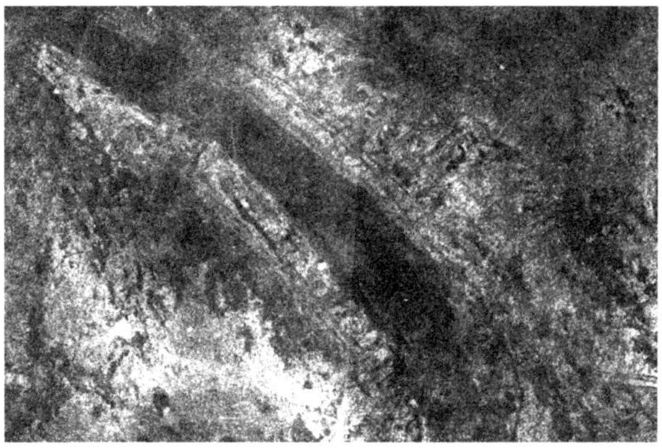

Figure 7. An ancient river channel showing as a discolored area in a 1968 Corona Image.

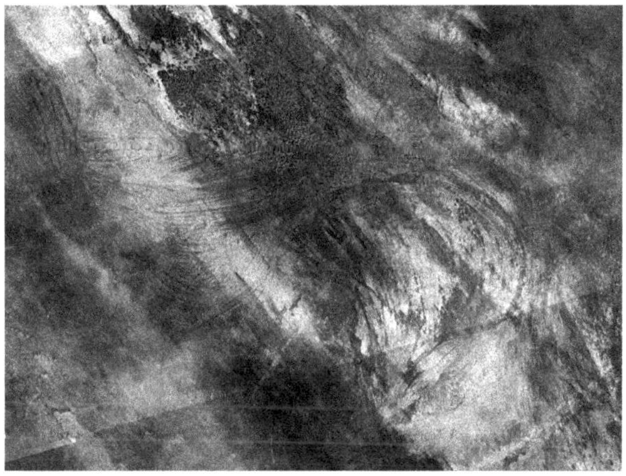

Figure 8. Large meander scars visible in a 1968 Corona Image.

A more difficult question is how we date the traces that we see. We are fortunate that much, but by no means all, of the area of interest was subjected to surface surveys conducted by Robert Adams (Adams 1965, 1972, 1981, Adams and Nissen 1972, Gibson 1972, Wright 1981) and his students and colleagues in the 1950s, 60s and 70s, of which the most important for this project are the Nippur (Adams 1981) and Uruk surveys (Adams and Nissen 1972). Where sites line up with the traces of watercourses seen in the imagery, we can make a reasonable case that they were contemporary. There are, however, some difficulties with the dating of the sites. This is the result of two factors. First, good stratigraphic ceramic sequences were rare at that time, and even where present they tended to be based on whole pots–most of which are found in graves–rather than the potsherds which better reflect living debris. Thus, sites dating to the period in which we are interested--the transition from the Ur III period (2112-2004 B.C.) to the Isin-Larsa period (2004-1763 B.C.) and the early years of the Old Babylonian period (1763-1720 B.C.)–fall into only one chronological category in the survey data. For this reason, the only way to determine if there were changes in settlement within this 450 year period is to look at the dates of excavated sites. Since these tend to be the larger cities, it seems reasonable to assume that if we find evidence for their abandonment, then it is highly likely that the smaller sites in their vicinity known only from survey were abandoned at the same time.

A second and related problem with the survey data is that when the work was carried out it had not yet been appreciated that southern and central Babylonia experienced a complete or almost complete loss of permanent settlement for the nearly four centuries between around 1720 and 1350 B.C. (Stone 1977, Armstrong and Brandt 1994). Thus neither the type fossils used in the survey work nor the excavated sequences on which these were based reflect this long gap. As a consequence, the data collected during the Nippur and Uruk surveys grouped chronological periods that we would prefer to be discrete–and grouped them in different ways between surveys. The result of these vicissitudes is that while the coincidence of occupied sites and traces of ancient watercourses can provide a rough indication of the probable date of the river in question, more nuanced questions regarding the differential development of the Tigris and Euphrates systems are less easy to answer definitively. In addition, much of the eastern portion of the plain remains unsurveyed, leaving the date of the traces seen in the imagery very much open to question.

Nevertheless, the use of the combination of remotely sensing and the survey data allow the reconstruction of both Tigris and Euphrates channels for the late third and early second millennia B.C. Here I provide a description of how I now see the ancient watercourses in the southern plain. I cannot claim at this time to have examined in detail every possible piece of imagery pertaining to the area–but the general outlines seem to be clear (Fig. 9).

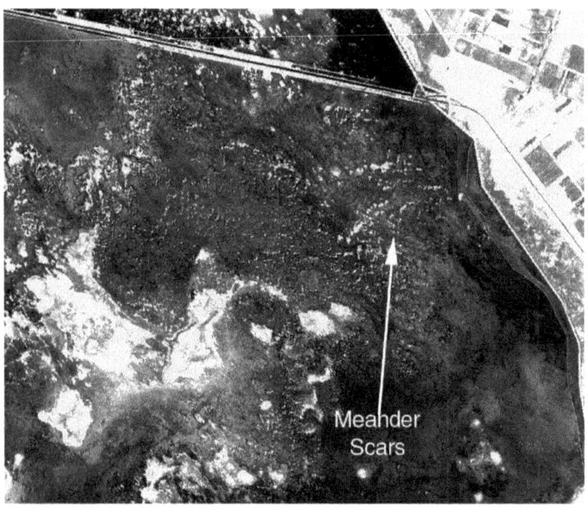

Figure 9. Meander scars just visible within lake deposits seen in a 1988 Spot Panchromatic Image.

I will begin in the north–merely because this is the source of all the water to be found in southern and central Babylonia. Here Gasche and Cole (1998) have argued convincingly that in the early second millennium–i.e. in the Isin-Larsa and Old Babylonian periods–northern Babylonia was characterized by a typical braided stream with both the Tigris and Euphrates flowing in multiple branches, and where some of the Euphrates branches flowed into the Tigris–an indication that the Tigris flowed at a somewhat lower elevation than the Euphrates. At the southern margins of the area pursued by Gasche and Cole, they identified a bifurcation of the Euphrates at Kish with one branch heading southeast towards Nippur and Adab, and the other going more directly south, perhaps to Isin. Two other branches must also be postulated one passing through Babylon and possibly continuing south to flow to the west of Ur, and another (not on our map) flowing still further west through Borsippa and Dilbat. Most traces of these western branches have long been obliterated by the modern Euphrates so they must remain speculative. Gasche and Cole (1998) also see two branches of the Tigris taking off to the southeast.

The easternmost branch of the Euphrates can be traced to a point a little north of Nippur, where it splits into two branches, one passing through Nippur and on to Kisurra, Shuruppak, Uruk and Ur, and the other through Adab and on nearly to Umma, where it turns back towards the southwest and heads towards Uruk. Adams postulated that a further branch of the Nippur channel fed the important city of Isin, but the Spot image of the area suggests instead that the watercourse associated with Isin came from the northwest–perhaps the western branch from Kish. Since the area between Isin and Kish remains unsurveyed–and indeed would be very difficult to survey due to the intense irrigation in the area–we have no evidence in the form of contemporary sites that could help us here.

We see not two but four branches of the Tigris as we move south from the area focused on by Gasche and Cole. The easternmost lies completely outside the survey area and flows back into the central branch. The two central branches reunite a little further south, flow past Mashkan-shapir, birfurcating a little further southeast with the eastern of the two branches flowing past Wilaya before rejoining the more western branch.

The most important branch for our purposes is the western branch, whose large meander scars in the area between Mashkan-shapir and Nippur were noted by Adams at the time of his last survey. He noted that late fourth millennium B.C. sites lined up with these scars and, as a consequence, dated these traces to that period (Adams 1981, 62). However, although this branch was certainly not in existence for most of the third millennium–as demonstrated by the complete absence of settlement in the area (Fig. 10), a number of sites dated to the Ur III through Isin-Larsa periods line up equally well with these traces. It may therefore be postulated that this western Tigris levee was dry in the third millennium, but received renewed flow in the period in which we are most interested. This should not be surprising since heavily meandering rivers frequently shift their beds from levee to levee. The bulk of the flow then seems to have rejoined the western of the other two branches of the Tigris, but we also have clear traces–with appropriately dated sites associated with them, of a channel that parallels the Adab branch of the Euprates before turning towards the east just before Zabalam. From this point water is taken off to water Umma, Bad Tibera, Larsa and Girsu. Indeed, meander traces can also be seen linking this part of the Tigris to the point where the Adab branch of the Euphrates turns southwest to flow towards Uruk, an indication that at this part of the valley, the Tigris was probably at a slightly higher elevation than the Euphrates.

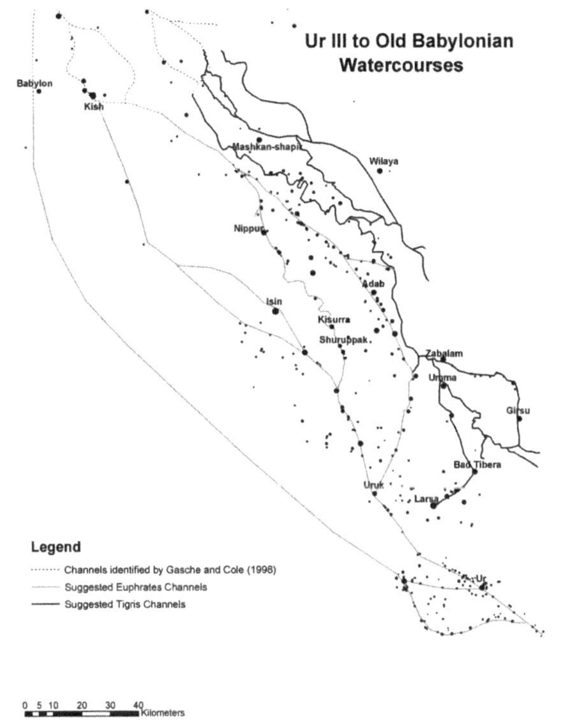

Figure 10. Map of sites dating to the late third to early second millennium together with a reconstruction of the major channels of the Tigris and Euphrates rivers of that time.

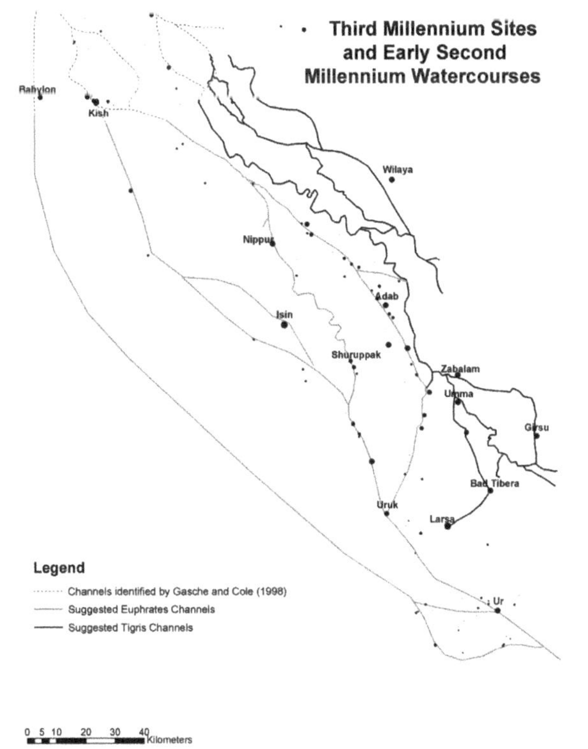

Figure 11. Map showing mid-third millennium sites with the reconstruction of the major channels of the Tigris and Euphrates thought to date to the late third to early second millennia B.C. Note that there are no sites associated with the western branches of the Tigris, an indication that this bed was dry at that time.

This is as far as the use of remote sensing and survey data can take us in determining the role of the ancient Tigris in ancient Mesopotamian settlement. These data make it clear that although the Tigris may have played only a minor role for most of the third millennium, by the beginning of the second millennium its waters were critical to life in the southern part of the alluvium. Indeed, these data suggest that Tigris water, sometimes admixed with water from the Euphrates, was critical to the survival–and perhaps even the growth–of the important cities of Uruk, Ur, Larsa, and Umma, as well as those sites that were actually on the Tigris itself, like Zabalam and Girsu. Understanding how the interplay of the Tigris and Euphrates systems affected Mesopotamian history, however, must rely on a mixture of precision dating based on excavated sites and the interpretation of the available historical data–a project beyond the scope of this paper.

References

ADAMS, ROBERT MCC., 1965. *Land behind Baghdad.* Chicago: University of Chicago Press.

ADAMS, ROBERT MCC , 1966. *The Evolution of Urban Society.* Chicago: Aldine Publishing Company.

ADAMS, ROBERT MCC, 1972. "Settlement and irrigation patterns in ancient Akkad," Appendix V of M. Gibson, *The city and area of Kish.* Coconut Grove: Field Research Projects, 182-208.

ADAMS, ROBERT MCC , 1981. *Heartland of cities: Surveys of ancient settlement and land use on the central floodplain of the Euphrates.* Chicago: University of Chicago Press.

ADAMS, ROBERT MCC. and Hans Nissen, 1972. *The Uruk countryside: The natural setting of urban societies.* Chicago: University of Chicago Press.

ARMSTRONG, J. A. AND M. C. BRANDT, 1994. *Ancient Dunes at Nippur,* in Cinquante-deux reflexions sur le Proche-Orient ancien: offertes en hommage à Léon de Meyer, H. Gasche, eds. Mesopotamian History and Environment, Occasional Publications 2. Leuven: Peeters.

GASCHE, HERMANN, AND STEPHEN COLE, 1998. *Second and First-Millennium B.C. Rivers in Northern Babylonia,* in Changing Watercourses in Babylonia: Towards a Reconstruction of the Ancient Environment in Lower Mesopotamia, H. Gasche and M. Tanret, eds. Mesopotamian History and Environment II, Memoires 5. Leuven: Peeters, 1-64.

GIBSON, MCGUIRE, 1972 *The city and area of Kish.* Miami: Field Research Projects.

HEIMPEL, WOLFGANG, 1990. *Ein zweiter Schritt zur Rehabiliteirung der Rolle des Tigris in Sumer,* Zietschrift für Assyriologie 80, 204-213.

STEINKELLER, PIOTR, 2001. *New Light on the Hydrology and Topography of Southern Babylonia in the Third Millennium,*

Zietschrift für Assyriologie 91, 22-84.

STONE, ELIZABETH C., 1977. *Economic crisis and social upheaval in Old Babylonian Nippur,* in Mountains and lowlands, Louis D. Levine and T. Cuyler Young, Jr., eds. Bibliotheca Mesopotamica 7 Malibu: Undena Press. Pp. 267-289.

STONE, ELIZABETH C. AND PAUL ZIMANSKY, 1995. *The Tapestry of Power in a Mesopotamian City,* Scientific American 272/4: 92-97.

WRIGHT, HENRY T., 1981. *The southern margins of Sumer,* Appendix in R. Adams, Heartland of cities. Chicago: University of Chicago Press, 295-345.

Hydraulic Landscapes and Social Relations in the Middle Horizon Andes

Patrick Ryan Williams [1]

[1] Field Museum of Natural History - Department of Anthropology
1400 S. Lake Shore Dr. - Chicago, IL 60605
rwilliams@fmnh.org

Abstract. Between A. D. 600 and 1000, two great states dominated the Peruvian Andes. Archaeological investigations have revealed their direct overlap in only one place: the Moquegua Valley of Southern Peru. Satellite imaging and GIS models are used to recreate the hydraulic landscape of the valley during this time period, a landscape which represents the economic basis for the survival of each colony. A network analysis of water distribution systems and paleoclimate records provide an intriguing assessment of the changing dynamics of imperial interaction over the 400 years of co-occupation of the valley. Results of the analysis of digital landscape models are used to assess the charismatic nature of social exchange during periods of cooperation and conflict based on water availability in the driest desert in the world.

1 Introduction

A seminal problem in understanding the relationship between prehistoric societies in the Peruvian Andes is the nature of the differential access to that most precious of resources: water. Water feeds the rivers and canals that make agriculture possible on the dry mountain slopes. Without irrigation, only the hardiest of desert cactus grows, and no grasses or domesticated crops can survive for long in the desert without water. Once agriculture became the predominant means of subsistence in the sierra ecozones, water became the limiting resource that curtailed population growth and served to define productive success. There is an ample supply of desert land that with sufficient labor can be turned into productive terrace systems, yet without water nothing will grow.

Thus, the nature of social relations among irrigation societies within a bounded hydrological system is predicated on the distribution of water. If environmental or social variables change to alter the availability or distribution of that most precious resource, it upsets the water balance and forces cultural agrarian productive systems to change. In this study, I present a model of water availability and social distribution mechanisms for the Moquegua Valley, Peru during the Middle Horizon (A. D. 500 - 1000). Moquegua is the only valley in the Andes in which colonies of both Wari and Tiwanaku, the two great polities of this era, reside within the same hydrological unit (Figure 1). The model assesses the impact of social water distribution practices on the agrarian systems of the two colonies throughout the four hundred year duration of their co-occupation of the Moquegua sierra.

2 Creating the Database

The first step in creating a model of landscape use in Middle Horizon Moquegua involved the acquisition of topographical, hydrological, and archaeological data. Although significant alterations to the landscape have occurred over the past 1500 years, the general topographic and hydrological characteristics of the Moquegua sierra have not been substantially altered.

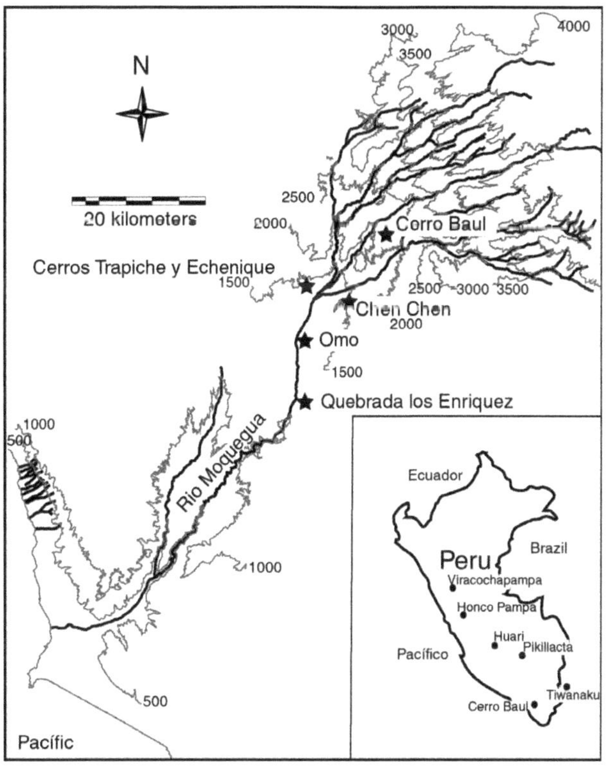

Fig. 1. The Moquegua Valley of Southern Peru, showing prominent Middle Horizon settlements

The presence of archaeological remains, including standing architecture, dating to this time period on the modern surface attests to the age of those surfaces. The natural hydrology of the valley includes steeply entrenched tributaries that merge into a single river below the city of Moquegua. Fluctuations in river discharge have substantially affected the valley bottoms, but these areas are circumscribed by the steep topography of the region, and unlike the flat plains of a meandering river basin, the reworking of sediments by hydraulic resources is limited to a narrow, entrenched canyon. There have been significant tectonic and climatic events including volcanic eruptions, earthquakes, and landslides that have substantially effected ancient human systems and have altered the surface geography of specific locales in the valley, but at a regional scale, the topography and natural hydrology of the Moquegua Valley has been fairly consistent over the past millenia. Thus, the topographic and natural hydrologic databases for the study region can be approximated by a modern digital elevation data. In this case, the source for the DEM is a photogrammetrically created 30 m resolution DEM based on the ASTER image on the Terra Platform acquired on October 12, 2000 and identified by DEM file number 20011025110437 (Figure 2).

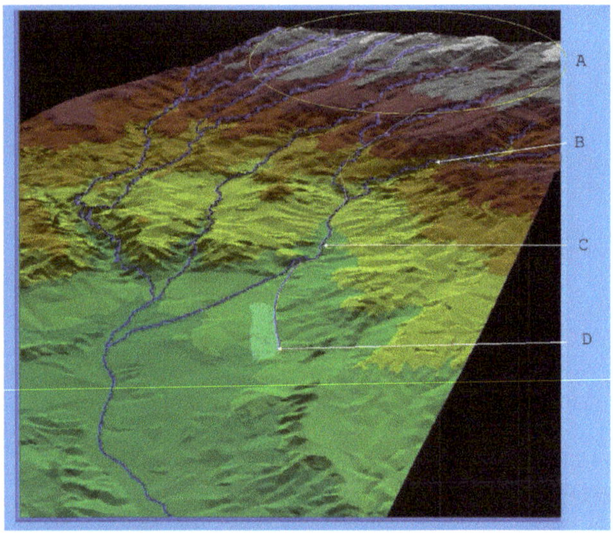

Fig. 2. Digital Elevation Model (orthographic perspective) of the Moquegua Valley derived from ASTER data. Water flows from the humid basin (A) through the natural hydrological network (B), which is assigned network friction values based on water loss rates. The best route through the network of man-made and natural water systems (C) is calculated using network analysis methods, and the derived water transport cost is assigned to the agrarian plot (D) based on its network travel costs and irrigable area

However, due to the extreme topographic variability in one portion the upper sierra of the drainage where slopes actually reach 90 degrees vertical, the photogrammetric DEM proves inadequate. In this area, around the unique geological mesa formation known as Cerro Baúl, shadow created by the vertical cliff faces make the DEM algorithms unresolvable. Since this mesa was critical to the development of Middle Horizon agriculture, a second DEM was created for this area based on a 1:5000 scale topographic map created by the Lockwood Survey company for the Southern Peru Copper Corporation. This high resolution map with a 5 meter contour interval was digitized and a 10 m DEM interpolated from the resultant data set. I used Arc/Info's Topogrid command to complete the interpolation, but the results did not accurately reflect the topography, instead providing a stepped contour model instead of the more smoothly differentiated surface. However, by applying a 5 x 5 kernel mean filter to the dataset, the steeped edges were removed, and the surface conformed to topographic data points collected with a total station within tolerable limits. It should be noted that this 5x5 kernel, which covered 50 meters on the ground, represented the average distance between adjacent major 25 m contour lines on the image (Figure 3).

After much experimenting, it was determined that this distance removed the stepped effect without compromising the integrity of the DEM. Thus , when choosing a mean filter to apply to a DEM derived from topographic contours, the best result might be obtained by using a kernel which will span the average inter-contour distance, but no greater.

The ancient cultural features of the valley were incorporated into the geographic information system based on archaeological survey and site specific studies over the past 15 years by researchers affiliated with the Contisuyo research program. The location and cultural context of major archaeological settlements and their associated agricultural systems was recorded from previous publications and from the author's own infrastructure reconnaissance in the valley. Positional data was obtained using a Topcon ET-1 total station with bench marks located by differentially corrected global positioning systems data. Systematic archaeological survey of the coastal section

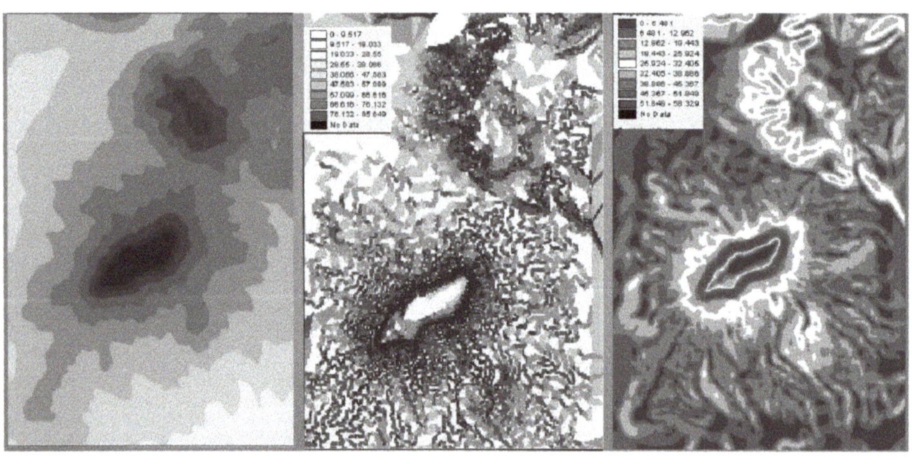

Fig. 3. DEM derived from topographic interpolation of contour lines around Cerro Baúl (left) using Arc/Info's Topogrid function. A slope derivative of the DEM reveals the flaws in the interpolation method (center), resulting in a "stairstepped pattern". After applying a mean filter to the data, slope derivates are more representative of the land surface (right).

and Otora tributary of the valley has been published (Owen 1993; Stanish 1985) and systematic archaeological survey of the middle valley and Torata and Tumilaca tributaries was completed (Goldstein 1993; Owen 1993). Although the analysis of the latter is still in process and significant new data has resulted from these studies, the preliminary results fall in accord with the data included in this analysis.

The author's archaeological infrastructure reconnaissance was undertaken coincident with the latter systematic survey projects and was designed to obtain specific location data on ancient agricultural systems and major archaeological settlements. Since systematic surveys were already in progress, the purpose of the reconnaissance was to document permanent infrastructure remains associated with agricultural production. The methodology for this reconnaissance involved the identification of all archaeological features identifiable from three aerial data sources including a SPOT satellite image, a SIR-C synthetic aperture radar image, and a set of black and white aerial photographs that were converted to digital products and orthographically corrected. Infrastructure features from these sources were confirmed by pedestrian survey and mapped in from the georeferenced data source. In order to document major infrastructure features that were not identified from remotely sensed data, several pedestrian transects were completed between each river tributary in the upper valley and along either side of the river in the middle and lower valley. While these transects do not represent a complete and systematic archaeological survey of the region, they did serve to identify all major archaeological settlements and all extensive ancient irrigation systems.

Three primary sources of cultural data were extracted from this study and incorporated into the geographic information system. These data layers included polygons representing archaeological settlements, polygons representing areas modified by human agriculture past and present (Figure 4), and a polyline coverage of the modern canal systems of the valley. The last data layer is incomplete for ancient time periods because entire ancient canal courses are rarely completely preserved. However, fragments of these canal systems do exist, and their locations were noted in the reconnaissance phase as point data. In order to complete the cultural hydrology data layer, principal canal courses were reconstructed from the fragmentary evidence using ARC/INFO's GRID module.

The reconstruction of canal courses is a complex process that involves a number of steps. First, a digital elevation model of the valley with a resampled 5 meter spatial resolution was created based on the topographic data layers. The ancient canal fragment point data was input into an equivalent resolution GRID file. Each point was then assigned to an ancient canal whose course would be reconstructed in the subsequent step. These assignments can be made because there is a recognizable canal typology in the Moquegua sierra and because the number of principal canals on a hillside rarely exceeds four stratified systems. Canal types include ridge top canals that follow the crest of a hill, contour canals that follow the contour of a slope with a very slight declination, and drop canals that run perpendicular to the contour (Stanish 1985).

The first two canal types represent the principal canal course that draws directly from the natural water source: a river or a spring. The drop canal is most often found as a secondary distribution mechanism that brings water to the farming surfaces and is rarely part of the principal canal line (canal madre). By examining the canal fragment's orientation to the slope in the field, one can distinguish the canal type. Associated fragment's can then be located by following the contour towards the water source for contour canals, following the ridge crest upslope for ridge top canals, and following the perpendicular to the slope for drop canals.

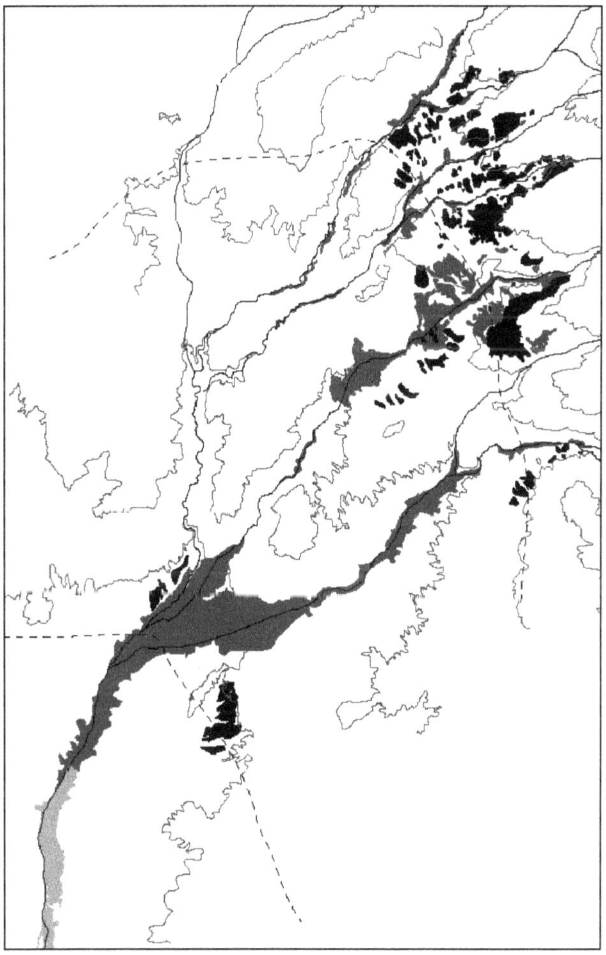

Fig. 4. Imagery classification and survey results for agricultural land use in the Moquegua Valley. Dark gray represents abandoned terracing, medium gray represents modern irrigation fed agriculture, and light gray represents areas irrigated today by wells.

Once canal fragments have been assigned to a canal system, the next step is the interpolation of the ancient canal course. This is accomplished by modifying the digital elevation model (DEM) to reflect the slope of the canal between two adjacent fragments. The first step involves the calculation of the average slope between the two points, which is defined as the elevation difference ex

Fig. 5. Methodology for reconstructing a canal course based on preserved fragments and a DEM. Here, an orthographically corrected aerial photo is draped over the DEM to provide a visual context from an orthographic perspective.

tracted from the DEM divided by the length of the course between them. The latter measurement is estimated by calculating the length of the least cost path between the two points on a slope model derived from the DEM. This path is close in length to the actual canal course, but does not necessarily follow a steadily declining canal course, so a second step must be undertaken to create a new cost surface. Once the average slope between the two points is calculated, the DEM is effectively tilted by that same slope so that the two points on the modified DEM will have the same elevation. The DEM tilting is accomplished by running a Euclidean distance operand on the entire DEM, adding the product of the average canal slope and the Euclidean distance from the higher canal point to the elevation of each cell in the DEM. A least cost path along the slope derived from the modified DEM provides the model for the actual path of the ancient canal between the two points. In effect, the contour between the two points of the modified DEM represents the course of the canal (Figure 5).

This procedure is repeated for each set of adjacent canal point fragments within each canal system until there is a complete canal course created. The final step involves undertaking the procedure between the highest canal point and the nearest water source, usually a linear river. Since a linear feature does not have a single elevation equivalent on the DEM, an alternative estimate of the canal slope between the river and the highest canal point is needed. The best estimate is the slope between the highest preserved canal remnant and the next highest canal point. Thus, one can use the slope model from the same modified DEM, but the least cost path in this case will be between the linear river and the highest canal point. The raster based canal courses are then converted to a vector canal system and imported into the vector GIS hydrology layer. At this point, a systemic perspective of ancient cultural water distribution networks is in place. However, the hydraulic characteristics of these ancient canal systems and the temporal extents of their utility still need to be addressed. Furthermore, areas of ancient agricultural production, partially accounted for by the agriculturally modified polygon layer, need to be associated with each canal system. In order to obtain additional data to address these issues, samples of preserved canal fragments for each ancient system were excavated.

3 Canal Excavation Methods

Ancient preserved canal fragments were bisected by a 50 cm wide excavation trench perpendicular to the course of the canal. These trenches 1) revealed information on construction techniques, 2) allowed examination of the post-abandonment depositional patterns, and 3) permitted the reconstruction of maximum discharge capacity. Excavation proceeded stratigraphically by natural levels with a 40 g soil sample from each level taken as sedimentological evidence and for hydraulic analysis. In addition, a one liter sample of the bed material which constituted the base of the canal was collected in order to evaluate bed material composition. A second trench was cut approximately ten meters upstream or downstream of the first, depending on preservation, in order to estimate instantaneous slope of the canal at the primary trench. Instantaneous slope differs from the average slope along the canal course in that it represents the exact slope at the point of the cross section. The elevation difference between the mean base of the canal in the two trenches was divided by the exact horizontal distance to obtain an estimate of the instantaneous slope, although it is more accurately described as the average slope over a short distance in the proximity of the trench. A profile was drawn for all primary trenches and basal views were drafted when canals were lined with stones or uncommon construction techniques were encountered. Canals that were too poorly preserved, had been re-excavated for failed revitalization attempts, or were in modern use were not excavated, although some modern canals were profiled to confirm the Ministry of Agriculture's maximum capacity measurements (Moquegua Ministry of Agriculture, 1983).

In the lab, the 40-gram samples were evaluated on color and size to identify water deposited layers. Bed material samples were passed through a set of nested sieves in order to calculate D84 values (diameter which exceeds 84 percent of the particles). These numbers are used in the Chezy, Manning, and Darcy-Weisbach equations for calculation of maximum discharge. Based on measured discharge rates from modern operating canals and calculations of their estimated discharge rates using the discharge equations, the Darcy-Weisbach equation best estimates the actual flow regimes of these canal systems. The Darcy-Weisbach equation is defined by ff = 8gRs / v2 where 1/(ff)½ = c log (aR/D84). In these equations, ff is the resistance coefficient; c and a are constants; R is the hydraulic radius of the channel, defined as the cross-sectional area divided by the wetted perimeter; s is the slope of the energy gradient; and g is the gravity constant. An estimate of mean water velocity (v) can be calculated for an ancient channel. Multiplying v by the channel cross-sectional area (A) yields the discharge capacity (Q) of the ancient channel (Knighton, 1984). Each canal trench that was excavated was subjected to hydraulic discharge analyses in order to better understand the amount of land under cultivation and to account for differences in preservation and cultural practices. By reconstructing discharge capacities of ancient canals, we can approach a more complete understanding of the relationship between hydraulics and cultivation.

The estimated bankfull discharge, the maximum carrying discharge of the canal, is then compared to a regression equation created by comparing bankfull discharge to irrigation area for the 77 modern canals in the Moquegua Valley today, including several that were probably originally built by prehistoric farmers. Bankfull discharge represents the maximum discharge capacity of the canal before it begins to overflow its banks. This regression analysis yields a Pearson's R of 0.689, indicating that a majority of the variance in irrigation area is attributable to discharge (Williams 1997). Furthermore, an F-statistic test at the .01 level allows us to easily reject the null hypothesis that no correlation exists between canal discharge and irrigation area. Excavation evidence for bankfull discharge analysis of abandoned canals is coordinated with survey data to determine the reliability of agricultural area interpretations based on this relationship. Canal systems whose bankfull discharges do not correlate with the polygonal areas of land modified by human agriculture are examined in detail. Lack of preservation of agricultural infrastructure or the use of lands for agricultural purposes without investing in substantial infrastructure can lead to underestimation of the irrigated area of a canal system. These "lost parcels" can be accounted for by the bankfull discharge equation. New agricultural polygons representing the lost agrarian parcels are added to the agricultural infrastructure coverage so that a complete, systemic perspective is achieved. The exact locations of the lost parcels could not be modeled for this analysis, and polygons were randomly located within the potentially irrigated area of the canal that was not occupied by other agrarian infrastructure. This method does not significantly affect water distribution models at the regional scale, such as the valley wide analysis this study undertakes. Attempts to determine exact locations for detailed analyses of intra-system hydraulic dynamics by soil chemical analysis and remote sensing methods are underway.

4 Assigning Agricultural Land Use to Time Periods

The most difficult task in reconstructing the evolution of agriculture is dating agricultural systems. One of the problems is that land use in the Moquegua valley is extremely dynamic. Over the centuries, canals are constructed, used, abandoned, revitalized, and reused over and over again. Within a functioning canal system, certain parcels may only be used in the wet season, while over the decades, distal parcels may be abandoned in favor of parcels more proximal to the water source or vice versa. The dating method used in this analysis is the association of principal canal and terraced areas with archaeological sites, accompanied by superposition analysis and dating criteria derived from canal excavations.

I date sites and the use of their associated agricultural fields using radiocarbon dates from published excavation reports, major temporal ceramic styles associated with radiometric dates in other areas, and topological arguments. Superposition of one canal system over another, or of a site over a canal, augments the direct association method of dating by providing information on relative chronology within a small area. Differences in surface patination and preservation are also used to understand the relative chronology within a local area of the drainage. Finally, canal deposits can be used to estimate the period of last use of a canal. The volcanic eruption of Huaynaputina leaves a thick deposit of volcanic ash in those canals abandoned before but close to 1600 A.D. The depth and purity of the ash deposit can be used to provide a relative age of the abandonment of the canal. Those canals associated with late prehispanic sites tend to have large deposits of ash while canals associated with earlier sites (circa 1000 A.D. and before) have little to no ash deposits. This is due to the fact that earlier abandoned canals had an opportunity to fill in with eroded sediment before the ash fall.

The association of agricultural areas with archaeological habitation sites is based on the concept of the site catchement area (Vita-Finzi and Higgs 1970), which argues that the most intensively accessed resources will be located within a certain proximity to the site. The spatial relationship between habitation sites and agricultural fields is especially strong in highland irrigation communities for several reasons. Irrigated agricultural fields require a great deal of monitoring. They need to be protected from animals, theft, or destruction. The feeder canals must be opened and closed during the designated cyclical watering cycles. The terraces and canals must be regularly inspected for damages and repaired if necessary. Most modern farmers live on or adjacent to their agricultural lands, and their community centers are located within the agricultural system or near the principal canal that feeds it as well. Those few landowners who are not resident near their agricultural fields must rely on a neighbor to keep vigilance over their fields. There is also a strong correlation between certain isolated archaeological sites and irrigation systems that indicates that this resident habitation pattern is characteristic of prehispanic societies as well. Thus, one factor in dating a canal system is to look at site distribution patterns for each archaeological time period in the valley.

Defining the actual agricultural area for a specific time period involves further considerations, including the association of architectural features of known age with the canal and irrigation infrastructure. For example, an aqueduct at the Wari site of El Paso channeled water across a divide. This same aqueduct formed part of the wall of a group of rooms at the site, and was covered by a later period plaza, indicating its temporal contemporaneity with the rooms, and its abandonment by the time of the construction of the plaza. At the site of Camata, the terraces associated with an Inca agricultural system (ca. A. D. 1500) cover the principal canal associated with the Late Intermediate (ca. A. D. 1300) village of Cerro Huayco. Thus, local features of architectural compatibility and superposition can refine the functioning of irrigation system components and bring entire canal systems into chronological focus.

The attribute table associated with the agricultural polygons to this point only contains the standard information: an id and an area. The table is updated with a series of temporal boolean fields that define whether or not a plot was under cultivation during a specific time period. The chronological resolution of the current model is on the order of a century, so fields for each century were added to the attribute table and a value of True or False is coded for each agricultural polygon for each century since A. D. 500. While AMS radiocarbon dates on organics within the construction matrices of irrigation works could provide additional chronological information, a substantial number from each irrigation system would need to be analyzed in order to further refine the irrigation chronology. Bulk soil dates from the final silts on one canal had a high error range (a one sigma range of 200 years), and the date was incompatible (too recent) with architectural associations and depositional characteristics. Bulk soil dates do not seem to be an effective means of dating irrigation works in Moquegua. Future refinement of the agrarian chronology may be realized through the application of AMS dates to organic materials in construction contexts, but the present methods do provide an assessment of patterns of agricultural use at a broad regional scale. In discussing changes over a 500 year period, the present dating methods permit us to compare the long term effects of highland water use and drought on down river irrigation systems. The dynamics of water availability are modeled through an additional set of hydraulic analysis methods that are used to reconstruct the availability and distribution of water throughout the time periods represented in this analysis.

5 Hydraulic Analysis Methods

The hydraulic analysis is basically a water cost/availability function and is based on two principal data sources. The first is a record of precipitation over the past 1500 years based on glacial ice cores from the Quelccaya Ice Cap near Cuzco (Thompson, 1992). While year to year fluctuations will change from region to region, long-term general trends are characteristic of Peruvian climate as a whole. It is these decadal trends that will also most profoundly influence major changes in agricultural production strategies (Shimada et al., 1991). The ice cap as a proxy for long term precipitation trends in the South-Central Andes is confirmed by recent lake core evidence from Lake Titicaca (Binford et al., 1997). The securely dated portions of lake level fluctuations from the Titicaca cores over the past 1500 years parallel long term ice accumulation trends. An examination of fifty-year averages of Quelccaya accumulation converges with securely dated lake core records in all cases, while the variation between the two records exists only for the dates for which lake level measurements are less secure (Fig. 6).

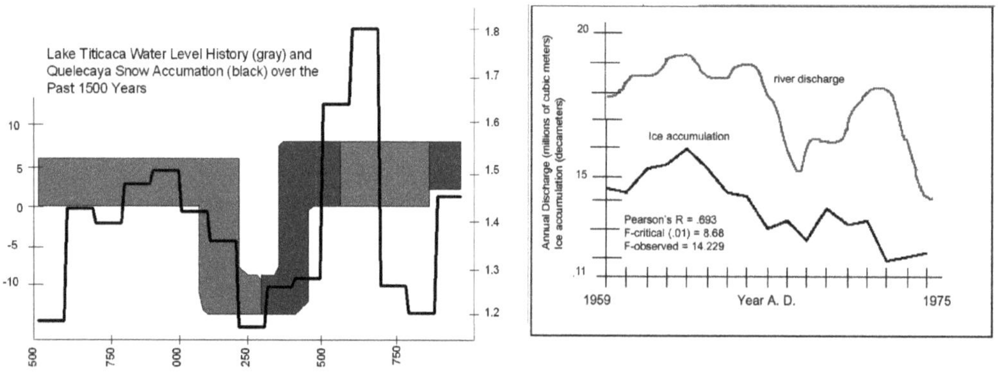

Fig. 6. Correlation of long term paleoclimate records (left) and short term river discharge of the Moquegua Valley (right) with Quelccaya ice accumulation (Thompson 1992). Dark gray shading represents secure interpretation of lake levels, while light gray is representative of less secure interpretations (after Binford et al 1997). Ice data is shown on the left in half century averages for ease of water balance comparisons. River discharge and ice accumulation graphs are depicted using decadal running averages (Williams 1997)

cords as models of past precipitation values (Williams 1997).

The second data source in the water availability model is the geography of precipitation and runoff in the Moquegua drainage. Only the area of the drainage above 3900 meters, referred to as the humid basin, actually contributes water to the river system (Rice, 1989). Therefore, the farther water must travel from the 3900 meter contour, the more water is lost to evaporation and seepage.

Each plot of river irrigated agricultural land in the valley, past and present, can be assigned a water transport cost that accounts for water lost to evaporation and seepage per unit area (Fig. 2). This cost is calculated using the ArcView Network Analyst applied to the hydrology coverage (rivers and canal systems) for Moquegua. The Network Analyst is employed to calculate the best route from the humid basin to each field plot along the natural and human made hydrological system. The terminal locations of the route are specified by choosing the point at which the canal first intersects the field plot and the point at which the river that feeds that canal enters the humid basin. The best route is calculated, and the linear route distance along the hydraulic path is derived. Since water loss is a relatively consistent four percent in unlined canal beds and river channels in highland areas between 1500 and 3000 meters above sea level, the route length is the principal factor in determining the evaporation and seepage cost of bringing water to a field system. A new field named hydraulic cost ratio is added to the attribute table for the field plot and is defined as 0.96 x, where x is the linear distance in kilometers along the best route. Based on discharge measurements and water loss calculations made by the author, an irrigated unit's hydraulic cost ratio can be calculated as HCR = 1 / (.96 D), where .96 represents the amount of water conserved per kilometer traveled and D is the distance in kilometers from the humid basin above 3900 meters via the hydraulic route (Williams 1997).

The hydraulic cost for the plot is defined as the product of the area of the field plot, the hydraulic cost ratio, and a constant that represents the conversion from agricultural area (square meters) to the water volume needed to sustain that area (cubic meters). The constant is derived by solving the equation for the sum of all agricultural plots of the modern time period, which has a known average discharge of the humid basin. Thus, the constant can be calculated in the following equation: Σ (HC) / Σ (HCR x A) = C where HC is the hydraulic cost and its sum is the total modern discharge of the humid basin, HCR is the hydraulic cost ratio for each plot of modern cultivated land, A is the area of each plot, and C is the constant. In this manner, each plot of land within the drainage is assigned a hydraulic cost ratio that represents its relative efficiency in the use of water in comparison to other plots within the system and a hydraulic cost that represents the amount of water from the humid basin needed to travel the specified distance and irrigate the plot with an equivalent amount of water to all plots in the basin. Likewise, the hydraulic cost field is added to the agricultural polygon attribute table. The water requirements of an entire production system at a specific period of time can be calculated by summing the hydraulic cost fields in the attribute table for all plots within the system that are coded as operating in that century.

The analysis assumes that each plot of land requires the same amount of water to be productive, and this is problematic as an exact reconstruction of each parcel's water requirements. Both local filtration rates (which is affected by soil type and human investment in terracing) and crop types will affect the rate of water consumption by a plot of land. Since these factors cannot be controlled for at this point, the analysis assumes that agricultural plots of equal area require similar amounts of water. Also, a large proportion of the water lost to seepage enters the groundwater system and feeds the aquifers and springs of the lower and coastal valley. This does not effect agricultural systems dependent on river fed irrigation, but it is possible that some of that seepage re-enters the surface hydrological system and contributes to water available for river fed irrigation agriculture at lower elevations. Without additional hydrological fieldwork, it is not possible to model seepage contribution to superficial hydrology. Once again, this limits the ability of the model to characterize differences on a highly local scale, but in the regional scale of the basin at large, we can still characterize changes in water availability between broad areas from time A to time B.

In order to compare water requirements with water availability and the social distribution of water, it is necessary to estimate average discharges of the humid basin for decades over the past 1500 years. Modern discharge values for the humid basin are divided by amount of modern Quelccaya ice core accumulations, and this ratio is applied to Quelccaya ice core accumulations of past decades to produce Moquegua humid basin discharge rates for the same time frame (Fig. 7).

Based on the earliest radiocarbon dates from a Tiwanaku context in Moquegua, the first Tiwanaku colonies appear ca. AD 550 at the site of Omo (Goldstein 1989). This colonization represents a major change in political dynamics in the valley. Although these early Tiwanaku settlements seem to be autonomous entities that did not necessarily exert binding political control over the valley inhabitants, their intrusion into the Huaracane production niche represents a usurpation of resources, especially water and irrigable land, that must have impacted the local populations. These settlements, characterized by imported Tiwanaku IV ceramics, intrusive altiplano house forms, and distinct communities organized around public plazas, focused on middle valley floodplain farming (Goldstein 1989).

By AD 650, a new imperial order had joined Tiwanaku in the settlement of Moquegua. The Wari colony on and around Cerro Baúl was established early in the expansion

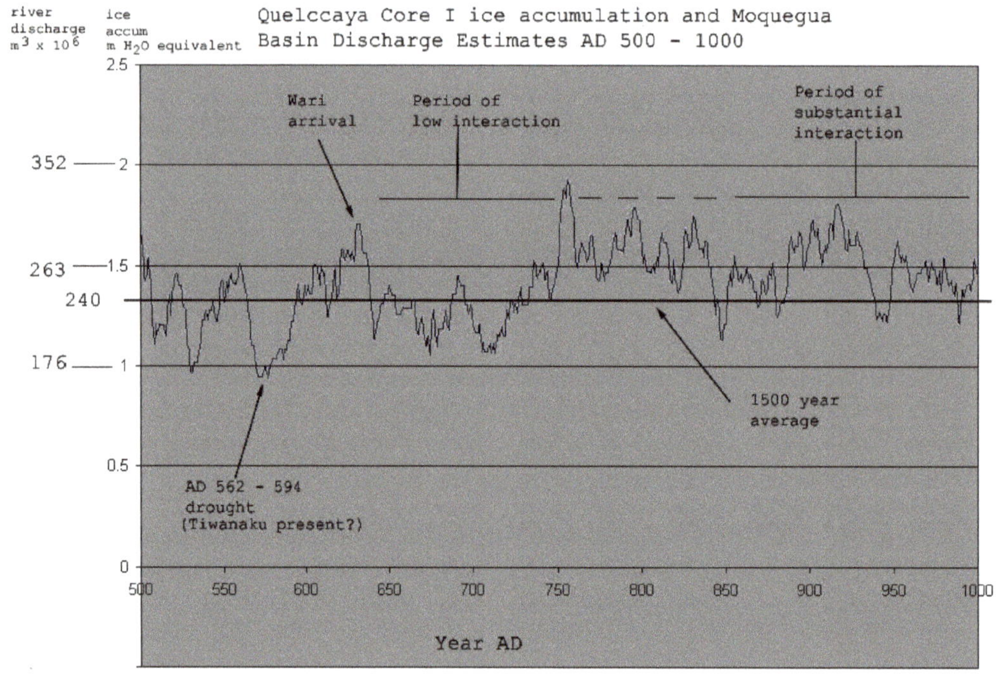

Fig. 7. Estimates of Moqegua River discharge based on Quelccaya ice accumulation data from Thompson (1992). Data is displayed annually for decadal running average of meters of water equivalence. Correlation between annual discharge of the Moquegua Basin is made by assessing discharge measured in recent decades to recent ice accumulation values. Periods of low and increased social interaction are also indicated.

of the first South American empire (Williams 2001). The Wari introduced the new agricultural technology of expansive high sierra terraced agriculture to the upper Moquegua Valley between 2000 and 2500 meters above sea level. This area was only sparsely occupied, if occupied at all, prior to Wari's arrival (Owen 1993), and thus Wari opened a new ecological niche to irrigation agriculture (Fig. 8).

An assessment of water use and water availability using the GIS network analysis model from AD 500 to AD 750 provides a provocative reason for hostilities between the two states early in their expansive periods.

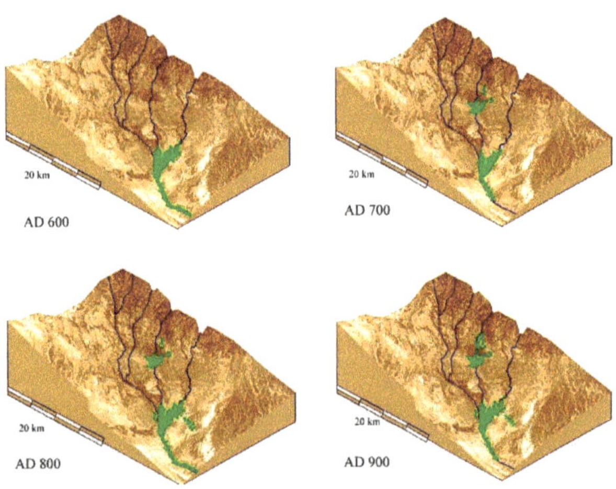

Fig. 8. Landscape models of agrarian land use during the centuries of Middle Horizon occupation in the Moquegua Valley.

During the sixth century AD, water availability is estimated to have averaged 220 million cubic meters annually, with significant deficit fluctuations between AD 525 and 550 and AD 560 to 590 to decadal averages as low as 175 million cubic meters. Huaracane fields could have been irrigated with 185 million cubic meters of annual discharge, and likely would have suffered some hardship during these environmental downturns. With the addition of Tiwanaku agricultural production to the system, the total water needs of the system would likely have grown slightly, to around 200 million cubic meters annually by AD 600.

The Wari expansion around AD 600 occurred after 30 years of 20-30% below normal precipitation (Shimada et al., 1991). High sierra farming radically improved the efficient use of water by decreasing the amount of water lost to evaporation and seepage in transport from the rainfall areas (Moseley 1992). In Moquegua, I estimate that the Wari canals could irrigate 2.5 times the area irrigated by the Tiwanaku lands, due to their closer proximity to the highland rains above 3500 meters above sea level and the use of stone masonry terracing. That is, Wari agriculture was more than twice as efficient in the use of water. However, the addition of a new zone of agricultural production in the upper sierra, despite its efficiency in water use, would still drain resources from the total hydrological system. Wari irrigation needed 25 million cubic meters of annual discharge to support itself. While less than 15 percent of the overall hydrological output, this became increasingly significant in times of drought.

Despite above normal levels of water availability between

AD 600 and 640 on the tails of the 3 decade long drought, the ice core data indicate fluctuating levels of water supply between AD 650 and 750, with pronounced deficits from the long term average between approximately AD 660 to 680 and AD 700 to 720. The water availability model suggests needs outstripped supply during these two intervals and conflict between the two groups over water usage is likely during these decades. In fact, the average annual discharge is still just over 220 million cubic meters, but the new Wari fields upstream of the middle valley reduce the average available discharge of the humid basin to 195 million cubic meters annually. Tiwanaku water needs of 200 million cubic meters annually were not being met.

By AD 750, however, water levels had recovered and were well above average (Figure 7). They would remain so with periods of substantial excess except for brief downturns around AD 850 and AD 950 until after AD 1000. Between AD 750 and 1000, average discharge is almost 270 million cubic meters annually, and decadal averages exceed 240 million cubic meters annually 9 out of 10 decades. Thus, after AD 750 the constraints on middle valley agriculture imposed by upland irrigation and below average rainfall are withdrawn. The result is an environment of decreasing potential conflict and détente between the superpowers.

Around AD 850, the Tiwanaku colony witnessed a political reorganization that resulted in the establishment of a demographic center at the site of Chen Chen and the construction of a Tiwanaku style temple at the site of Omo. This temple reflected the three-tiered platform of altiplano temple centers, and is the only such complex constructed outside the Titicaca Basin (Goldstein 1993). This phase of settlement also reflects the development of a three tier site size hierarchy in the Moquegua Valley, which mirrors the new political organization of provincial administration of the middle valley. Chen Chen, with thousands of burials, is a vast necropolis as well. Located at the neck of the valley where the river narrows significantly, the Pampa Chen Chen and its canal intake represent an important geographical nexus for the control of water flowing to the middle valley. Chen Chen, with thousands of burials, is a vast necropolis as well. Located at the neck of the valley where the river narrows significantly, the Pampa Chen Chen and its canal intake represent an important geographical nexus for the control of water flowing to the middle valley. The Wari colony in the high sierra continued to thrive, however, and may have even experienced agricultural expansion.

Around AD 900, the changes that were taking place in the Tiwanaku colony were also complemented in the Wari colony with a massive programs of urban renewal and institutional restructuring at the Wari provincial capital of Cerro Baúl (Williams 2001). This period of time also witnesses the intrusion of Tiwanaku settlers into settlements around the Wari administrative center and intermingled with Wari domestic settlements (Williams 2002). Furthermore, three rustic tripartite structures of public architecture, one of which Owen (1999) argues to reflect temple architecture in the Tiwanaku style, are established on the slopes of Cerro Baúl. These changes in the settlement data reflect the substantially increased interaction between the two groups. While they strictly maintain their separate identities in terms of material culture, their interaction in lithic exchange (see Burger et al 2001), in the sharing of water resources (Williams 2002), and in maintaining settlements necessitating confrontation with the other population in the course of day to day activities reflects a substantial change in the pattern of daily life for the two groups. This interaction is likely witnessed late in the Middle Horizon for several reasons, one of the most important of which is the reduction in potential hostilities afforded by the increased availability of water resources that are no longer stressed by agricultural expansion. This interaction, in fact, may have paved the way for the development and further expansion of the agricultural system of the upper sierra by local Tiwanaku peoples called Tumilaca in the tenth century AD. By AD 1050, both colonies had collapsed, perhaps survived slightly by the Tumilaca tradition. Collapse may also be assessed from the water availability model, which sees decadal rainfall levels drop to below 230 million cubic meters annually in the eleventh century AD for the first time in 3 centuries. Ironically, the interaction and expansion of upland agricultural fields may have been instrumental in lowland agricultural collapse that fed back into the social system and resulted in colonial abandonment on both sides (Williams 2002).

References

BINFORD, M.W., A.L. KOLATA, M. BRENNER, J. JANUSEK, M.T. SEDDON, M. B. ABBOTT, AND J. CURTIS, 1997 *Climate variation and the rise and fall of an Andean civilization.* Quaternary Research 47:235-248.

BURGER RICHARD L., KAREN L. MOHR CHAVEZ AND SERGIO J. CHAVEZ, 2001 *Through the glass darkly: Prehispanic obsidian procurement and exchange in southern Peru and northern Bolivia.* Journal of World Prehistory 14(3).

GOLDSTEIN, P., 1989, *Omo, a Tiwanaku Provincial Center in Moquegua,* Peru. Ph.D. dissertation, University of Chicago.

GOLDSTEIN, P., 1993 *Tiwanaku Temples and State Expansion: A Tiwanaku Sunken-Court Temple in Moquegua*, Peru. Latin American Antiquity 4(1): 22-47.

GOLDSTEIN, P., 2000 *Exotic Goods and Everyday Chiefs: Long-Distance Exchange and Indigenous Sociopolitical Development in the South Central Andes.* Latin American Antiquity. 11(4) : 335-61.

KNIGHTON, D., 1984, *Fluvial Forms and Processes.* London: Edward Arnold.

KOLATA, A., 1994, The *Tiwanaku: Portrait of an Andean Civilization.* Cambridge: Blackwell.

MOQUEGUA MINISTRY OF AGRICULTURE IX AGRARIAN REGION, 1983. *Diagnostico del Distrito de Riego Moquegua y Sector de Riego Carumas*. Lima: Ministerio de Agricultura, Convenio Direccion General de Aguas Suelo e Irrigaciones, Programa Sectorial Agropecuario.

MOSELEY, M., 1992, *The Incas and Their Ancestors*. Thames and Hudson, New York.

MOSELEY, M., R. FELDMAN, P. GOLDSTEIN, AND L. WATANABE, 1991- *Colonies and Conquest: Tiwanaku and Wari in Moquegua. In Wari Administrative Structure: Prehistoric and Monumental Architecture and State Government.* W. Isbell & G. McEwan, eds. Washington, D. C.:Dumbarton Oaks.

NASH, D., 1996 *Cerro Petroglifo: Settlement Pattern and Social Organization of a Residential Wari Community*. M. A. thesis, University of Florida.

ORTLOFF, C. & A. KOLATA, 1993 *Climate and Collapse: Agro-ecological Perspectives on the Drought of the Tiwanaku State*. Journal of Archaeological Science 20:195-221.

OWEN, B., 1993 *A Model of Multiethnicity: State Collapse, Competition, and Social Complexity from Tiwanaku to Chiribaya in the Osmore Valley*, Peru. Ph.D. dissertation, UCLA.

Inventario Arqueológico del Drenaje Superior del Río Osmore: *Informe del Campo e Informe Final.* Informe entregado al INC-Lima y INC-Moquegua.

1997 *Informe de Excavaciones en los Sectores Mortuorios de Chen Chen, Parte del Proyecto Rescate de Chen Chen,* Temporada de 1995. Informe entregado al INC-Lima y INC-Moquegua.

PROYECTO *Vecinos de Cerro Baúl*, 1997, Informe del Campo e Informe Final. Informe entregado al INC-Lima y INC-Moquegua.

RICE, DON , 1989 *Osmore Drainage, Peru: The Ecological Setting. in Ecology, Settlement, and History in the Osmore Drainage*. D. Rice, C. Stanish, and P. R. Scarr, Eds. Oxford: BAR International Series 545.

SHIMADA, I., C. SCHAAF, L. THOMPSON, AND E. MOSLEY-THOMPSON, 1991 - *Cultural Impacts of Severe Droughts in the Prehistoric Andes: Application of a 1,500-year Ice Core Precipitation Record.* World Archaeology 22(3):247-271.

STANISH, C., 1985, *Post-Tiwanaku Regional Economies in the Otora Valley, Southern Peru*. Ph.D. dissertation, University of Chicago.

THOMPSON, L., 1992. *Quelccaya Ice Core Database.* IGBP PAGES/World Data Center-A for

Paleoclimatology *Data Contribution Series* # 92-008. NOAA/NGDC Paleoclimatology Program, Boulder CO, USA.

VITA-FINZI, C., AND E.S. HIGGS, 1970, *Prehistoric Economy in the Mount Carmel Area of Palestine: Site Catchment Analysis.* In Proceedings of the Prehistoric Society, edited by J. Coles. Vol. XXXVI, December 1970:1-37

WILLIAMS, P. R., 1997, *The Role of Disaster in the Evolution of Agriculture and the Development of Social Complexity in the South Central Andean Sierra.* Ph. D. Dissertation, University of Florida.

WILLIAMS, P. R., 2001, *Cerro Baúl: A Wari Administrative Center on the Tiwanaku Frontier.* Latin American Antiquity 12(1):67-83.

WILLIAMS, P. R., 2002, *A Re-examination of Disaster-Induced Collpase* in the Case of the Andean Highland States: Wari and Tiwanaku. World Archaeology 33(3):361-374

WILLIAMS, P. R., D. J. NASH, AND M. MOSELEY, 2000 *Empires of the Andes: A Majestic Frontier Outpost Chose Cooperation Over War.* Discovering Archaeology 8: 68-73.

The Archaeologist, the Neural Network, and the Random Pattern: Problems in Spatial and Cultural Cognition of Landscapes

Ezra Zubrow

National Center for Geographic Information
Anthropology, University at Buffalo

The stones themselves are thick with history, and those cats that dash through the alleyways must surely be the ghosts of the famous dead in feline disguise...Erica Jong [Jong, 23 Mar 86 #1]

1 Introduction

Archaeologists want to study landscapes. In landscape nature and society as well as time and space are organized perfectly. Archaeologists find contemporary landscapes fascinating for they fervently hope they will provide insights into the prehistoric environment and into the processes by which the ancient surroundings change.

When analyzing landscapes there are numerous crosscutting issues. Initially, I wish to focus on two – legibility and modification (Figure 1). Some contemporary landscapes are more relevant than others to their ancestral predecessors. Why is this case? Some landscapes are easier to understand or "read" than others. Some landscapes maintain more of the prehistoric features, processes, and meaning than others.

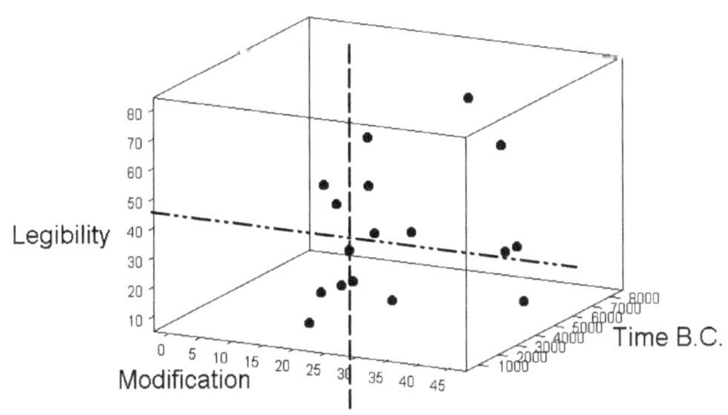

Figure 1. Classifying prehistoric landscapes by legibility and modification. (Dots are prehistoric landscapes).

2. Thick and thin landscapes

Archaeologists need to be "able to read" the landscape. Similar to children archaeologists prefer to read easily legible landscapes that have large print and easy to read fonts. In other words, their goal is to find landscapes with high legibility and low modification. In other words, the upper left hand quadrant of figure 1 as indicated by the black box is preferable. By legibility, I am referring to Scott's (Scott 1998)concept. To be able to read the landscape is to be able to abstract the complexity of the landscape into those variables or quantities that are needed. For archaeologists, these variables frequently are concerned with "ancient features" and "social memory". The landscapes that possess this legibility are frequently called "thin." They can be read and used by all and are as Lillesand claims the reflection of a consistent vision. (Lillesand and Kiefer 2000)

When archaeologists have "thin landscapes" they are able to argue for centralization and the uniformity of the spatial data. They are asserting the constancy of definitions, representations, and formats that makes the landscape legible to many archaeologists, anthropologists and even diverse societies. The latter might even include ethnographic and even prehistoric societies who lived in the area. This uniformity of thin maps resonates with the often-lauded appeal of archaeological survey. Archaeological survey provides a consistent vision. It helps us see large swaths of the land virtually simultaneously.[1] It augments our vision[2] and facilitates our ability to discern spatial patterns over large regions. This legibility establishes its authority through representation. All landscapes establish their own authority for the landscape is self-evident and was self-evident. Classifications of present or even prehistoric landscapes seem to speak to us directly without the aid of mediating text. There is a "naturalness" of the landscape. It is seemingly translated directly from nature. Even when one knows that people have modified it substantially, there is a tendency to suggest that there is no other author than nature itself. This "naturalizing" of the landscape further substantiates the landscape's authority as a comprehensive depiction of reality. This reality is

[1] It assumes the perspective from above—almost as though it is from space. Though this is definitely only one of a variety of potentially different perspectives, the idea of an unvarying vision from above, independent of place on the earth or specialized context is reminiscent of critiques of the "perspectiveless gaze", the "god-trick." Similar to aerial photographs and images from space, it has inherent and unquestioned authority since it is similar to what "god does".

[2] They augment our vision for they allow us to examine what is actually thought to be below the surface and what existed in the dim mists of time past.

uniform, "thin," abstracted, and highly legible through the computer aided environment. .

Conversely, landscapes that are "thick" have multiple layers of meaning that can only be deciphered by small groups of "contextual zed players." These are indicated in figure 1 in the lower right quadrant and are marked by the open circle. Based upon the work of anthropologist Clifford Geertz's The Interpretation of Cultures (Geertz 1973) the focus is to capture the symbolic meaning of culturally charged phenomena. From this perspective when an individual acts within the landscape they are "...suspended in webs of (cultural) significance (they have) spun..." (p. 5). Even the most basic undertaking can denote different things depending on the cultural codes at work. A marked building in a contextualized landscape may be simultaneously a landmark for navigation, a shelter from the elements, the residence of a "god", the family homestead, the location for a mythological event, and a business and administrative center to name only a few functions and significata. The implications of Geertz's argument for understanding the landscape is most basically that one should never be satisfied with a "thin description" of a landscape that tells little more than that there are mountains, streams, fields and villages. .

It is essential to know many specifics before one can decide what specific features of the landscape means. Were the actions that took place there generic or did they focus on a particular issue? Were they of recent origin or long ago? Do they last generation after generation or are they immediate and only interim? Answers to questions such as these can make a world of difference. One needs to remember that individual members of the same society can imbue identical acts with somewhat different meanings, while observers can attribute to them ones that are completely different. These variations may make all the difference in their future relationship (or lack thereof).

3. Cognition, Environment, Genetics, and Learning:

Underlying some views of cognitive archaeology is the assumption that cognition is embodied. It arises from bodily interactions with the world and that it is continuously meshed with them (Lakoff 1987) (Johnson, 1987) (Varela, et al. 1991) et al., 1991). It is not "I think therefore I am" rather the causal arrow is reversed- "I am therefore I think". For proponents of this view, thinking emerges from environmental experiences, from sensory-motor coordinated interactions into the landscape, and from exploration of the surroundings. Human infants, for instance, exhibit a wide range of exploration strategies: from mouthing, through touching to listening (Kellman and Arterberry 1998). Concept formation, the prerequisite for high level cognition, is the result of infants' actions to generate stable sensory patterns in different sensory channels that may be exploited to form cross-modal associations.(Thelen and Smith 1994)

The twentieth century may be characterized as the century that accepted the view that human cognitive content was derived from individual experience with the social and physical world. It was believed that the learning apparatus imparted no content of its own to the outcome. Namely, human nature was a "tabula rasa" with a capacity to learn culture. Recently, however, a series of findings from archaeology, artificial intelligence, cognitive science, evolutionary biology, linguistics, neuroscience, physical anthropology, and psychology have emerged that confront this position.

This new view suggests that cognition results from general learning systems as well as functionally specialized computational equipment analogous to expert systems in artificial intelligence. One may think of these learning systems as analogous to factory "preinstalled" programs. From them may come the building blocks of ideas, cultural products and institutions. For example, all normally developing humans from every culture appear to have what has been called an intuitive physics or object mechanics. They date at least as far back as hominids. We know that early hominids at Olduvai and other prehistoric African sites from millions of years ago knew how to recognize and obtain flint nodules from riverbeds; to have the spatial knowledge and dexterity to strike them with other stones in order to make pre-conceived tools. They also had the knowledge and ability to use them for predetermined functions in new and different places in the landscape. Their landscapes must have been divided by a variety of recognized functions (resource location areas –flint deposits from which they obtained flint; construction and modification sites –where tool construction took place, and tool use areas where tools were used for scavenging, hunting and butchering).

Some of the clearest traces of the impact of evolutionary processes on the mind are apparent not in encultured adult behaviors but in the capacities of the infant mind. (Cross, et al. 2000) There is evidence (Keil 1994) (Spelke 1999) that very young infants develop certain abilities too rapidly to be explained as the outcome of learning processes that involve interaction with the environment. "Very young infants quickly cue in to behaviour that can be interpreted as "goal-directed". This suggests that they are primed to learn a sort of intuitive biology, physics, linguistics and psychology. (Cross, et al. 2000) In fact, almost all human infants acquire language quickly and expertly though they require continuous linguistic interaction with other humans in order for their language capacity to be fully expressed (Pinker 1994)

Evolution and culture are complementary. Evolution can be thought of as acting on the mind in terms of shaping predispositions. There are clear adaptive advantages for children to be primed to deal with certain types of information rapidly and expertly without being taught to do so. Culture, in the form of human interactions that are shaped by common ways of understanding, particularizes the developmental trajectory of those predispositions through implicit and explicit education. [3]. If evolution has shaped the human mind, it has probably selected at the level of predispositions, and culture can be thought of as particularizing – shaping into specific and distinct forms - the expression of those predispositions.

How does culture particularize? There is considerable literature. It ranges from the Sapir –Whorf hypothesis through Frake's and Berlin's ethnoscience. Even our own work on this subject has been somewhat contradictory in terms of the implications. For example the Hunt-Zubrow study of shape showed that cultural concepts of even such basic geometric notions as round/ not-round vary by culture and temporal period. Thus, even describing the shape of the environment may be problematic (Zubrow, 1986) On the other hand, the Zubrow/ Daly study of hunting/ gathering/ fishing representations of space showed that more than 70 plus hunting gathering and fishing societies as well as many prehistoric societies make representations of space and that there is some degree of semiotic consistency using points, lines, and polygons which crosses both cultural and temporal boundaries. Thus, there may be consistent use of "squiggly line" for a river and either a circle or a rectangle for a habitation etc. thousands of miles and thousands of years apart (Zubrow and Daly 1998).

4. Conceiving and Learning the Landscape

The goal of this paper is to explore how the landscape is conceived and learned. As discussed above there are thick and thin conceptions of the landscape as well as evolutionary predispositions and systems of learning. Presently, there is a considerable literature on understanding the landscape both from a "western perspective" and from an "ethnographic perspective". Archaeologists have been trying to determine what the prehistoric landscape would look like and what it meant to prehistoric cultures. Methodologically, this paper tries to use a testable processual design more than an interpretationist "post processual design" while trying to be cognizant of the post-processual issues.

The research design consists of two parts. Part one is trying to extract the "Western" from the western analysts consideration of landscapes. (See figure 2). Part two is an ethnographic experiment about the organization of landscapes.

There are a variety of methods to "emaciate" a thick cultural description of a landscape. Rather than use ethno science or any of the more "interpretative" post processual techniques this paper will focus on the more "cognitive" archaeological methods that are more processual and derived from artificial intelligence. In particular, the focus will be to use neural networks. There are numerous reasons to use neural networks for this task. First, they have a remarkable ability to derive meaning from complicated or imprecise data. Second, they may be used to extract patterns that are too complex to be noticed by either humans or other computer techniques. Third, they can be "trained" either by people or by themselves and once trained they become "expert" in the categories of information needed to analyze. Fourth, they are capable of adaptive learning. Fifth, they are self-organizing and can improve their organization. Sixth, they are capable of real time operation. Seventh and finally they very fault tolerant because of the possibility of redundant information coding.
In this paper a "neural network" is an "artificial neural network" (ANN). Biological neural networks are much more complicated in their elementary structures than the mathematical models we use for ANNs. A neural network of many very simple processors ("units"), each is having a (small amount of) local memory. The units are connected by unidirectional communication channels ("connections"), which carry numeric (as opposed to symbolic) data. The units operate only on their local data and on the inputs they receive via the connections.

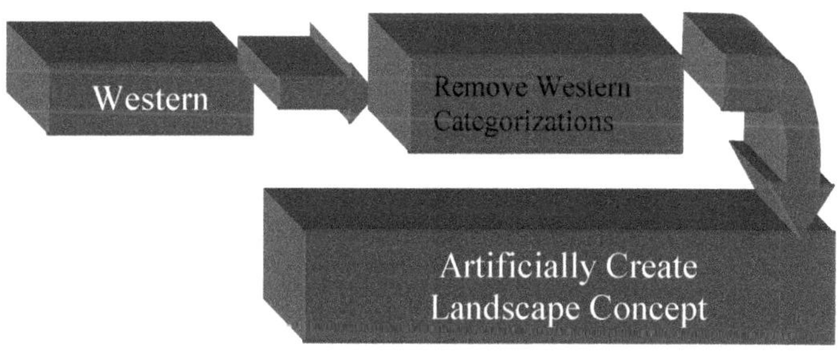

Figure 2. Research Design Part 1

Unlike other mathematical and statistical techniques neural networks have a motivation designed in them. Most neural networks have some sort of "training" rule whereby the weights of connections are adjusted on the basis of presented patterns. In other words, neural networks "learn" from examples, just like children learn to recognize dogs from examples of dogs, and exhibit some structural capability for generalization. (Prechelt 1995)

One of the characteristics that is particularly relevant because of the ongoing work that Zubrow, Branting, Tilson and Miller are doing at the Center for Computational Research is examining the great potential neural networks have for parallelism since the computations for each component is independent of each other.

[3] For example, in the case of language, human interactions lead to the acquisition of mature competence in a particular language with a specific syntactic structure, lexicon, etc.

Name and type of Neural Network	Function	Brief Description
ADALINE Adaline Network	Pattern Recognition	This is essentially a single-layer backpropagation network. It is trained on a pattern recognition task, where the aim is to classify a bitmap representation into the corresponding classes. Due to the limited capabilities it only recognizes the exact training patterns.
BPN Back Propagation Machine	Time Series Forecasting	This program implements the now classic multi-layer back propagation network with bias terms and momentum. It is used to detect structures in a time-series, which is presented to the network using a simple tapped delay-line memory. To avoid over fitting, termination is achieved by the stopped training method
HOPFIELD Hopfield Model	Autoassociative recall	The Hopfield model is used as an autoassociative memory to store and recall a set of bitmap images stored in a corresponding weight matrix. Thereafter, starting from an arbitrary configuration, the memory will settle on the stored image nearest to the starting configuration in terms of Hamming distance. This gives an incomplete or corrupted version of a stored image.
BAM Bidirectional Associative Memory	Heteroassociative recall	The bidirectional associative memory can be viewed as a generalization of the Hopfield model, to allow for a heteroassociative memory to be implemented.. The memory even shows a limited degree of fault-tolerance in case of corrupted input patterns.
BOLTZMAN Boltzmann Machine	Optimization	The Boltzmann machine is a stochastic version of the Hopfield model, whose network dynamics incorporate a random component. Starting with a high value and gradually decreasing, it allows the network to reach equilibrium at any step; chances are good, that the network will settle a global minimum. The weight matrix is chosen such that the global minimum of the energy function corresponds to a solution of a particular instance of the traveling salesman problem.
CPN Counter propagation Network	Landscapes-Vision	The counter propagation network is a competitive network, designed to function as a self-programming lookup table with the additional ability to interpolate between entries.
SOM Self-Organizing Map	Control	The self-organizing map is a competitive network with the ability to form topology-preserving mappings between its input and output spaces. The task of the network is to establish a mapping between the state variables and forces. This is done using a reinforcement learning approach: For any given state the network tries a slight variation of the mapped force. If the new force results in better control, the map is modified, using the current state variables and the new force as a training vector.
ART1 Adaptive Resonance Theory	Brain Modeling	This program is mainly a demonstration of the basic features of the adaptive resonance theory network, namely the ability to plastically adapt when presented with new input patterns while remaining stable at previously seen input patterns.

Furthermore, there are a wide range of neural network architectures. A few of them are listed below in table 1. (Kutza 2001)[4].

Results Research Design 1:

The process has been to create a set of Western landscapes and then use neural networks to classify the features and draw out the relationships as independently as possible of the analyst (in other words the archaeologist). In each of the following diagrams, one sees an increasingly complicated landscape represented figuratively on the left and the neural network results on the right. Figure 3 is an undifferentiated landscape. One might consider a "lake plain" or a "steppe" as the real life analogues. The single feature is located in the neural network space. One should note that the feature is located very close to the northwestern vertex of fourth grid of the northern most line of squares.[5] Generally, the number of objects is indicated by the distance the feature or "term" is from the plane. The relationship of objects is indicated by the distance the object is from other "features", "terms." or "objects" on the planar space. This convention will be common to all the following pictures.

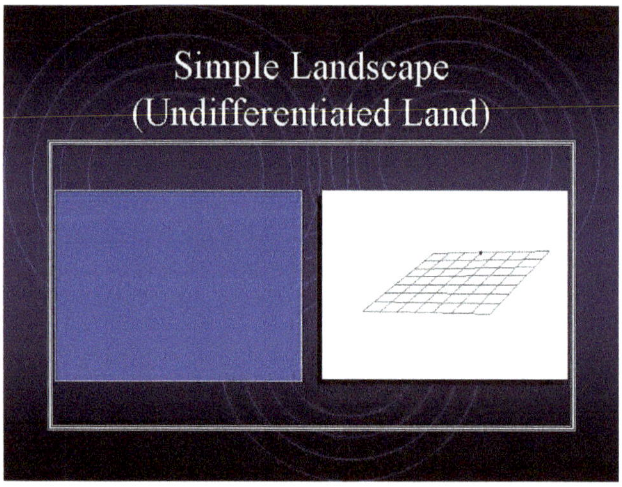

Figure 3. A simple undifferentiated landscape and its neural network analysis.

Into this environment, one inserts a new feature a road. Figure 4 shows the road and the undifferentiated landscape. In this landscape the road is central and divides the environment into two parts. The neural network shows the simple undifferentiated landscape now differentiated by the road and places the two categories relatively far apart. Note that the undifferentiated landscape category has moved from the northwestern vertex of the fourth grid of the northernmost line of grids to the northwestern vertex of the sixth grid of the northernmost line of grids. The road is located on the northeastern vertex of the eighth grid of the second line of grids. This is quite revealing. One should note that they are not maximally differentiated. If they were, the location of the two features would be at extreme ends of

[4] This table is based upon the work of Karsten Kutza of Chausseestraße 1 D-14109 Berlin Germany and is broadly taken, paraphrased, and quoted from his work.

[5] These geographic labels "north, northeastern etc." in the neural network space does not have any geographic meanings. They only are being used to help the reader find the location of the feature being discussed.

the diagonal across the neural network space. The reasons that they are not at the opposite ends are a bit complicated. However, the easy non-mathematical interpretation would be that the road and the undifferentiated landscape are not the "opposites" of each other. The opposite of "undifferentiated" would be "differentiated" and the "opposite' of road would be "non road". Furthermore, they are spatially not unrelated. One bisects the other. This is pointed ou t here only because the "naïve" archaeologist (naïve in the sense of categorization) might initially divide the landscape into two parts and suggest they are complete complements of each other.

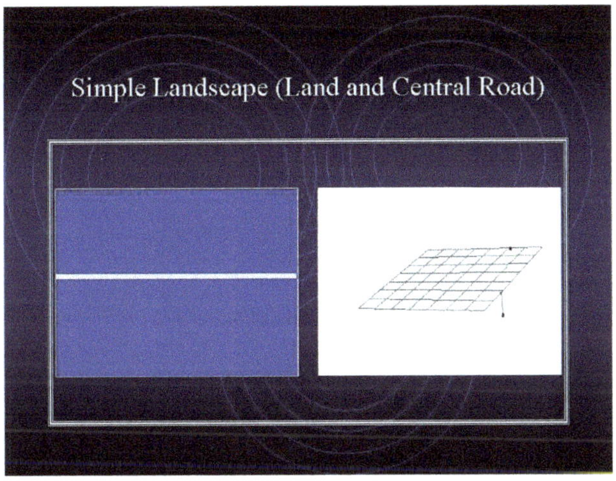

Figure 4 . Simple undifferentiated land and with east west road on third line and its neural network analysis

What happens if the location of the road in the landscape changes? Figure 5 shows a road on the northern periphery of the landscape rather than a single road in the center of the environment. It now is on the northern periphery. Although this is a single case, it is meant to be representative of a wide variety of cases where the roads and paths are not bisecting the landscape equally. This is, of course, the more frequent situation in prehistoric as well as modern landscapes. There is one change that is noteworthy as a result of the relocation of the road. The degree of separation increases as indicated by the change in the spatial location of the road relative to the undifferentiated landscape. The undifferentiated landscape has moved back toward its original position (the northwestern vertex of the fourth grid) prior to there being a road while the road feature had moved forward toward the front of the neural space. This increased distance indicates a greater degree of contrast. The more general point is that location matters. The same feature located differently will increase or decrease the amount of perceived difference.

The next step in making the landscape more complex is to add more roads. Figure 6 shows two roads and the accompanying neural network analysis. Adding two roads rather than one makes a major and a subtle difference. The major shift is the shift of the undifferentiated landscape feature and the road feature towards where they were with a single central road (Figure 4). The landscape feature has moved back toward the northwestern vertex of the sixth grid of the northernmost line and the road feature has moved back to the northeastern vertex of the second line of grids. The subtle shift is that of the road feature moving slightly south in the neural network space. The amount of relationship and contrast for two roads and a simple undifferentiated landscape is about the same as a single central road. There is slightly more contrast for two roads than for one. But both "one central road" and "two roads" are less contrastive than a single road on the periphery.

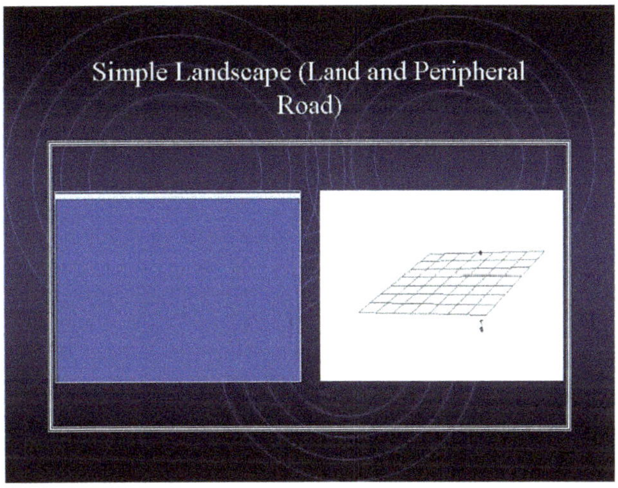

Figure 5. A simple undifferentiated landscape with peripheral road and its neural network analysis

Once again one may ask does the placement of the roads makes a difference. As one would expect it does. Figure 7 shows the same undifferentiated landscape with two roads crossing which allows comparison between two "parallel" and two "crossing roads. The undifferentiated landscape feature is now located in the northwestern vertex of the fifth grid of the northernmost grids. It is lies between all the previous figures. The road feature is now in the northeastern vertex of the eighth grid of the southernmost line. Crossing the roads adds contrast.

The next stage in the research is to make the landscape more complex. In these figures, figures 8-11, the left hand image is a simplified caricature of how the landscape actually appears. Thus, there are actually far more houses along the single road in figure 8. When one analyzes the neural network space for the three categories of land, house and road a new pattern emerges.

The three categories are in contrastive triangle with each other that does not have equal distance between the categories. The neural network diagram is a bit difficult to read because road and land are above the neural network space and the houses are below. There also has been a shift in the positions. The road feature is located in the third grid of the next to northern most line of squares. The house feature is located very close to the road (as one would expect) in the fourth grid of the next line south. Finally, the land feature has moved a long distance to the seventh grid of the same line as the house. Thus, the contrastive elements are the houses and the

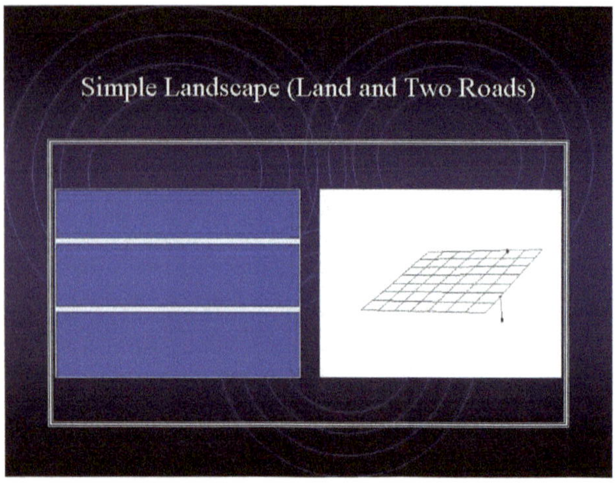

Figure 6. A simple undifferentiated landscape with two roads and its neural network analysis.

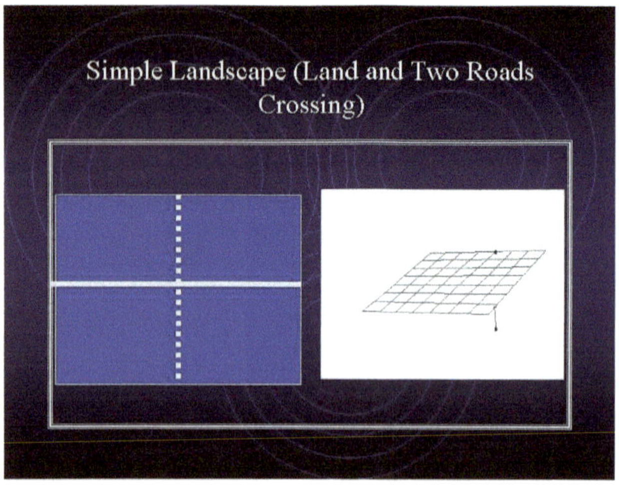

Figure 7. A simple undifferentiated landscape with two crossing roads and its neural network analysis.

roads are relatively close, the houses and the land are closer than the road to the land. This would make some cognitive sense.

Figure 9 presents a complicated landscape. It is complicating the landscape by several orders of magnitude. If the simplified environment had less than 10 features numerically, the moderate environment less than 100 features numerically, the complicated environment is five orders of magnitude greater than simplified environment but less than a million features. Although significantly more sophisticated, it is still simplified when compared to real landscapes. There are rivers, streams, lakes, pastures, houses, roads, mountains, woods, hills, paths, and a few other features repeated thousands of times to make up the landscape. There are as one might expect closely related features-for example river, stream and lake are clustered very closely in the neural space and field and pasture are moderately close. However, houses and roads or houses and paths or paths and roads are unexpectedly not particularly close.

Additional analyses were run showing the importance of particular routings through the landscape and having the neural network use non-western ethnographic categories as its beginning classification scheme. Both made significant differences in the neural spaces.

What is clear from the above is that neural networks provide an independent means to analyze and classify the landscape. It will provide a mechanism to help "thin" the "thick landscape".

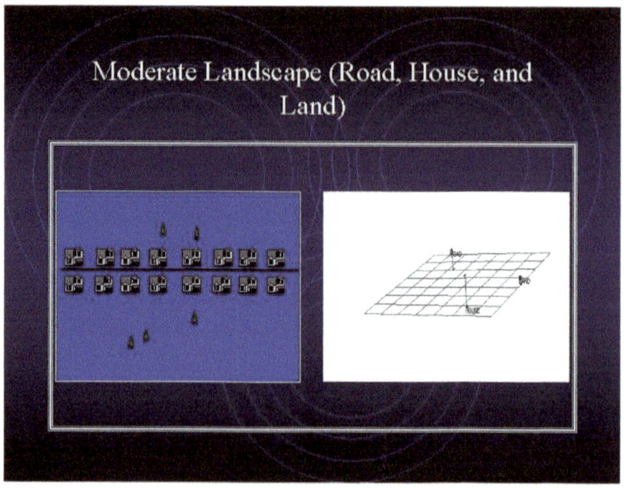

Figure 8. A moderate landscape with land , a road, and numerous houses and its neural network analysis.

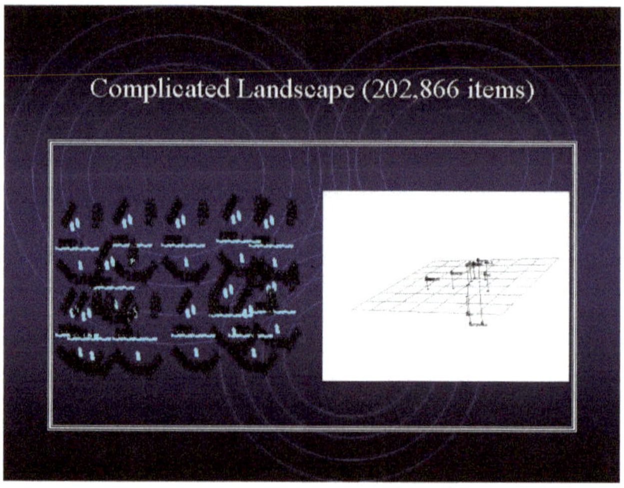

Figure 9. A complicated landscape and its neural network analysis.

5. Results Research Design 2:

The second part of the research on landscapes consisted of a small ethnographic experiment. The author requested twenty archaeologists to analyze and reconstruct a prehistoric landscape and its embedded settlement pattern from a set of provided data. They were encouraged to use any technique that they wished. However, GIS and geostatistics were specifically mentioned as potential methods.

The landscape was simpler than the complicated landscapes but more complicated than moderate landscapes discussed above. There were five types of landscape features with one thousand instances that were laid down upon the space. There were eight types of site features (each type a different type of site) with two hundred instances being laid down across the space.

In the first trial, eighteen archaeologists returned the experiment. Two did not do it. Four more were sent out with exactly the same instructions. The first two that were returned were included to make up the twenty reconstructed landscapes and settlement patterns.

No two archaeologists had the same landscape reconstruction, nor the same settlement pattern. This was both surprising and not surprising. All the data was created using a double random algorithm. What was not surprising, therefore, that they reconstructed different landscapes and different settlement patterns. What was surprising was that every archaeologist did construct a landscape and a settlement pattern. It seems to be clear that there is a deep seated drive in archaeologists to create order. One might argue that it is what archaeologists do best. They are trained to create order and patterns out of fragmentary and incompletely preserved data. They are masters of seeing patterns and making discoveries where others "ignore the clues". Clearly, it is possible to artificially construct a landscape and a settlement pattern without cultural content –or at least too much cultural content. Perhaps, it is a characteristic of Western culture. One will always find a pattern whether it is there or not.

Conclusions

The archaeologist is tied to the landscape both present and past. Some landscapes are more legible than others and some have been more modified than others. Ideally, archaeologists want legible non-modified landscapes. There is a problem for many landscapes are "thick" having multiple layers of meaning that can only be deciphered by small groups of "contextualized players." Archaeologists need techniques that will "decontextualize", make them "thin" and thus make legible past landscapes so that they can be understood in the present.

This paper has shown that neural networks provide a mechanism for "thinning" the landscape. It operates in the manner of a non-culturally specific human mind. Neural networks derive meaning from imprecise data., extract hidden complicated patterns, train themselves to become expert, are capable of adaptive learning and self-organization, and are fault tolerant. Secondarily, it has been shown that archaeologists are capable of reconstructing landscapes and settlement patterns from random data. There is something about an archaeologist that needs patterning. Neural networks may help them resist the temptation of "over ordering".

The understanding of landscapes by archaeologists is a process that has just begun. Whether they are thick or thin and whether they are analyzed using remote sensing, physical reconstruction, neural networks or virtual reconstruction, ultimately, it is the same process of understanding their diversity and structure. . Unfortunately, it is not easy and there is a significant problem. The serious archaeologist must be sufficiently reflexive as to determine what is theory driven, what is methodology driven, and what is data driven. Perhaps, one needs a reality "visualizer". When one examines a reconstruction of the landscape, the settlement pattern or the site, it would show one how much and what parts of the landscape is derived from the theory, and what parts are derived from the methodology, and what parts reflect the actual data. However, this is a discussion for another paper.

There are not only "thick and thin" landscapes, there are "thick and thin" archaeologists. In either case one must reconstruct the domain of the ancient times including the landscape. If we fail, the past ....

...the best I've known,
Stays here, and changes, breaks, grows old, is blown
About the winds of the world, and fades from brains
Of living men, and dies.
Nothing remains." ...Rupert Brooke (1887-1915)
(Brooke 1915)

References

BROOKE, R., 1915 - 1914 and other poems. Sidgwick and Jackson, London.

CROSS, I., F. COWAN AND E. B. W. ZUBROW, 2000 *Music and Prehistoric Behavior: A Preliminary Analysis.* Paper presented at the Origins of Music - Musical Intersections, Toronto.

GEERTZ, C., 1973. *The interpretation of cultures*; selected essays. Basic Books, New York.

JOHNSON, M., 1987. *The body in the mind : the bodily basis of meaning, imagination, and reason.* University of Chicago Press, Chicago.

KEIL, F. C., 1994. *The birth and nurturance of concepts by domains: the origins of concepts of living things. I. In Mapping the mind: domain specificity in cognition and culture*, edited by L. A. Hirsh and S. A. Gelman, pp. 234-252 C.U.P, Cambridge.

KELLMAN, P. J. AND M. E. ARTERBERRY, 1998. *The Cradle of Knowledge.* MIT Press. A Bradford Book., Cambridge, MA.

KUTZA, K., 2001 *Neural Networks at your Fingertips* in http://www.geocities.com/CapeCanaveral/1624/

LAKOFF, G., 1987. *Women, fire, and dangerous things: what categories reveal about the mind.* University of Chicago Press, Chicago.

LILLESAND, T. M. AND R. W. KIEFER, 2000. *Remote sensing*

and image interpretation. 4th ed. John Wiley & Sons, New York.

PINKER, S., 1994. *The language instinct.* Allen Lane, London.

PRECHELT, L., 1995. *ai-faq/neural-nets/part1*. vol. 2002.

SCOTT, J. C., 1998. *Seeing like a state: how certain schemes to improve the human condition have failed.* Yale University Press, New Haven.

SPELKE, E., 1999. *Infant cognition.* In MIT encyclopedia of cognitive sciences,, edited by R. A. R. A. Wilson and F. C. Keil. MIT Press., Cambridge, MA:.

THELEN, E. AND L. SMITH, 1994. *A Dynamic Systems Approach to the Development of Cognition and Action.* MITPress. A Bradford Book., Cambridge, MA.

VARELA, F. J., E. THOMPSON and E. ROSCH, 1991. *The Embodied Mind: Cognitive Science and Human Experience.* MIT Press, Cambridge, MA.

ZUBROW, E. B. W., 1986. *The Nature of Prehistoric Shape.* Paper presented at the Symposium Form and Design in Archaeology: A Grammatical Approach. Society of American Archaeology Meetings, New Orleans.

ZUBROW, E. B. W. and P. DALY, 1998. *Symbolic Behaviour: the Origin of a Spatial Perspective. In Cognition and Material Culture: the Archaeology of Symbolic Storage*, edited by C. Renfrew and C. Scarre, pp. 157-174. McDonald Institute Monographs. McDonald Institute for Archaeological Research, Cambridge UK.

www.ingramcontent.com/pod-product-compliance
Lightning Source LLC
LaVergne TN
LVHW071210210926
848482LV00015B/770